Noteables™

Interactive Study Notebook with Foldables™

Mathematics

Applications and Concepts

Course 3

Contributing Author
Dinah Zike

Foldables™

Consultant
Douglas Fisher, PhD
Director of Professional Development
San Diego State University
San Diego, CA

New York, New York   Columbus, Ohio   Chicago, Illinois   Peoria, Illinois   Woodland Hills, California

The McGraw·Hill Companies

Copyright © by The McGraw-Hill Companies, Inc. All rights reserved. Printed in the United States of America. Except as permitted under the United States Copyright Act, no part of this book may be reproduced in any form, electronic or mechanical, including photocopy, recording, or any information storage or retrieval system, without prior written permission of the publisher.

Send all inquiries to:
The McGraw-Hill Companies
8787 Orion Place
Columbus, OH 43240-4027

ISBN: 0-07-868216-9

Mathematics: Applications and Concepts, Course 3 *(Student Edition)*
Noteables™: Interactive Study Notebook with Foldables™

7 8 9 10 009 09 08 07 06

Contents

Contents

Organizing Your Foldables

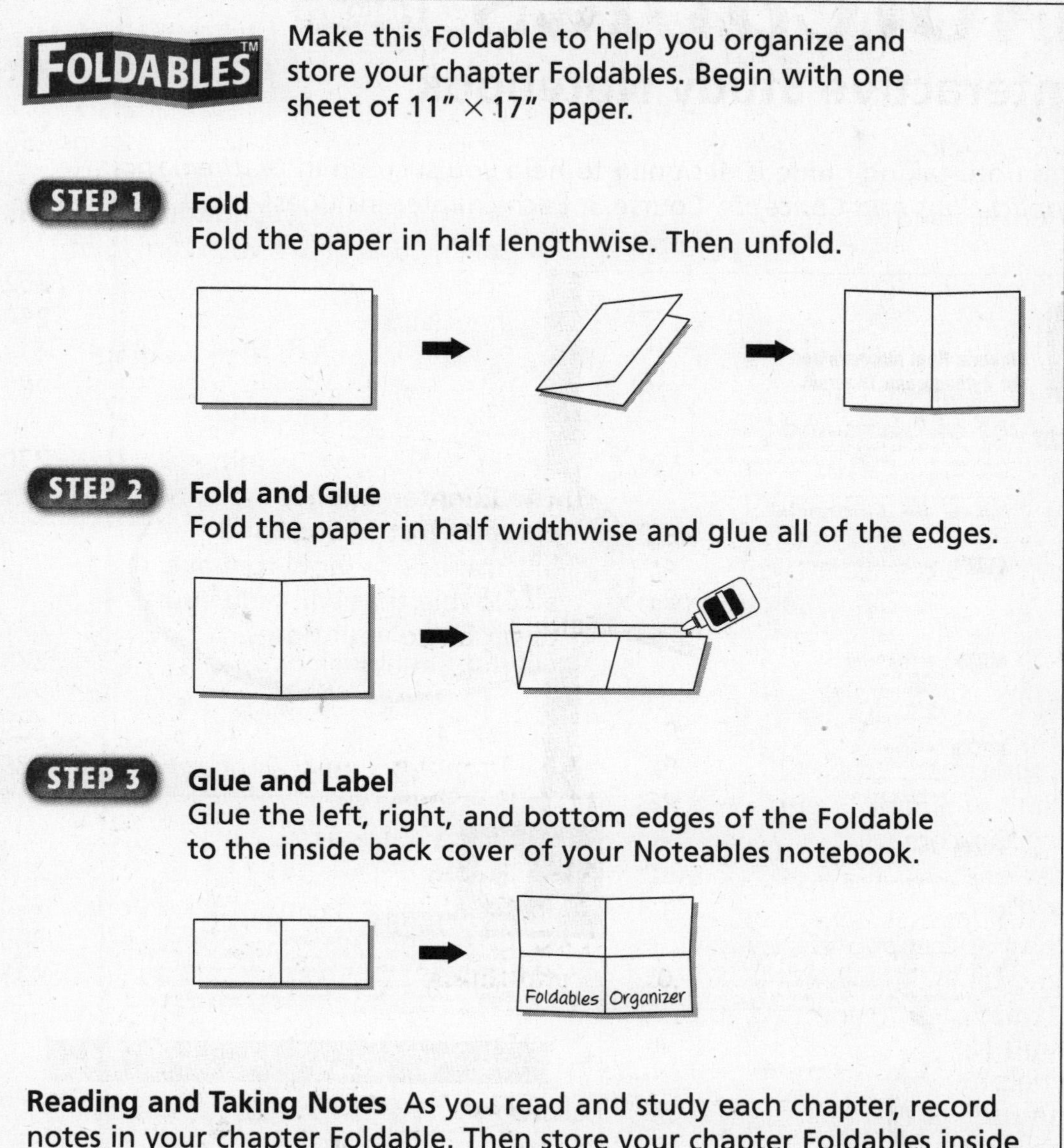

Make this Foldable to help you organize and store your chapter Foldables. Begin with one sheet of 11″ × 17″ paper.

STEP 1 Fold
Fold the paper in half lengthwise. Then unfold.

STEP 2 Fold and Glue
Fold the paper in half widthwise and glue all of the edges.

STEP 3 Glue and Label
Glue the left, right, and bottom edges of the Foldable to the inside back cover of your Noteables notebook.

Reading and Taking Notes As you read and study each chapter, record notes in your chapter Foldable. Then store your chapter Foldables inside this Foldable organizer.

Copyright © Glencoe/McGraw-Hill

Using Your Noteables™ with FOLDABLES
Interactive Study Notebook

This note-taking guide is designed to help you succeed in *Mathematics: Applications and Concepts,* Course 3. Each chapter includes:

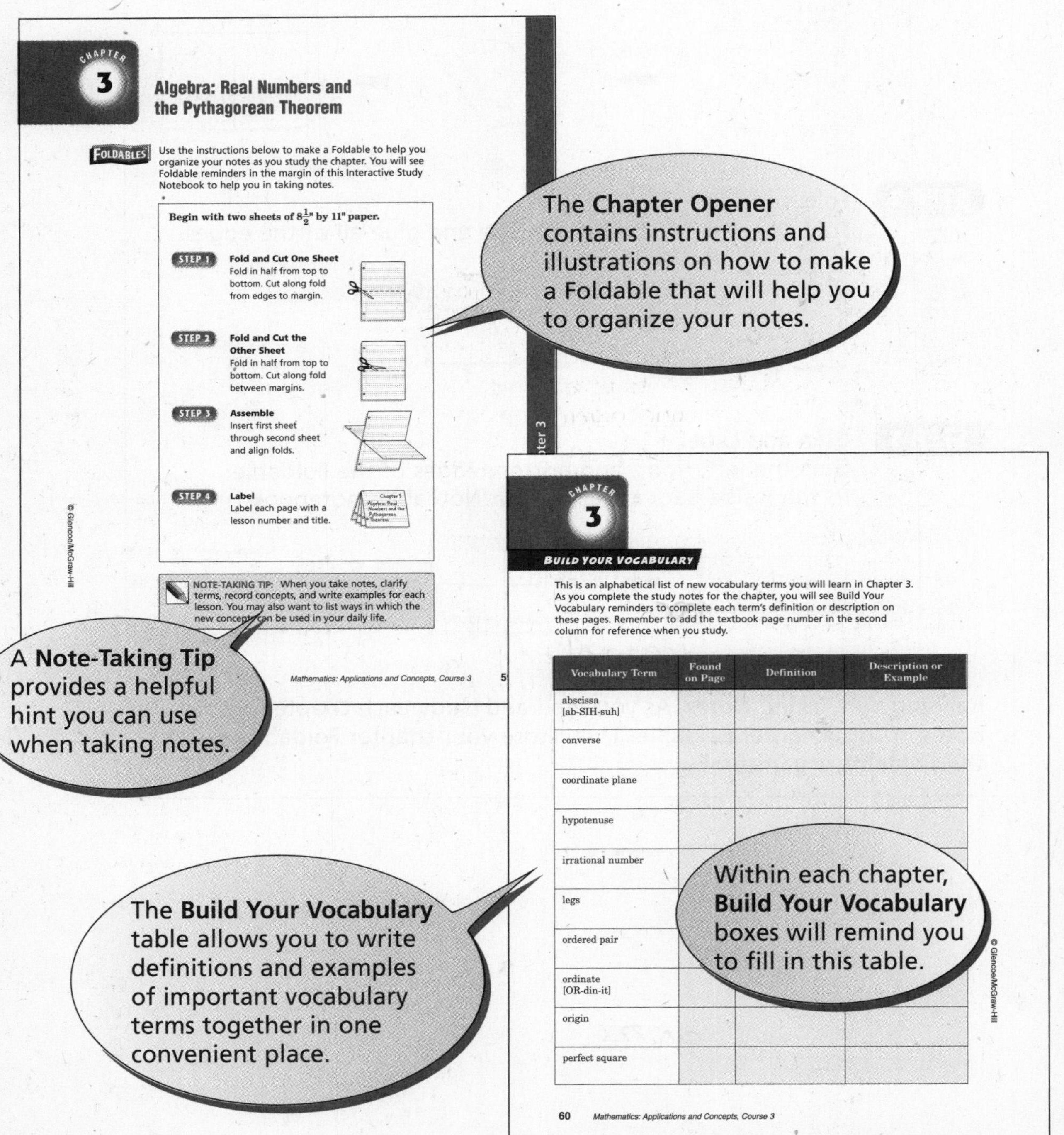

Copyright © Glencoe/McGraw-Hill

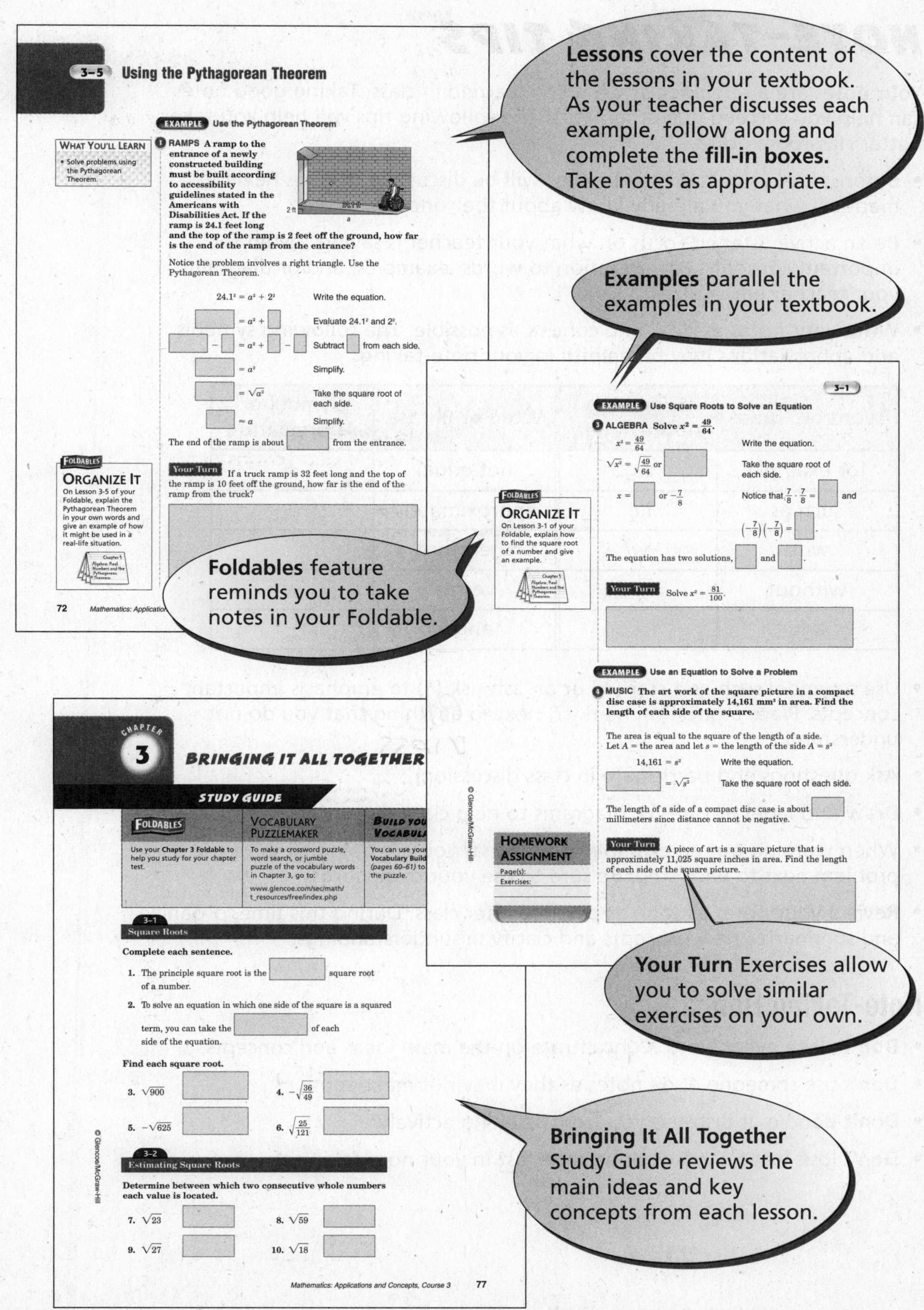

Lessons cover the content of the lessons in your textbook. As your teacher discusses each example, follow along and complete the **fill-in boxes.** Take notes as appropriate.
Examples parallel the examples in your textbook.
Foldables feature reminds you to take notes in your Foldable.
Your Turn Exercises allow you to solve similar exercises on your own.
Bringing It All Together Study Guide reviews the main ideas and key concepts from each lesson.
3–5 Using the Pythagorean Theorem
EXAMPLE Use the Pythagorean Theorem
WHAT YOU'LL LEARN
• Solve problems using the Pythagorean Theorem.
1 RAMPS A ramp to the entrance of a newly constructed building must be built according to accessibility guidelines stated in the Americans with Disabilities Act. If the ramp is 24.1 feet long and the top of the ramp is 2 feet off the ground, how far is the end of the ramp from the entrance?
Notice the problem involves a right triangle. Use the Pythagorean Theorem.
$24.1^2 = a^2 + 2^2$ Write the equation.
Evaluate 24.1^2 and 2^2.
Subtract from each side.
Simplify.
Take the square root of each side.
Simplify.
The end of the ramp is about from the entrance.
FOLDABLES ORGANIZE IT
On Lesson 3-5 of your Foldable, explain the Pythagorean Theorem in your own words and give an example of how it might be used in a real-life situation.
Your Turn If a truck ramp is 32 feet long and the top of the ramp is 10 feet off the ground, how far is the end of the ramp from the truck?
72 Mathematics: Applications
3–1
EXAMPLE Use Square Roots to Solve an Equation
3 ALGEBRA Solve $x^2 = \frac{49}{64}$.
$x^2 = \frac{49}{64}$ Write the equation.
$\sqrt{x^2} = \sqrt{\frac{49}{64}}$ or Take the square root of each side.
$x =$ or $-\frac{7}{8}$ Notice that $\frac{7}{8} \cdot \frac{7}{8} =$ and $\left(-\frac{7}{8}\right)\left(-\frac{7}{8}\right) =$
The equation has two solutions, and .
FOLDABLES ORGANIZE IT
On Lesson 3-1 of your Foldable, explain how to find the square root of a number and give an example.
Your Turn Solve $x^2 = \frac{81}{100}$.
EXAMPLE Use an Equation to Solve a Problem
4 MUSIC The art work of the square picture in a compact disc case is approximately 14,161 mm² in area. Find the length of each side of the square.
The area is equal to the square of the length of a side. Let A = the area and let s = the length of the side $A = s^2$
$14{,}161 = s^2$ Write the equation.
$= \sqrt{s^2}$ Take the square root of each side.
The length of a side of a compact disc case is about millimeters since distance cannot be negative.
HOMEWORK ASSIGNMENT
Page(s):
Exercises:
Your Turn A piece of art is a square picture that is approximately 11,025 square inches in area. Find the length of each side of the square picture.
© Glencoe/McGraw-Hill
CHAPTER 3 BRINGING IT ALL TOGETHER
STUDY GUIDE
FOLDABLES Use your Chapter 3 Foldable to help you study for your chapter test.
VOCABULARY PUZZLEMAKER To make a crossword puzzle, word search, or jumble puzzle of the vocabulary words in Chapter 3, go to: www.glencoe.com/sec/math/t_resources/free/index.php
BUILD YOUR VOCABULARY You can use your Vocabulary Builder (pages 60–61) to complete the puzzle.
3–1 Square Roots
Complete each sentence.
1. The principle square root is the square root of a number.
2. To solve an equation in which one side of the square is a squared term, you can take the of each side of the equation.
Find each square root.
3. $\sqrt{900}$
4. $-\sqrt{\frac{36}{49}}$
5. $-\sqrt{625}$
6. $\sqrt{\frac{25}{121}}$
3–2 Estimating Square Roots
Determine between which two consecutive whole numbers each value is located.
7. $\sqrt{23}$
8. $\sqrt{59}$
9. $\sqrt{27}$
10. $\sqrt{18}$
© Glencoe/McGraw-Hill
Mathematics: Applications and Concepts, Course 3 77

Copyright © Glencoe/McGraw-Hill

NOTE-TAKING TIPS

Your notes are a reminder of what you learned in class. Taking good notes can help you succeed in mathematics. The following tips will help you take better classroom notes.

- Before class, ask what your teacher will be discussing in class. Review mentally what you already know about the concept.
- Be an active listener. Focus on what your teacher is saying. Listen for important concepts. Pay attention to words, examples, and/or diagrams your teacher emphasizes.
- Write your notes as clear and concise as possible. The following symbols and abbreviations may be helpful in your note-taking.

Word or Phrase	Symbol or Abbreviation	Word or Phrase	Symbol or Abbreviation
for example	e.g.	not equal	$\neq$
such as	i.e.	approximately	$\approx$
with	w/	therefore	$\therefore$
without	w/o	versus	vs
and	+	angle	$\angle$

- Use a symbol such as a star (★) or an asterisk (*) to emphasis important concepts. Place a question mark (?) next to anything that you do not understand.
- Ask questions and participate in class discussion.
- Draw and label pictures or diagrams to help clarify a concept.
- When working out an example, write what you are doing to solve the problem next to each step. Be sure to use your own words.
- Review your notes as soon as possible after class. During this time, organize and summarize new concepts and clarify misunderstandings.

Note-Taking Don'ts

- **Don't** write every word. Concentrate on the main ideas and concepts.
- **Don't** use someone else's notes as they may not make sense.
- **Don't** doodle. It distracts you from listening actively.
- **Don't** lose focus or you will become lost in your note-taking.

Copyright © Glencoe/McGraw-Hill

Algebra: Integers

Use the instructions below to make a Foldable to help you organize your notes as you study the chapter. You will see Foldable reminders in the margin of this Interactive Study Notebook to help you in taking notes.

Begin with a plain piece of 11" × 17" paper.

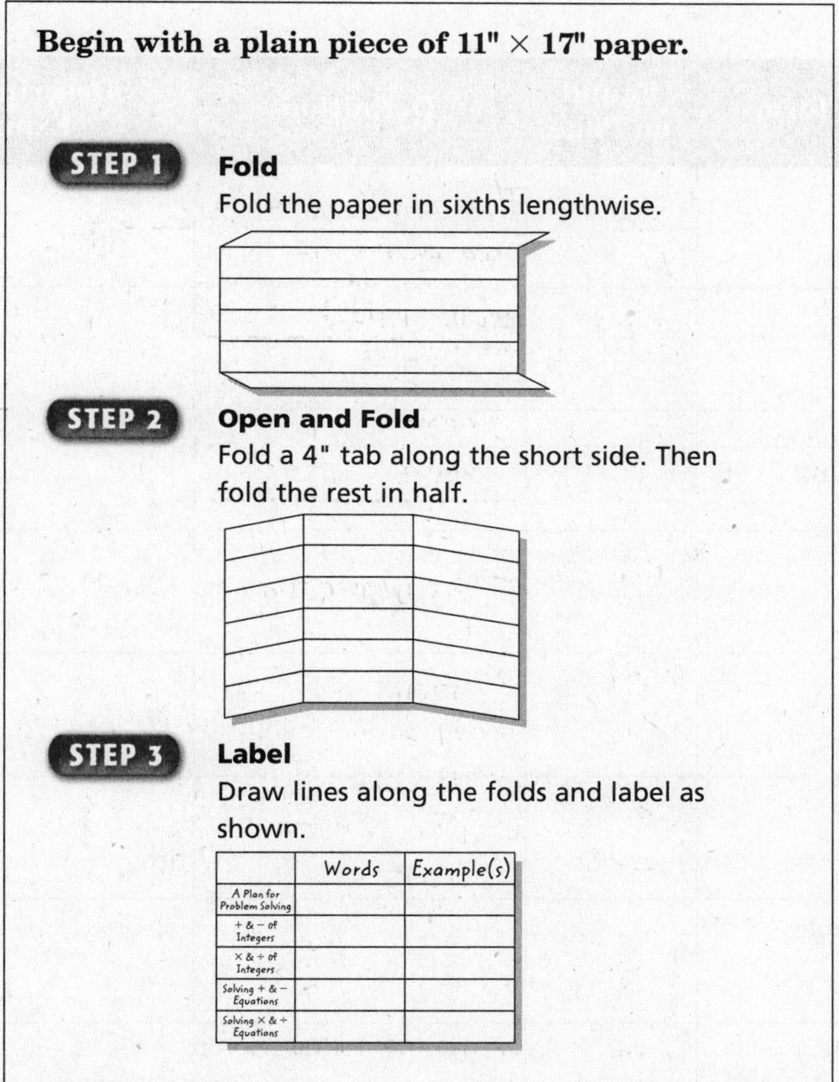

Fold
Fold the paper in sixths lengthwise.

Open and Fold
Fold a 4" tab along the short side. Then fold the rest in half.

Label
Draw lines along the folds and label as shown.

NOTE-TAKING TIP: When taking notes, it may be helpful to explain each idea in words and give one or more examples.

© Glencoe/McGraw-Hill

BUILD YOUR VOCABULARY

This is an alphabetical list of new vocabulary terms you will learn in Chapter 1. As you complete the study notes for the chapter, you will see Build Your Vocabulary reminders to complete each term's definition or description on these pages. Remember to add the textbook page number in the second column for reference when you study.

Vocabulary Term	Found on Page	Definition	Description or Example
absolute value	19	The distance a number is from zero on the number line	5 4 3 2 1 0 -1 -2 -3 -4 -4
additive inverse	25	Two integers that are opposite of each other are called additive inverse.	
algebraic expression [AL-juh-BRAY-ihk]	11	A combination of variales number and at least one operation.	
conjecture	7	An educated guess	
coordinate	18	A number associat with a point on the number line	
counterexample	13	A statement or example that show a conjecture is false.	
defining a variable	39	Choosing a varible and a quantity variable to represent in an expression or equation	
equation [ih-KWAY-zhuhn]	13	A mathematical sentence that contains an equal sign.	
evaluate	11	To find the value of an expression by replacing the variable with numerals	
inequality	18	A mathematical sentence that	

© Glencoe\McGraw-Hill

Vocabulary Term	Found on Page	Definition	Description or Example
integer [IHN-tih-juhr]	17	The set of whole number and their opposites	
inverse operations	46	Pairs of operations that undo each other.	
negative number	17	A number that is less than zero.	
numerical expression	11	A mathematical expreeion that has combination of number and at least one operation.	
open sentence	13	An equation that contains a variable.	
opposites	25	Two number with the same absolute value but different signs.	
order of operations	11	The rulse to follow when more than one operation is used in one expression.	
powers	12	Number written using exponents.	
property	13	An open sentense that is true for any	
solution	45	The value for the variable that makes an equation true	
solve	45	Find the value of the variable that makes the equation true.	
variable	11	A symbol usually a letter used to represent a number in mathematical expressons	

© Glencoe\McGraw-Hill

1–1 A Plan for Problem Solving

What You'll Learn

- Solve problems by using the four-step plan.

BUILD YOUR VOCABULARY (pages 2–3)

Some problem solving strategies require you to make an ______ or **conjecture**.

FOLDABLES

ORGANIZE IT

Summarize the four-step problem-solving plan in words and symbols. Include an example of how you have used this plan to solve a problem.

	Words	Example(s)
A Plan for Problem Solving		
+ & − of Integers		
× & ÷ of Integers		
Solving + & − Equations		
Solving × & ÷ Equations		

EXAMPLES Use the Four-Step Plan

1 **HOME IMPROVEMENT** **The Vorhees family plans to paint the wall in their family room. They need to cover 512 square feet with two coats of paint. If a one-gallon can of paint covers 220 square feet, how many one-gallon cans of paint should they purchase?**

EXPLORE Since they will be using ______ coats of paint, we must ______ the area to be painted.

PLAN They will be covering ______ × ______ square feet or ______ square feet. Next, divide ______ by ______ to determine how many cans of paint are needed.

SOLVE ______ ÷ ______ = ______

EXAMINE Since they will purchase a whole number of cans of paint, round ______ to ______.

They will need to purchase ______ cans of paint.

Your Turn Jocelyn plans to paint her bedroom. She needs to cover 400 square feet with three coats of paint. If a one-gallon can of paint covers 350 square feet, how many one-gallon cans of paint should she purchase?

______

© Glencoe/McGraw-Hill

REMEMBER IT

Always check to make sure your answer is reasonable. You can solve the problem again if you think your answer is not correct.

2 **GEOGRAPHY** **The five largest states in total area, which includes land and water, are shown. Of the five states shown, which one has the smallest area of water?**

Largest States in Area		
State	**Land Area (mi²)**	**Total Area (mi²)**
Alaska	570,374	615,230
Texas	261,914	267,277
California	155,973	158,869
Montana	145,556	147,046
New Mexico	121,364	121,598

Source: U.S. Census Bureau

EXPLORE You are given the total area and the land area for five states. You need to find the water area.

PLAN To determine the water area, ______ the ______ from the ______ for each state.

SOLVE Alaska = 615,230 − 570,374 = ______

Texas = 267,277 − 261,914 = ______

California = 158,869 − 155,973 = ______

Montana = 147,046 − 145,556 = ______

New Mexico = 121,598 − 121,364 = ______

EXAMINE Compare the water area for each state to determine which state has the least water area.

______ has the least water area with ______ square miles.

HOMEWORK ASSIGNMENT

Page(s):

Exercises:

Your Turn Refer to Example 2. How many times larger is the land area of Alaska than the land area of Montana?

© Glencoe/McGraw-Hill

1–2 Variables, Expressions, and Properties

WHAT YOU'LL LEARN

- Evaluate expressions and identify properties.

BUILD YOUR VOCABULARY (pages 2–3)

A **variable** is a ______, usually a letter, used to represent a ______.

An **algebraic expression** contains a ______, a number, and at least one ______ symbol.

When you substitute a number for the ______, an algebraic expression becomes a **numeric expression**.

To **evaluate** an expression means to find its ______ value.

To avoid confusion, mathematicians have agreed on a ______ called the **order of operations**.

KEY CONCEPT

Order of Operations

1. Do all operations within grouping symbols first; start with the innermost grouping symbols.
2. Evaluate all powers before other operations.
3. Multiply and divide in order from left to right.
4. Add and subtract in order from left to right.

EXAMPLE Evaluate a Numerical Expression

1 **Evaluate $3 + 9 - 2 \times (8 \div 2)$.**

$3 + 9 - 2 \times (8 \div 2)$

$= 3 + 9 - 2 \times$ ______ (4) ______ inside parentheses first.

$= 3 + 9 -$ ______ (8) ______ next.

$=$ ______ (12) $- 8$ or ______ (4) Add and subtract in order from left to right.

Your Turn Evaluate $2 + 6 - 3 \times (6 \div 3)$.

______

© Glencoe/McGraw-Hill

BUILD YOUR VOCABULARY (pages 2–3)

Expressions such as 7^2 and 2^3 are called **powers** and represent repeated ______.

EXAMPLES Evaluate Algebraic Expressions

2 Evaluate the expression $3r + 2s - 4$ if $r = 6$ and $s = 3$.

$3r + 2s - 4 = 3(6) + 2(3) - 4$ $r = 6$, $s = 3$

$= \square + \square - 4$ Multiply.

$= \square$ Add and subtract in order from left to right.

3 Evaluate the expression $\frac{6q}{3r - 3}$ if $q = 5$ and $r = 6$.

The fraction bar is a grouping symbol. Evaluate the expressions in the numerator and denominator separately before dividing.

$\frac{6q}{3r-3} = \frac{6\square}{3\square - 3}$ $q = \square$, $\square = 6$

$= \frac{\square}{\square - 3}$ Multiply.

$= \square$ or $\square$ Subtract in the denominator. Then divide.

Your Turn **Evaluate each expression.**

a. $2r + 4s - 2$ if $r = 5$ and $s = 4$

b. $\frac{3s}{q + 4}$ if $q = 2$ and $s = 4$

© Glencoe/McGraw-Hill

BUILD YOUR VOCABULARY (pages 2–3)

A mathematical sentence that contains an ______ sign (=) is called an **equation**.

An equation that contains a ______ is an **open sentence**.

Properties are ______ sentences that are true for any numbers.

A **counterexample** is an example that shows that a conjecture is ______.

REMEMBER IT

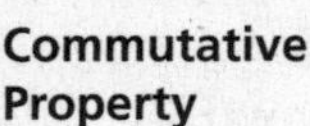

Commutative Property

$a + b = b + a$
$a \cdot b = b \cdot a$

Associative Property

$a + (b + c) = (a + b) + c$
$a \cdot (b \cdot c) = (a \cdot b) \cdot c$

Distributive Property

$a(b + c) = ab + ac$
$a(b - c) = ab - ac$

Identity Property

$a + 0 = a$
$a \cdot 1 = a$

EXAMPLE Identify Properties

4 Name the property shown by $12 \cdot 1 = 12$.

Multiplying by 1 does not change the number.

This is the ______ Property.

Your Turn Name the property shown by $3 \cdot 2 = 2 \cdot 3$.

______

EXAMPLE Find a Counterexample

5 State whether the following conjecture is *true* or *false*.

The sum of an odd number and an even number is always odd.

This conjecture is ______.

Your Turn State whether the following conjecture is *true* or *false*. If false, provide a counterexample.

Division of whole numbers is associative.

______

HOMEWORK ASSIGNMENT

Page(s):

Exercises:

© Glencoe/McGraw-Hill

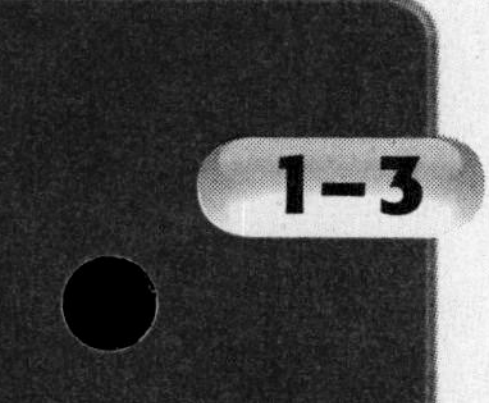

1–3 Integers and Absolute Values

WHAT YOU'LL LEARN

- Graph integers on a number line and find absolute value.

BUILD YOUR VOCABULARY (pages 2–3)

A **negative number** is a number ______ than zero.

______ numbers like −86, positive numbers like +125, and ______ are members of the set of **integers**.

EXAMPLES Write Integers for Real-Life Situations

1 **Write an integer to represent a 35¢ discount.**

The integer is ______.

2 **Write an integer for the expression *a fever of 4 degrees above normal.***

The integer is ______.

Your Turn **Write an integer for each situation.**

a. a 25¢ tax ______

b. a loss of 10 yards in football ______

BUILD YOUR VOCABULARY (pages 2–3)

The ______ that corresponds to a ______ on a number line is called the **coordinate** of that point.

A sentence that ______ two different numbers of quantities is called an **inequality**.

© Glencoe/McGraw-Hill

EXAMPLES Compare Two Integers

3 **Replace ● in −3 ● 3 with <, >, or = to make a true sentence.**

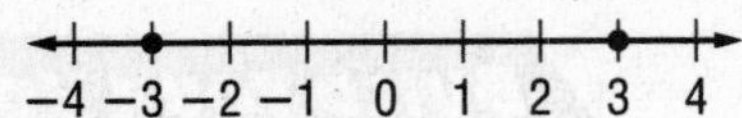

The number line shows that −3 is ______ than 3, since it lies to the ______ of 3. So, write −3 ______ 3.

Your Turn **Replace ● with <, >, or = to make a true sentence.**

a. −2 ● 2

b. −4 ● −6

EXAMPLE Order Integers

4 **FOOTBALL** **Order the statistics in the table from least to greatest.**

Player	Yards Rushing
Marty	8
Autry	4
Shane	5
Brad	−10
Glyn	3
Jim	−19

Graph each integer on a number line.

The order from least to greatest is ______.

Your Turn Order the temperatures 83°, 81°, −54°, −30° from least to greatest.

© Glencoe/McGraw-Hill

BUILD YOUR VOCABULARY (pages 2–3)

The **absolute value** of a number is the distance the number is from ____ on the number line.

REMEMBER IT

The absolute value of a number is *not* the same as the opposite of a number. Remember that the absolute value of a number cannot be negative.

EXAMPLES Expressions with Absolute Value

Evaluate each expression.

5 $|5|$

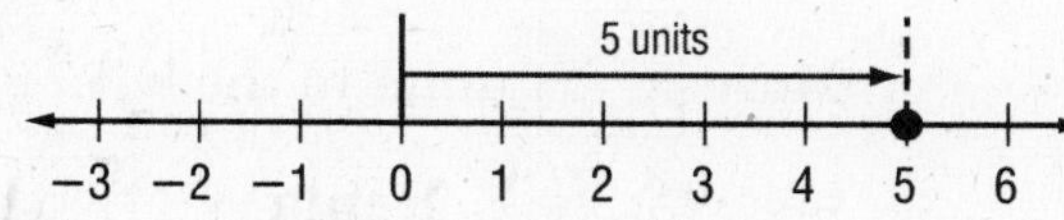

The graph of 5 is ____ units from 0 on the number line.

So, $|5| =$ ____.

6 $|6| - |-5|$

$\|6\| - \|-5\| =$ ____ $- \|-5\|$	The absolute value of 6 is ____.
$= 6 -$ ____	$\|-5\| =$ ____
$=$ ____	Simplify.

7 **Evaluate $|x| + 13$ if $x = -4$.**

$\|x\| + 13 = \|$ ____ $\| + 13$	Replace x with ____.
$=$ ____ $+ 13$	$\|-4\| =$ ____
$=$ ____	Simplify.

Your Turn

Evaluate each expression.

a. $|-3|$

b. $|9| - |-6|$

c. Evaluate $|x| + 7$ if $x = -2$.

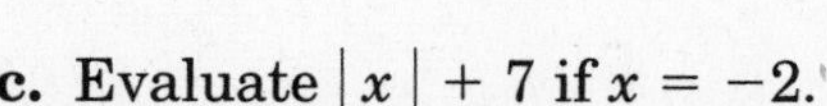

HOMEWORK ASSIGNMENT

Page(s):

Exercises:

© Glencoe/McGraw-Hill

1–4 Adding Integers

EXAMPLES **Add Integers with the Same Sign**

WHAT YOU'LL LEARN

- Add integers.

1 **Find −8 + (−4).**

METHOD 1 Use a number line.

Start at zero.

Move ______ units to the left.

From there, move 4 units ______.

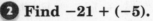

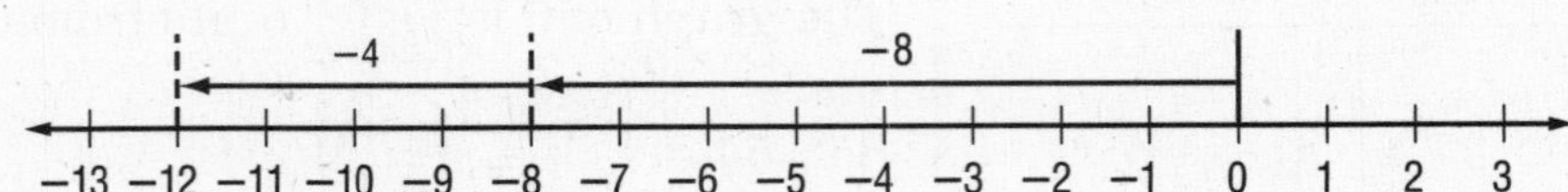

METHOD 2 Use counters.

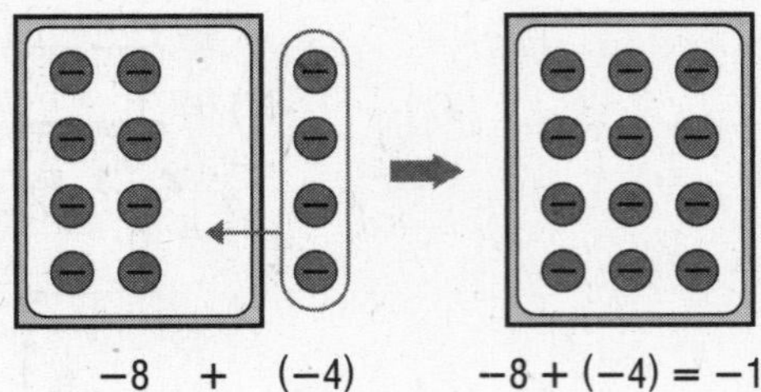

So, −8 + (−4) = ______.

KEY CONCEPT

Adding Integers with the Same Sign To add integers with the same sign, add their absolute values. Give the result the same sign as the integers.

2 **Find −21 + (−5).**

−21 + (−5) = ______

Add |−21| and |−5|. Both numbers are negative, so the sum is negative.

Your Turn **Add using a number line or counters.**

a. −3 + (−6)

b. −13 + (−12)

© Glencoe/McGraw-Hill

EXAMPLES Add Integers with Different Signs

3 **Find 4 + (−6).**

You can use a number line or counters to find the sum. Use a number line.

Start at ______.

Move 4 units ______.

From there, move ______ units left.

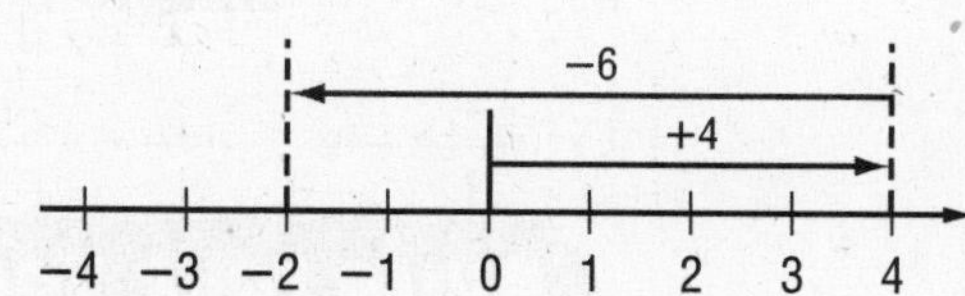

So, 4 + (−6) = ______.

4 **Find −5 + 9.**

Use Counters to find the sum.

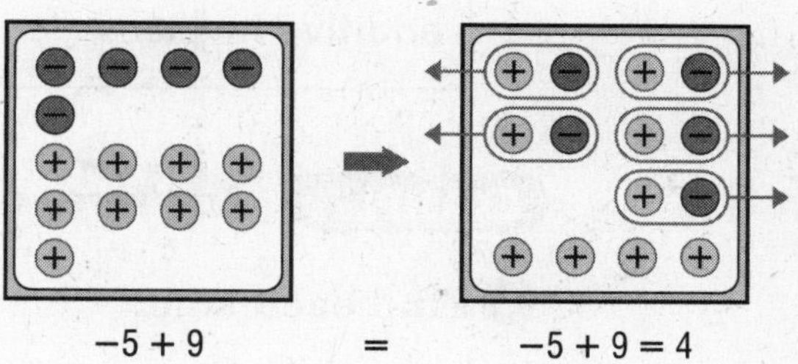

So, −5 + 9 = ______.

EXAMPLE Add Integers with Different Signs

5 **Find 33 + (−16).**

33 + (−16) = ______

To find 33 + (−16), subtract |−16| from |33|.

The sum is ______ because |33| > |−16|.

FOLDABLES

ORGANIZE IT

Explain and give examples of how to add integers with the same sign and how to add integers with a different signs.

	Words	Example(s)
A Plan for Problem Solving		
+ & − of Integers		
× & ÷ of Integers		
Solving + & − Equations		
Solving × & ÷ Equations		

KEY CONCEPT

Adding Integers with Different Signs To add integers with different signs, subtract their absolute values. Give the result the same sign as the integer with the greater absolute value.

© Glencoe/McGraw-Hill

Your Turn

Add.

a. $3 + (-5)$

b. $-6 + 8$

c. $25 + (-15)$.

BUILD YOUR VOCABULARY (pages 2–3)

Two numbers with the same ________ but different signs are called **opposites**.

An integer and its ________ are also called **additive inverses**.

EXAMPLES Add Three or More Integers

6 **Find each sum.**

$2 + (-5) + (-3)$

$2 + (-5) + (-3)$	$= 2 + [\ \_\_\_\_ + (-3)]$	Associative Property
	$= 2 + \_\_\_\_$	Order of operations.
	$= \_\_\_\_$	Simplify.

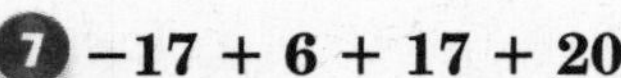

7 **$-17 + 6 + 17 + 20$**

$-7 + 6 + 17 + 20$	
$= -17 + 17 + \_\_\_\_ + 20$	Commutative Property
$= (-17 + 17) + (6 + 20)$	Associative Property
$= \_\_\_\_ + \_\_\_\_$	Additive Inverse Property
$= \_\_\_\_$	Simplify.

KEY CONCEPT

Additive Inverse Property The sum of any number and its additive inverse is zero.

© Glencoe\McGraw-Hill

Your Turn **Find each sum.**

a. $3 + (-6) + (-2)$

b. $-10 + 5 + 10 + 7$

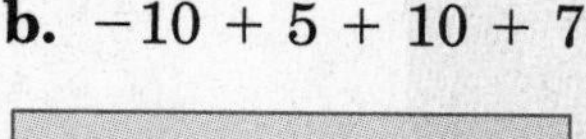

8 **STOCKS An investor owns 50 shares in a video game manufacturer. A broker purchases 30 shares more for the client on Tuesday. On Friday, the investor asks the broker to sell 65 shares. How many shares of this stock will the client own after these trades are completed?**

Selling a stock decreases the number of shares, so the integer for selling is ____.

Purchasing new stock increases the number of shares, so the integer for buying is ____.

Add these integers to the starting number of shares to find the new number of shares.

$50 + \_\_\_ + (\_\_\_)$

$= (50 + \_\_\_) + (\_\_\_)$ Associative Property

$= \_\_\_ + (-65)$ $50 + \_\_\_ = \_\_\_$

$= \_\_\_$ Simplify.

Your Turn Jaime gets an allowance of \$5. She spends \$2 on video games and \$1 on lunch. Her best friend repays a \$2 loan and she buys a \$3 pair of socks. How much money does Jaime have left?

© Glencoe\McGraw-Hill

HOMEWORK ASSIGNMENT

Page(s):

Exercises:

1–5 Subtracting Integers

WHAT YOU'LL LEARN

- Subtract integers.

KEY CONCEPT

Subtracting Integers

To subtract an integer, add its opposite or additive inverse.

FOLDABLES

ORGANIZE IT

Record in your Foldable how to subtract integers. Be sure to include examples.

	Words	Example(s)
A Plan for Problem Solving		
+ & − of Integers		
× & ÷ of Integers		
Solving + & − Equations		
Solving × & ÷ Equations		

EXAMPLES Subtract a Positive Integer

1 **Find 2 − 6.**

$2 - 6 = 2 + (-6)$ To subtract 6, add ______.

= ______ Add.

2 **Find −7 − 5.**

$-7 - 5 = -7$ ______ (-5) To subtract ______ add −5.

$= -12$ Add.

EXAMPLES Subtract a Negative Integer

3 **Find 11 − (−8).**

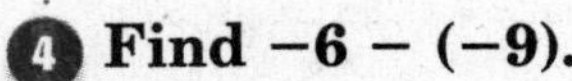

$11 - (-8) =$ ______ $+ 8$ To subtract −8, add ______.

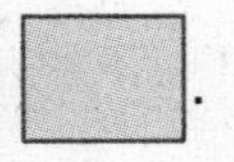

= ______ Add.

4 **Find −6 − (−9).**

$-6 - (-9) = -6$ ______ To subtract ______, add ______.

= ______ Add.

Your Turn **Subtract.**

a. $3 - 7$

b. $-6 - 2$

c. $15 - (-3)$

d. $-7 - (-11)$

© Glencoe/McGraw-Hill

WRITE IT

Explain why $-b$ does not necessarily mean that the value of $-b$ is negative.

____

____

____

____

EXAMPLES Evaluate Algebraic Expressions

5 **Evaluate $12 - r$ if $r = -7$.**

$12 - r = 12 -$ ____ Replace r with ____.

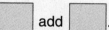

$= 12 +$ ____ To subtract ____ add ____.

$=$ ____ Add.

6 **Evaluate $q - p$ if $p = 6$ and $q = -3$.**

$q - p = -3 - 6$ Replace q with 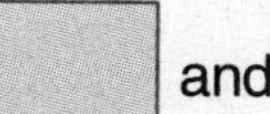 and p with ____.

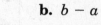

$= -3 +$ ____ To subtract 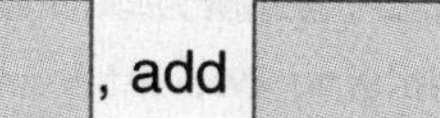, add ____.

$=$  Add.

Your Turn **Evaluate each expression if $a = 3$, $b = -6$, and $c = 2$.**

a. $10 - c$

b. $b - a$ 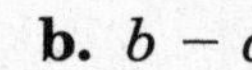

HOMEWORK ASSIGNMENT

Page(s):

Exercises:

© Glencoe\McGraw-Hill

1–6 Multiplying and Dividing Integers

WHAT YOU'LL LEARN

- Multiply and divide integers.

KEY CONCEPTS

Multiplying Two Integers The product of two integers with different signs is negative.

The product of two integers with the same sign is positive.

Dividing Integers The quotient of two integers with different signs is negative.

The quotient of two integers with the same sign is positive.

REMEMBER IT

Decide on the sign of the product before multiplying. If the number of negatives is *even* the product is positive. If the number of negatives is *odd* the product is negative.

EXAMPLE Multiply Integers with Different Signs

1 Find 8(−4).

8(−4) = ______

The factors have ______ signs. The product is ______.

EXAMPLE Multiply Integers with the Same Sign

2 Find −12(−12).

−12(−12) = ______

The factors have the ______ sign. The product is ______.

EXAMPLE Multiply More Than Two Integers

3 Find 6(−2)(−4).

6(−2)(−4) = [6(−2)] ______ ______ Property

= −12 ______ 6(−2) = ______

= ______ −12(−4) = ______

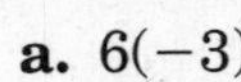

Your Turn **Multiply.**

a. 6(−3)

b. −2(6)

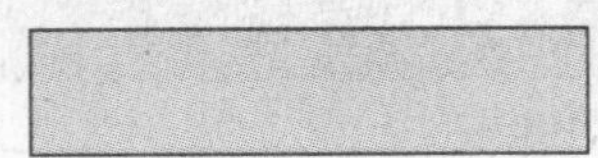

c. −8(−8)

d. 5(−3)(−2)

______

© Glencoe\McGraw-Hill

EXAMPLES Divide Integers

4 Find 30 ÷ −5.

$30 \div -5 =$ ______

The dividend and the divisor have ______ signs.

The quotient is ______.

FOLDABLES™

ORGANIZE IT

Describe why the product or quotient of two integers with the same sign is positive and the product or quotient of two integers with different signs is negative.

	Words	Example(s)
A Plan for Problem Solving		
+ & − of Integers		
× & ÷ of Integers		
Solving + & − Equations		
Solving × & ÷ Equations		

Your Turn **Divide.**

a. $36 \div (-6)$

b. $\frac{-30}{5}$

EXAMPLE Evaluate Algebraic Expressions

5 Evaluate $3x - (-4y)$ if $x = -10$ and $y = -4$.

$3x - (-4y)$

$= 3(\ \_\_\_\ ) - [-4(\ \_\_\_\ )]$ Replace *x* with ______ and *y* with ______.

$=$ ______ $-$ ______ $3(-10) =$ ______ $-4(-4) =$ ______

$= -30 +$ ______ To subtract ______, add  .

$=$ ______ Add.

Your Turn Evaluate $2a - (-3b)$ if $a = 26$ and $b = 24$.

© Glencoe/McGraw-Hill

EXAMPLE Find the Mean of a Set of Integers

6 WEATHER The table shows the low temperature for each month in McGrath, Alaska. Find the mean (average) of all 12 temperatures.

To find the mean of a set of numbers, find the sum of the numbers. Then divide the result by how many numbers there are in the set.

Average Low Temperatures	
Month	Temp. (°C)
Jan.	−27
Feb.	−26
March	−19
April	−9
May	1
June	7
July	9
Aug.	7
Sept.	2
Oct.	−8
Nov.	−19
Dec.	−26

Source: weather.com

$$\frac{-27 + (-26) + (-19) + (-9) + 1 + 7 + 9 + 7 + 2 + (-8) + (-19) + (-26)}{12}$$

$= \dfrac{\square}{12}$

$= \square$

Your Turn The table shows a set of record low temperatures. Find the mean (average) of all 12 temperatures.

Record Low Temperatures	
Month	Temp. (°C)
Jan.	−20
Feb.	−15
March	−5
April	10
May	25
June	31
July	41
Aug.	38
Sept.	34
Oct.	19
Nov.	3
Dec.	−15

Source: weather.com

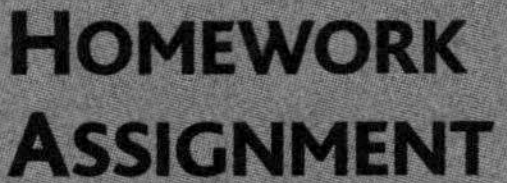

HOMEWORK ASSIGNMENT

Page(s):

Exercises:

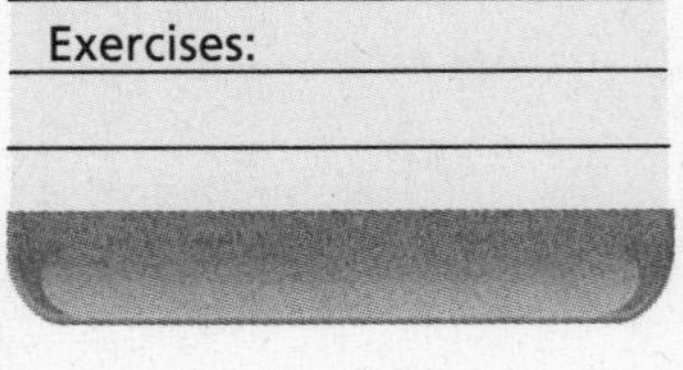

© Glencoe/McGraw-Hill

1–7 Writing Expressions and Equations

WHAT YOU'LL LEARN

- Write algebraic expressions and equations from verbal phrases and sentences.

BUILD YOUR VOCABULARY (pages 2–3)

When you choose a variable and an unknown quantity for the variable to represent, this is called **defining the variable.**

EXAMPLES Write an Algebraic Expression

1 Write *double the price of a loaf of bread* as an algebraic expression.

Words	double the price of a loaf of bread
Variable	Let p represent the price of a ______.
Expression	double the price of a loaf of bread
	______ $\cdot$ p

The expression is ______.

2 Write *stalks of celery are divided into 4 portions* as an algebraic expression.

Words	Stalks of celery are divided into 4 portions.
Variable	Let c represent the ______.
Expression	Stalks of celery are divided into 4 portions
	c ______ 4

The expression is ______.

REMEMBER IT

It is often helpful to select letters that can easily be connected to the quantity they represent. For example, age = a.

Your Turn Write each verbal sentence as an algebraic expression.

a. triple the cost of a cup of coffee

b. pies are divided into 8 portions

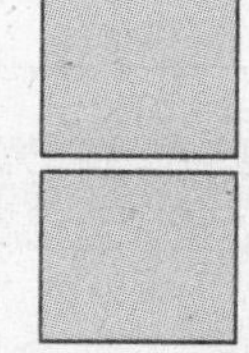

© Glencoe/McGraw-Hill

EXAMPLE Write an Algebraic Equation

REVIEW IT

Explain why it is important to read a word problem more than once before attempting to solve it.

______

______

______

______

______

______

3 **Write *the price of a book plus $5 shipping is $29* as an algebraic equation.**

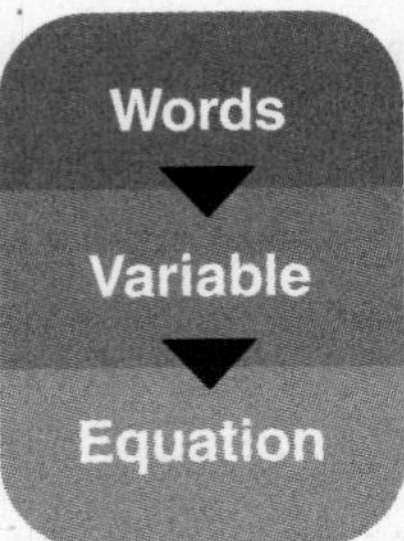

The price of a book plus $5 shipping is $29.

Let p represent the price of the book.

The price of a book	plus	$5 shipping	is $29.
______	+	______	= 29

The equation is ______.

Your Turn Write *the price of a toy plus $6 shipping is $35* as an algebraic equation.

______

EXAMPLE Write an Equation to Solve a Problem

4 **NUTRITION** **A particular box of oatmeal contains 10 individual packages. If the box contains 30 grams of fiber, write an equation to find the amount of fiber in one package of oatmeal.**

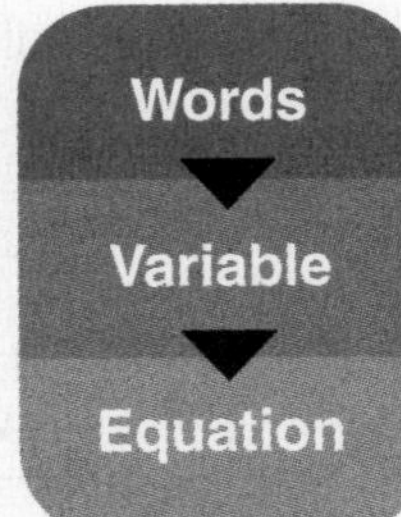

Ten packages of oatmeal contain 30 grams of fiber.

Let f represent the grams of fiber per package.

Ten packages of oatmeal	contain	30 grams of fiber.
______	=	30

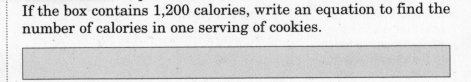

HOMEWORK ASSIGNMENT

Page(s): ______

Exercises: ______

Your Turn A particular box of cookies contains 10 servings. If the box contains 1,200 calories, write an equation to find the number of calories in one serving of cookies.

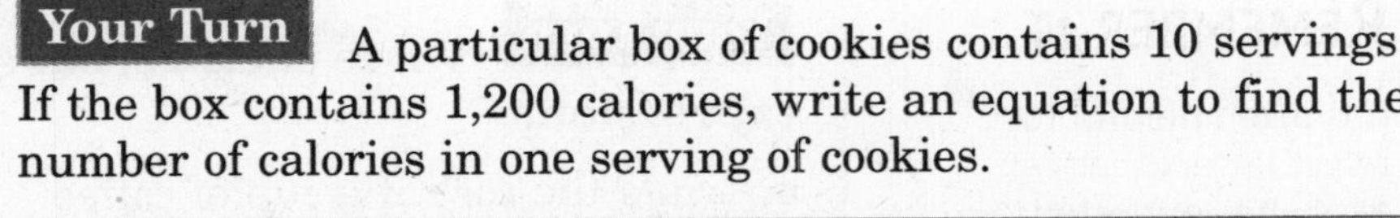

© Glencoe/McGraw-Hill

1–8 Solving Addition and Subtraction Equations

WHAT YOU'LL LEARN

- Solve equations using the Subtraction and Addition Properties of Equality.

BUILD YOUR VOCABULARY (pages 2–3)

When you **solve** an equation, you are trying to find the values of the variable that makes the equation ______.

A **solution** is the value of the variable that makes the variable ______.

KEY CONCEPTS

Subtraction Property of Equality If you subtract the same number from each side of an equation, the two sides remain equal.

Addition Property of Equality If you add the same number to each side of an equation, the two sides remain equal.

EXAMPLE Solve an Addition Equation

1 **Solve $7 = 15 + c$.**

METHOD 1 Vertical Method

$7 = 15 + c$ — Write the equation.

$7 = 15 + c$
$-15 = -15$ — Subtract ______ from each side.

______ $= c$ — $7 - 15 =$ ______; $15 - 15 =$ ______

METHOD 2 Horizontal Method

$7 = 15 + c$ — Write the equation.

$7 -$ ______ $= 15 + c -$ ______ — Subtract ______ from each side.

______ $= c$ — $7 - 15 =$ ______; and ______ $- 15 = 0$

Your Turn Solve $6 = 11 + a$.

© Glencoe/McGraw-Hill

BUILD YOUR VOCABULARY (pages 2–3)

Addition and subtraction are called **inverse operations** because they "undo" each other.

FOLDABLES ORGANIZE IT

Compare how to solve an equation involving whole numbers and an equation involving integers.

	Words	Example(s)
A Plan for Problem Solving		
+ & − of Integers		
× & ÷ of Integers		
Solving + & − Equations		
Solving × & ÷ Equations		

EXAMPLE Solve a Subtraction Equation

2 Solve $z - 16 = -5$.

Use the horizontal method.

$z - 16 = -5$	Write the equation.
$z - 16 + \_\_\_ = -5 + \_\_\_$	Add ___ to each side.
$z = \_\_\_$	$-16 + 16 = \_\_\_$ and
	$\_\_\_ + 16 = 11$.

Your Turn Solve $x - 12 = 26$.

EXAMPLE Write and Solve an Equation

3 What value of y makes the difference of 30 and y equal to 12?

$30 - y = 12$	Write the equation.
$30 - y + y = 12 + y$	___ y to each side.
$\_\_\_ = 12 + y$	___ $+ y = 0$
$30 - 12 = 12 - 12 + y$	___ 12 from each side.
$18 = \_\_\_$	$30 - 12 = \_\_\_$

Your Turn What value of x makes the difference of 30 and x equal to 5?

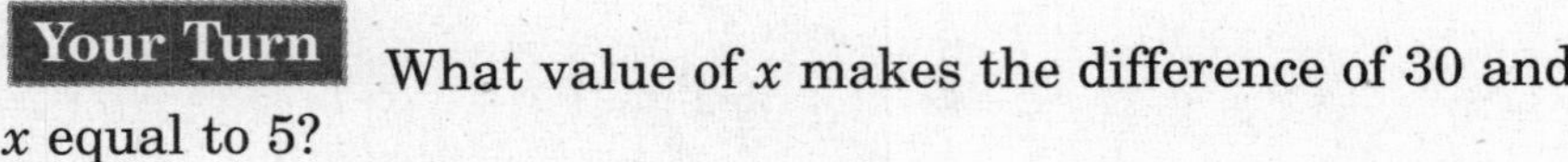

HOMEWORK ASSIGNMENT

Page(s):

Exercises:

© Glencoe/McGraw-Hill

1–9 Solving Multiplication and Division Equations

WHAT YOU'LL LEARN

- Solve equations by using the Division and Multiplication Properties of Equality.

KEY CONCEPTS

Division Property of Equality If you divide each side of an equation by the same nonzero number, the two sides remain equal.

Multiplication Property of Equality If you multiply each side of an equation by the same number, the two sides remain equal.

EXAMPLE Solve a Multiplication Equation

1 Solve $7z = -49$.

$7z = -49$ Write the equation.

$\frac{7z}{\square} = \frac{-49}{\square}$ 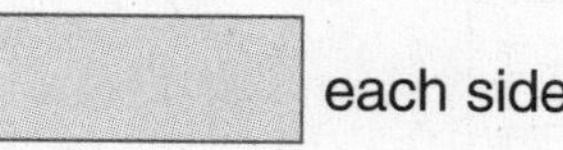 each side by ☐.

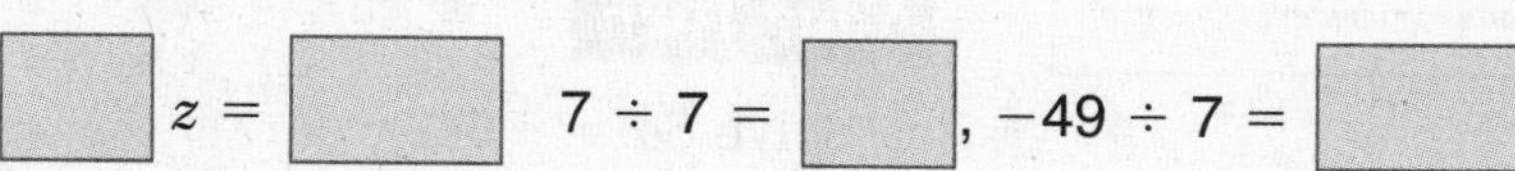

☐ $z =$ ☐ $7 \div 7 =$ ☐, $-49 \div 7 =$ ☐

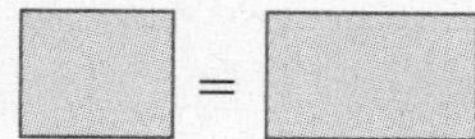

☐ $=$ ☐ Identity Property; $1z =$ ☐

EXAMPLE Solve a Division Equation

2 Solve $\frac{c}{9} = 26$.

$\frac{c}{9} = -6$ Write the equation.

$\frac{c}{9}$ ☐ $= -6$ ☐ Multiply each since by ☐.

$c =$ ☐ -6 ☐ $=$ ☐

EXAMPLE Use an Equation to Solve a Problem

3 SURVEYING English mathematician Edmund Gunter lived around 1600. He invented the *chain*, which was used to measure land for maps and deeds. One chain equals 66 feet. If the south side of a property measures 330 feet, how many chains long is it?

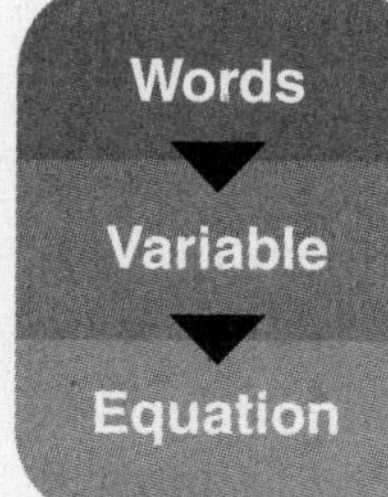

One chain equals 66 feet.

Let $c =$ the number of chains in  feet.

Measurement of property | is | 66 times the number of chains.

330 = ☐

© Glencoe/McGraw-Hill

FOLDABLES™

ORGANIZE IT

On your Foldable table, explain how to solve multiplication equations using the multiplication properties of equality.

	Words	Example(s)
A Plan for Problem Solving		
+ & − of Integers		
× & ÷ of Integers		
Solving + & − Equations		
Solving × & ÷ Equations		

Solve the equation.

$330 = 66c$ Write the equation.

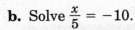

$\dfrac{330}{\square} = \dfrac{66c}{\square}$ Divide each side by ☐.

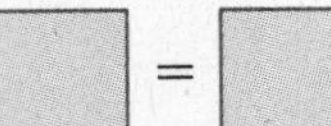

☐ = ☐ $330 \div$ ☐ = ☐

The number of chains in 330 feet is ☐.

Your Turn

a. Solve $8a = -64$.

b. Solve $\frac{x}{5} = -10$.

c. Most horses are measured in *hands*. One hand equals 4 inches. If a horse measures 60 inches, how many hands is it?

HOMEWORK ASSIGNMENT

Page(s):

Exercises:

© Glencoe/McGraw-Hill

BRINGING IT ALL TOGETHER

STUDY GUIDE

FOLDABLES™	VOCABULARY PUZZLEMAKER	BUILD YOUR VOCABULARY
Use your **Chapter 1 Foldable** to help you study for your chapter test.	To make a crossword puzzle, word search, or jumble puzzle of the vocabulary words in Chapter 1, go to: www.glencoe.com/sec/math/t_resources/free/index.php	You can use your completed **Vocabulary Builder** *(pages 2-3)* to help you solve the puzzle.

1-1 A Plan for Problem Solving

Use the four step plan to solve the problem.

1. Lisa plans to redecorate her bedroom. Each wall is 120 square feet. Three walls need a single coat of paint and the fourth wall needs a double coat. If each can of paint will cover 200 square feet, how many gallons of paint does Lisa need?

1-2 Variables, Expressions and Properties

2. Number the operations in the correct order for simplifying $2 + 4(9 - 6 \div 3)$.

☐ addition ☐ subtraction

☐ multiplication ☐ division

3. Describe how the expressions $2 + 5$ and $5 + 2$ are different. Then determine whether the two expressions are equal to each other. If the expressions are equal, name the property that says they are equal.

© Glencoe/McGraw-Hill

1-3

Integers and Absolute Values

Complete each sentence with either *left* or *right* to make a true sentence. Then write a statement comparing the two numbers with either < or >.

4. −45 lies to the ______ of 0 on a number line. ______

5. 72 lies to the ______ of 0 on a number line. ______

6. −3 lies to the ______ of −95 on a number line. ______

7. 6 lies to the ______ of −7 on a number line. ______

1-4

Adding Integers

Determine whether you *add* or *subtract* the absolute values of the numbers to find the sum. Give reasons for your answers.

8. 4 + 8

9. −3 + 5

10. 9 + (−12)

11. −23 + (−16)

1-5

Subtracting Integers

Rewrite each difference as a sum. Then find the sum.

12. 2 − 9

13. −3 − 8

14. 10 − (−12)

15. −5 − (−16)

© Glencoe/McGraw-Hill

1-6 Multiplying and Dividing Integers

Find each product or quotient.

16. $9(-2)$

17. $-6(-7)$

18. $12 \div (-4)$

19. $-35 \div (-7)$

1-7 Writing Expressions and Equations

Determine whether each situation requires *addition, subtraction, multiplication* or *division*.

20. Find the difference in the cost of a gallon of premium gasoline and the cost of a gallon of regular gasoline.

21. Find the flight time after the time has been increased by 15 minutes.

1-8 Solving Addition and Subtraction Equations

Solve each equation.

22. $x + 6 = 9$

23. $s - 5 = 14$

24. $4 = -3 + p$

25. $11 + m = 33$

1-9 Solving Multiplication and Division Equations

Solve each equation.

26. $8r = 32$

27. $3 = \frac{x}{7}$

28. $-9 = -9g$

© Glencoe/McGraw-Hill

CHAPTER 1

Checklist

ARE YOU READY FOR THE CHAPTER TEST?

Visit **msmath3.net** to access your textbook, more examples, self-check quizzes, and practice tests to help you study the concepts in Chapter 1.

Check the one that applies. Suggestions to help you study are given with each item.

- [ ] **I completed the review of all or most lessons without using my notes or asking for help.**
  - You are probably ready for the Chapter Test.
  - You may want take the Chapter 1 Practice Test on page 57 of your textbook as a final check.

- [ ] **I used my Foldable or Study Notebook to complete the review of all or most lessons.**
  - You should complete the Chapter 1 Study Guide and Review on pages 54–56 of your textbook.
  - If you are unsure of any concepts or skills, refer back to the specific lesson(s).
  - You may also want to take the Chapter 1 Practice Test on page 57 of your textbook.

- [ ] **I asked for help from someone else to complete the review of all or most lessons.**
  - You should review the examples and concepts in your Study Notebook and Chapter 1 Foldable.
  - Then complete the Chapter 1 Study Guide and Review on pages 54–56 of your textbook.
  - If you are unsure of any concepts or skills, refer back to the specific lesson(s).
  - You may also want to take the Chapter 1 Practice Test on page 57 of your textbook.

Student Signature

Parent/Guardian Signature

Teacher Signature

© Glencoe/McGraw-Hill

© Glencoe/McGraw-Hill

Algebra: Rational Numbers

Use the instructions below to make a Foldable to help you organize your notes as you study the chapter. You will see Foldable reminders in the margin of this Interactive Study Notebook to help you in taking notes.

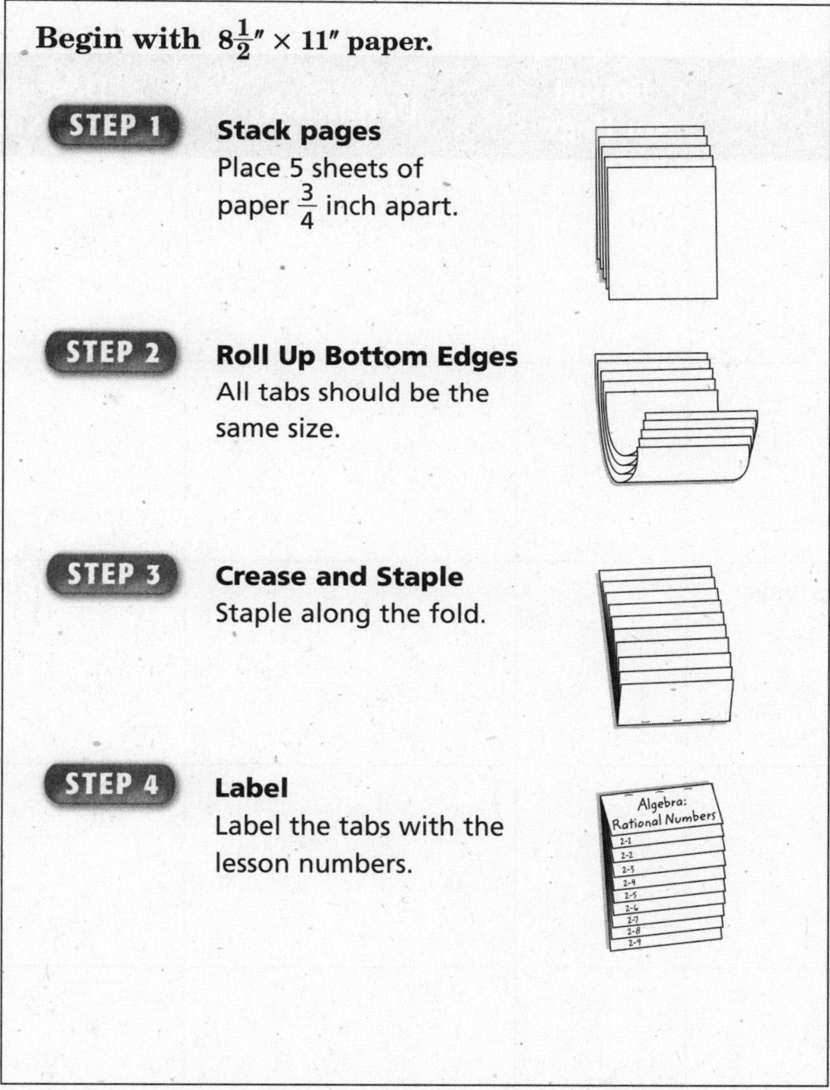

Begin with $8\frac{1}{2}'' \times 11''$ paper.

STEP 1 **Stack pages**
Place 5 sheets of paper $\frac{3}{4}$ inch apart.

STEP 2 **Roll Up Bottom Edges**
All tabs should be the same size.

STEP 3 **Crease and Staple**
Staple along the fold.

STEP 4 **Label**
Label the tabs with the lesson numbers.

NOTE-TAKING TIP: As you study a lesson, write down questions you have, comments and reactions, short summaries of the lesson, and key points that are highlighted and underlined.

Chapter 2

© Glencoe/McGraw-Hill

BUILD YOUR VOCABULARY

This is an alphabetical list of new vocabulary terms you will learn in Chapter 2. As you complete the study notes for the chapter, you will see Build Your Vocabulary reminders to complete each term's definition or description on these pages. Remember to add the textbook page number in the second column for reference when you study.

Vocabulary Term	Found on Page	Definition	Description or Example
bar notation			
base			
dimensional analysis			
exponent			
like fractions			
multiplicative inverses			

© Glencoe/McGraw-Hill

Vocabulary Term	Found on Page	Definition	Description or Example
power			
rational number			
reciprocals			
repeating decimal			
scientific notation			
terminating decimal			
unlike fractions			

© Glencoe/McGraw-Hill

2–1 Fractions and Decimals

BUILD YOUR VOCABULARY (pages 32–33)

A **rational number** is any number that can be expressed in the form $\frac{a}{b}$ where a and b are ______ and $b \neq 0$.

A decimal like 0.0625 is a **terminating** decimal because the division ends, or terminates, when the ______ is 0.

WHAT YOU'LL LEARN

- Express rational numbers as decimals and decimals as fractions.

KEY CONCEPT

Rational Numbers A rational number is any number that can be expressed in the form *a*/*b*, where *a* and *b* are integers and $b \neq 0$.

EXAMPLE Write a Fraction as a Decimal

1 Write $\frac{3}{16}$ as a decimal.

$\frac{3}{16}$ means 3 ______ 16.

$$\begin{array}{r} 0.1875 \\ 16\overline{)3.0000} \\ \underline{16} \\ 140 \\ \underline{128} \\ 120 \\ \underline{112} \\ 80 \\ \underline{80} \\ 0 \end{array}$$

Add a decimal point and zeroes to the dividend: 3 = 3.0000

Division ends when the ______ is 0.

You can also use a calculator.

The fraction $\frac{3}{16}$ can be written as ______.

Your Turn Write $\frac{1}{16}$ as a decimal.

© Glencoe/McGraw-Hill

BUILD YOUR VOCABULARY (pages 32–33)

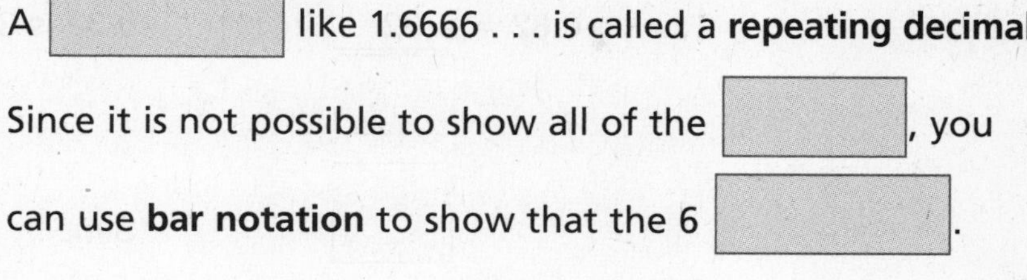

A ______ like 1.6666 . . . is called a **repeating decimal.**

Since it is not possible to show all of the ______, you can use **bar notation** to show that the 6 ______.

EXAMPLE Write a Mixed Number as a Decimal

2 Write $3\frac{2}{11}$ as a decimal.

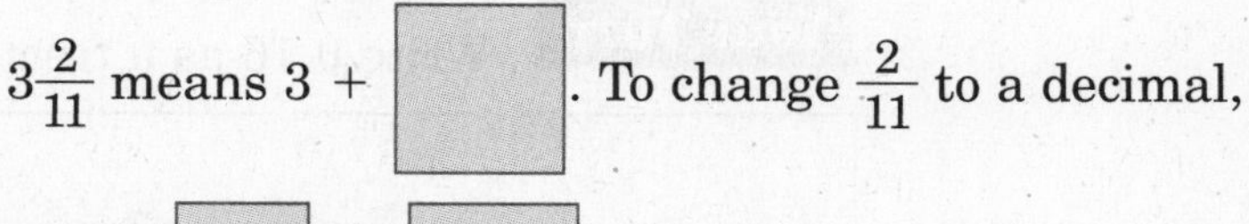

$3\frac{2}{11}$ means 3 + ______. To change $\frac{2}{11}$ to a decimal, divide ______ by ______.

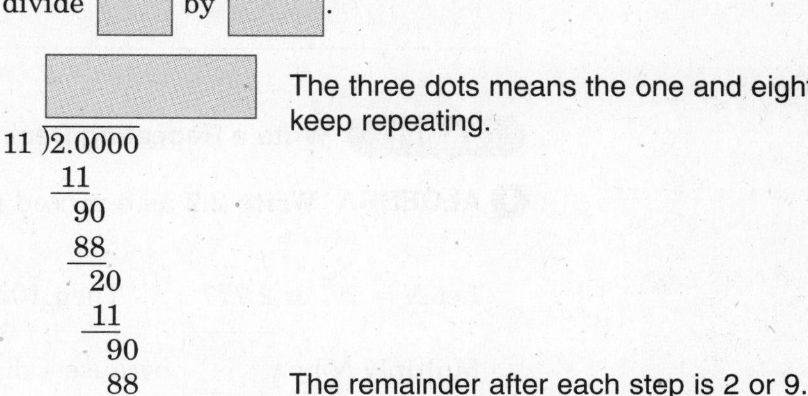

$$
\begin{array}{r}
 \\
11\,\overline{)2.0000} \\
\underline{11} \\
90 \\
\underline{88} \\
20 \\
\underline{11} \\
90 \\
\underline{88} \\
2
\end{array}
$$

The three dots means the one and eight keep repeating.

The remainder after each step is 2 or 9.

The mixed number $3\frac{2}{11}$ can be written as ______.

Your Turn Write $5\frac{1}{9}$ as a decimal.

WRITE IT

Explain how you decide where the bar is placed when you use bar notation for a repeating decimal.

______

______

______

______

______

______

FOLDABLES™ ORGANIZE IT

Under the tab for Lesson 2–1, explain in your own words how to express rational numbers as decimals and decimals as fractions.

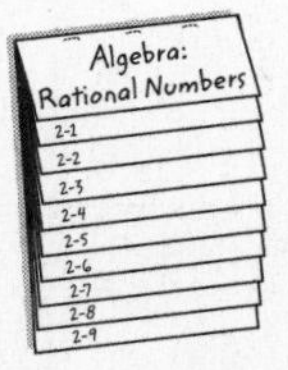

© Glencoe/McGraw-Hill

EXAMPLE Write a Terminating Decimal as a Fraction

3 Write 0.32 as a fraction.

$0.32 = \frac{32}{\square}$ 0.32 is 32 ______.

$= \square$ Simplify. Divide by the greatest common factor of 32 and 100, ______.

The decimal 0.32 can be written as ______.

Your Turn Write 0.16 as a fraction.

EXAMPLE Write a Repeating Decimal as a Fraction

4 ALGEBRA Write $2.\overline{7}$ as a mixed number.

Let $N = 2.\overline{7}$ or 2.777 . . . . Then $10N =$ ______.

Multiply N by ______ because 1 digit repeats.

Subtract $N = 2.777\ldots$ to eliminate the ______ part, 0.777 . . . .

$$10N = 27.777\ldots$$
$$-1N = \ \ 2.777\ldots \qquad N = 1N$$

$\square = 25$ $\qquad 10N - 1N = \square$

$\square = \square$ $\qquad$ Divide each side by ______.

$N = \square$ $\qquad$ Simplify.

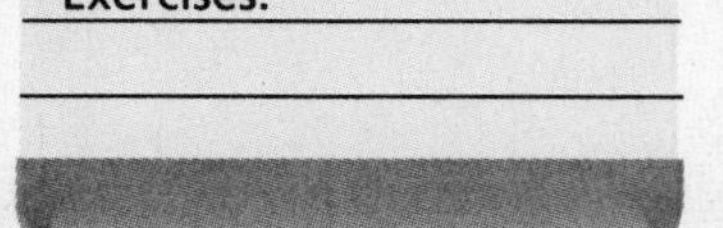

HOMEWORK ASSIGNMENT

Page(s):

Exercises:

Your Turn Write $1.\overline{7}$ as a mixed number.

© Glencoe/McGraw-Hill

2–2 Comparing and Ordering Rational Numbers

WHAT YOU'LL LEARN

- Compare and order rational numbers.

EXAMPLE Compare Rational Numbers

1 **Replace ● with <, >, or = to make $\frac{3}{7}$ ● $\frac{8}{13}$ a true sentence.**

METHOD 1 Write as fractions with the same denominator.

For $\frac{3}{7}$ and $\frac{8}{13}$, the least common denominator is 91.

$$\frac{3}{7} = \frac{3 \cdot \square}{7 \cdot \square} = \frac{\square}{91}$$

$$\frac{8}{13} = \frac{8 \cdot \square}{13 \cdot \square} = \frac{\square}{91}$$

Since $\frac{\square}{91} < \frac{\square}{91}$, $\frac{3}{7}$ $\square$ $\frac{8}{13}$.

METHOD 2 Write as decimals.

Write $\frac{3}{7}$ and $\frac{8}{13}$ as decimals. Use a calculator.

3 [÷] 7 [ENTER] 0.42857 . . .   8 [÷] 13 [ENTER] 0.61538 . . .

So, $\frac{3}{7}$ = 0.42857 . . . .   So, $\frac{8}{13}$ = 0.61538 . . . .

Since 0.42857 . . . < 0.61538 . . . , $\square$ < $\square$.

FOLDABLES

ORGANIZE IT

Under the tab for Lesson 2–2, explain how you can compare two numbers by expressing them as decimals and comparing the decimals.

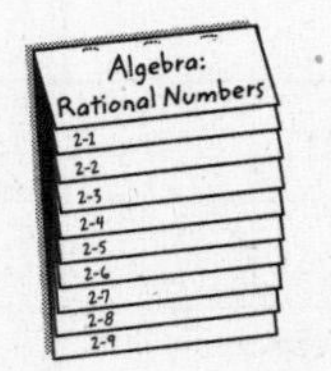

EXAMPLE Compare Negative Rational Numbers

2 **Replace ● with <, >, or = to make −6.7 ● $-6\frac{3}{4}$ a true sentence.**

Write $-6\frac{3}{4}$ as a decimal.

$\frac{3}{4}$ = $\square$, so $-6\frac{3}{4}$ = $\square$.

Since −6.7 > −6.75, −6.7 $\square$ $-6\frac{3}{4}$.

© Glencoe/McGraw-Hill

REMEMBER IT

On a number line, a number to the left is always less than a number to the right.

Your Turn Replace each ● with <, >, or = to make a true sentence.

a. $\frac{2}{3}$ ● $\frac{3}{5}$

b. -2.1 ● $-2\frac{1}{8}$

EXAMPLE Order Rational Numbers

3 **CHEMISTRY** **The values for the approximate densities of various substances are shown in the table. Order the densities from least to greatest.**

Substance	Density (g/cm3)
aluminum	2.7
beryllium	1.87
brick	$1\frac{4}{5}$
crown glass	$2\frac{1}{4}$
fused silica	$2.\overline{2}$
marble	$2\frac{3}{5}$
nylon	1.1
pyrex glass	2.32
rubber neoprene	$1.\overline{3}$

Source: *CRC Handbook of Chemistry and Physics*

Write each fraction as a decimal.

$1\frac{4}{5}$ =

$2\frac{1}{4}$ =

$2\frac{3}{5}$ =

From the least to the greatest, the densities are 1.1, $1.\overline{3}$, $1\frac{4}{5}$, 1.87, $2.\overline{2}$, $2\frac{1}{4}$, 2.32, $2\frac{3}{5}$, and 2.7. So, the nylon is the least dense, and aluminum is the most dense.

Your Turn The ride times for five amusement park attractions are shown in the table. Order the lengths from least to greatest.

Coaster	Ride Time (min)
Big Dipper	$1\frac{3}{4}$
Double Loop	1.5
Mind Eraser	1.8
Serial Thriller	$2\frac{1}{12}$
X–Flight	$2.\overline{3}$

Source: www.coasterglobe.com

HOMEWORK ASSIGNMENT

Page(s):

Exercises:

© Glencoe/McGraw-Hill

2–3 Multiplying Rational Numbers

BUILD YOUR VOCABULARY (pages 32–33)

Dimensional analysis is the process of including units of ________ when you ________.

WHAT YOU'LL LEARN

- Multiply fractions.

KEY CONCEPT

Multiply Fractions To multiply fractions, multiply the numerators and multiply the denominators.

EXAMPLE Multiply Fractions

1 **Find $\frac{3}{7} \cdot \frac{8}{9}$. Write in simplest form.**

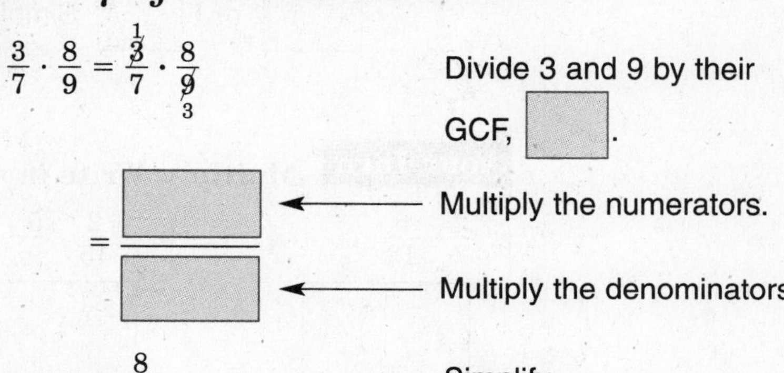

$\frac{3}{7} \cdot \frac{8}{9} = \frac{\overset{1}{\cancel{3}}}{7} \cdot \frac{8}{\underset{3}{\cancel{9}}}$ Divide 3 and 9 by their GCF, ____.

$= \frac{\square}{\square}$ ← Multiply the numerators. ← Multiply the denominators.

$= \frac{8}{21}$ Simplify.

EXAMPLE Multiply Negative Fractions

2 **Find $-\frac{3}{4} \cdot \frac{7}{12}$. Write in simplest form.**

$-\frac{3}{4} \cdot \frac{7}{12} = -\frac{\overset{-1}{\cancel{3}}}{4} \cdot \frac{7}{\underset{4}{\cancel{12}}}$ Divide −3 and 12 by their GCF, ____.

$= \frac{\square}{\square}$ ← Multiply the numerators. ← Multiply the denominators.

$= -\frac{\square}{\square}$ The factors have different signs, so the product is negative.

© Glencoe/McGraw-Hill

FOLDABLES™

ORGANIZE IT

Under the tab for Lesson 2–3, explain in your own words how to multiply rational numbers.

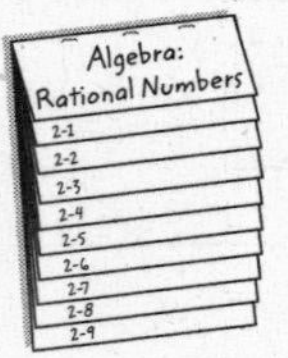

EXAMPLE Multiply Mixed Numbers

3 **Find $3\frac{1}{5} \cdot 1\frac{3}{4}$. Write in simplest form.**

$3\frac{1}{5} \cdot 1\frac{3}{4} = \square \cdot \square$ $\quad 3\frac{1}{5} = \square$, $1\frac{3}{4} = \square$

$= \frac{\overset{4}{\cancel{16}}}{5} \cdot \frac{7}{\underset{1}{\cancel{4}}}$ Divide 16 and 4 by their GCF, $\square$.

$= \frac{\square}{5 \cdot 1}$ ← Multiply the numerators. ← Multiply the denominators.

$= \square$, or $5\square$ Simplify.

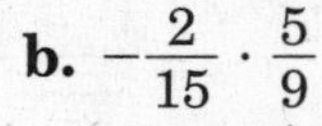

Your Turn **Multiply. Write in simplest form.**

a. $\frac{2}{7} \cdot \frac{5}{12}$

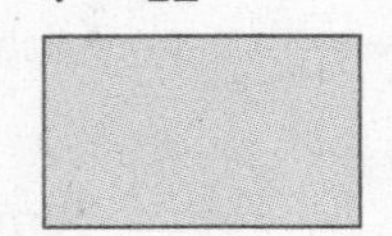

b. $-\frac{2}{15} \cdot \frac{5}{9}$

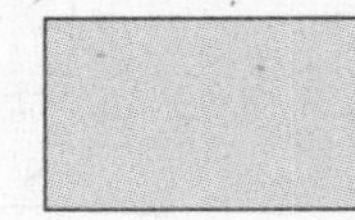

c. $3\frac{2}{5} \cdot 2\frac{2}{9}$

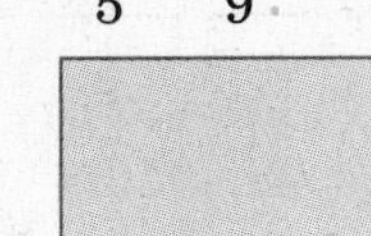

EXAMPLE Evaluate an Algebraic Expression

4 **ALGEBRA** **Evaluate *prq* if $p = \frac{2}{3}$, $r = -\frac{4}{5}$, and $q = \frac{7}{8}$.**

$prq = \square \cdot \square \cdot \square$ Replace p with $\frac{2}{3}$, r with $-\frac{4}{5}$, and q with $\frac{7}{8}$.

$= \frac{\overset{1}{\cancel{2}}}{3} \cdot -\frac{\overset{1}{\cancel{4}}}{5} \cdot \frac{7}{\underset{1}{\cancel{8}}}$ Divide out common factors.

$= \frac{1 \cdot (-1) \cdot 7}{3 \cdot 5 \cdot 1}$ or $\square$ Simplify.

HOMEWORK ASSIGNMENT

Page(s):

Exercises:

Your Turn Evaluate *prq* if $p = -\frac{3}{4}$, $r = \frac{8}{9}$, and $q = \frac{1}{2}$.

© Glencoe/McGraw-Hill

2–4

Dividing Rational Numbers

What You'll Learn

- Divide fractions.

Build Your Vocabulary (pages 32–33)

Two numbers whose product is one are **multiplicative inverses.**

The numbers 4 and $\frac{1}{4}$ are ________ or **reciprocals** of each other.

Key Concepts

Inverse Property of Multiplication The product of a rational number and its multiplicative inverse is 1.

Dividing Fractions To divide by a fraction, multiply by its multiplicative inverse.

EXAMPLE Find a Multiplicative Inverse

1 Write the multiplicative inverse of $-2\frac{4}{7}$.

$-2\frac{4}{7} =$ ________ Write $-2\frac{4}{7}$, as an improper fraction.

Since $-\frac{18}{7}\left(-\frac{7}{18}\right) =$ ____, the multiplicative inverse of $-2\frac{4}{7}$ is ________.

EXAMPLE Divide Fractions

2 Find $\frac{3}{10} \div \frac{2}{5}$. Write in simplest form.

$\frac{3}{10} \div \frac{2}{5} = \frac{3}{10} \cdot$ ____ Multiply by the multiplicative inverse of $\frac{2}{5}$.

$= \frac{3}{\overset{}{\cancel{10}}_{2}} \cdot \frac{\overset{1}{\cancel{5}}}{2}$ Divide 5 and 10 by their GCF, ____.

$=$ ____ Simplify.

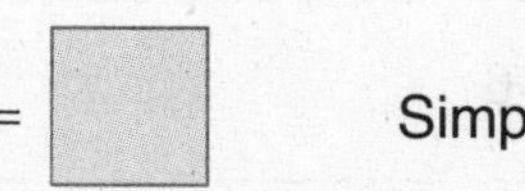

Your Turn

a. Write the multiplicative inverse of $-1\frac{5}{6}$.

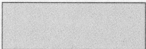

b. Find $\frac{4}{15} \div \frac{3}{5}$. Write in simplest form.

© Glencoe/McGraw-Hill

FOLDABLES

ORGANIZE IT

On the tab for Lesson 2–4, explain in your own words how to divide rational numbers.

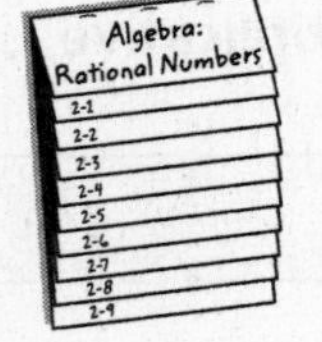

EXAMPLE Divide by a Whole Number

3 **Find $\frac{6}{7} \div 12$. Write in simplest form.**

$\frac{6}{7} \div 12 = \frac{6}{7} \div \frac{12}{1}$ — Write 12 as ______.

$= \frac{6}{7} \cdot$ ______ — Multiply by the multiplicative inverse of 12, which is .

$= \frac{\overset{1}{\cancel{6}}}{7} \cdot \frac{1}{\underset{2}{\cancel{12}}}$ — Divide 6 and 12 by their GCF, .

$=$ ______ — Simplify.

EXAMPLE Divide Negative Fractions

4 **Find $\frac{2}{7} \div -\frac{8}{9}$. Write in simplest form.**

$\frac{2}{7} \div -\frac{8}{9} = \frac{2}{7} \cdot$ ______ — Multiply by the multiplicative inverse of $-\frac{8}{9}$ which is 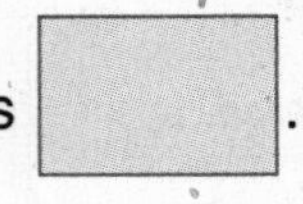.

$= \frac{\overset{1}{\cancel{2}}}{7} \cdot -\frac{9}{\underset{4}{\cancel{8}}}$ — Divide 2 and 8 by their GCF, .

$=$  — The fractions have different signs, so the quotient is negative.

WRITE IT

Explain how you would divide a fraction by a whole number.

Find each quotient. Write in simplest form.

a. $\frac{3}{4} \div 6$

b. $-\frac{3}{5} \div \frac{9}{10}$

© Glencoe/McGraw-Hill

EXAMPLE Divide Mixed Numbers

5 Find $3\frac{1}{4} \div \left(-2\frac{1}{8}\right)$. Write in simplest form.

$3\frac{1}{4} \div \left(-2\frac{1}{8}\right) = \square \div \left(\square\right)$ — $3\frac{1}{4} = \square$, $-2\frac{1}{8} = \square$

$= \square \cdot \left(-\frac{8}{17}\right)$ — The multiplicative inverse of $\square$ is $-\frac{8}{17}$.

$= \frac{13}{\cancel{4}_1} \cdot \left(-\frac{\cancel{8}^2}{17}\right)$ — Divide 4 and 8 by their GCF, $\square$.

$= -\frac{26}{17}$ or $\square$ — Simplify.

Your Turn Find $2\frac{1}{3} \div \left(-1\frac{1}{9}\right)$. Write in simplest form.

HOMEWORK ASSIGNMENT

Page(s):

Exercises:

2–5 Adding and Subtracting Like Fractions

What You'll Learn

- Add and subtract fractions with like denominators

BUILD YOUR VOCABULARY (pages 32–33)

Fractions with like ______ are called **like fractions.**

EXAMPLE Add Like Fractions

1 **Find $\frac{3}{16} + \left(-\frac{15}{16}\right)$. Write in simplest form.**

$\frac{3}{16} + \left(-\frac{15}{16}\right) = \frac{\square + (\square)}{16}$ ← Add the numerators. ← The denominators are the same.

$= \frac{-12}{16}$ or ______ Simplify.

EXAMPLE Subtract Like Fractions

Key Concepts

Adding Like Fractions To add fractions with like denominators, add the numerators and write the sum over the denominator.

Subtracting Like Fractions To subtract fractions with like denominators, subtract the numerators and write the difference over the denominator.

2 **Find $-\frac{7}{10} - \frac{9}{10}$. Write in simplest form.**

$-\frac{7}{10} - \frac{9}{10} = \frac{\square}{10}$ ← Subtract the numerators. ← The denominators are the same.

$= \frac{-16}{10}$ or ______ Rename $\frac{-16}{10}$ as $-1\frac{6}{10}$ or ______.

Your Turn **Find each difference. Write in simplest form.**

a. $\frac{2}{9} + \left(-\frac{8}{9}\right)$

b. $-\frac{7}{8} - \frac{5}{8}$

© Glencoe/McGraw-Hill

FOLDABLES™

ORGANIZE IT

Under the tab for Lesson 2–5, record models illustrating the addition and subtraction of like fractions.

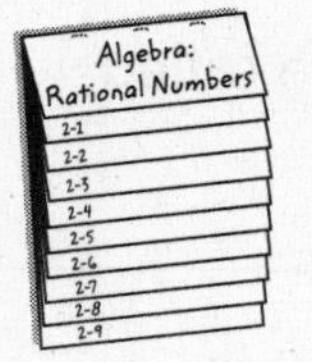

EXAMPLE Add Mixed Numbers

3 **Find $2\frac{5}{8} + 6\frac{1}{8}$. Write in simplest form.**

$2\frac{5}{8} + 6\frac{1}{8} = (\square + \square) + \left(\frac{5}{8} + \frac{1}{8}\right)$ — Add the whole numbers and fractions separately.

$= \square + \frac{5+1}{8}$ — Add the numerators.

$= \square$ or $\square$ — Simplify.

EXAMPLE Subtract Mixed Numbers

4 HEIGHTS **In the United States, the average height of a 9-year-old girl is $53\frac{4}{5}$ inches. The average height of a 16-year-old girl is $64\frac{1}{5}$ inches. How much does an average girl grow from age 9 to age 16?**

$64\frac{1}{5} - 53\frac{4}{5} = \frac{\square}{5} - \frac{\square}{5}$ — Write the mixed numbers as improper fractions.

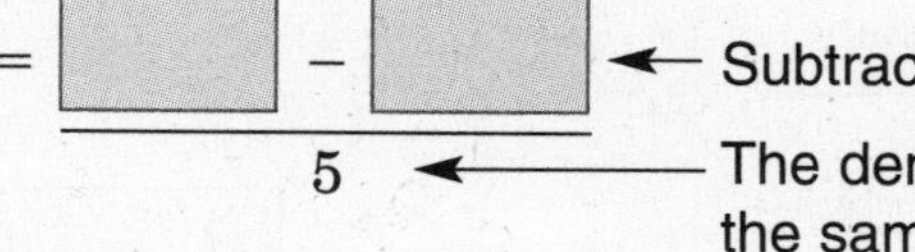

$= \frac{\square - \square}{5}$ ← Subtract the numerators. ← The denominators are the same.

$= \frac{52}{5}$ or $\square$ — Rename $\frac{52}{5}$ as  .

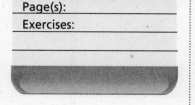

The average girl grows $\square$ inches from age 9 to age 16.

Your Turn

a. Find $3\frac{3}{10} + 4\frac{1}{10}$. Write in simplest form.

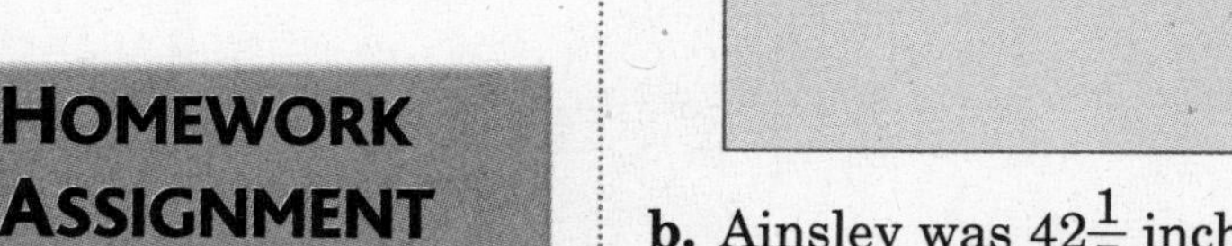

b. Ainsley was $42\frac{1}{7}$ inches tall when she was 4 years old. When she was 10 years old, she was $50\frac{3}{7}$ inches tall. How much did she grow between the ages of 4 and 10?

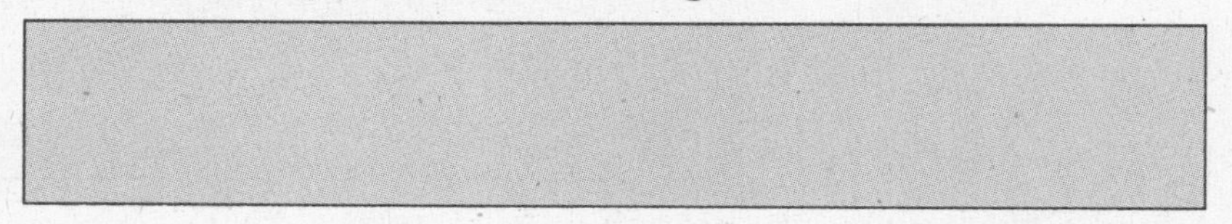

HOMEWORK ASSIGNMENT

Page(s):

Exercises:

© Glencoe/McGraw-Hill

2–6 Adding and Subtracting Unlike Fractions

WHAT YOU'LL LEARN

- Add and subtract fractions with unlike denominators.

BUILD YOUR VOCABULARY (pages 32–33)

Fractions with ______ denominators are called **unlike fractions.**

KEY CONCEPT

Adding and Subtracting Unlike Fractions To find the sum or difference of two fractions with unlike denominators, rename the fractions with a common denominator. Then add or subtract and simplify, if necessary.

EXAMPLE Subtract Unlike Fractions

1 **Find $\frac{1}{5} - \left(-\frac{2}{7}\right)$. Write in simplest form.**

$\frac{1}{5} - \left(-\frac{2}{7}\right) = \frac{1}{5} \cdot \frac{7}{7} - \left(-\frac{2}{7}\right) \cdot \frac{5}{5}$ — The LCD is $7 \cdot 5$ or ______.

$= \text{______} - \left(\text{______}\right)$ — Rename each fraction using the LCD.

$= \frac{7}{35} + \text{______}$ — Subtract $-\frac{10}{35}$ by adding its inverse, ______.

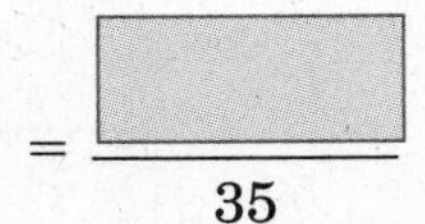

$= \frac{\text{______}}{35}$ — Add the numerators.

$= \text{______}$ — Simplify.

Your Turn Find $\frac{1}{3} - (-\frac{3}{5})$. Write in simplest form.

© Glencoe/McGraw-Hill

FOLDABLES™

ORGANIZE IT

Under the tab for Lesson 2–6, record the differences between adding and subtracting like and unlike fractions.

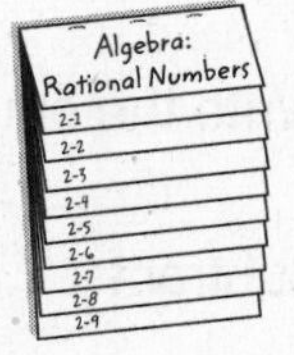

EXAMPLE Add Mixed Numbers

2 **Find $-4\frac{1}{8} + 2\frac{5}{12}$. Write in simplest form.**

$-4\frac{1}{8} + 2\frac{5}{12} = \square + \square$ Write the mixed numbers as fractions.

$= -\frac{33}{8} \cdot \frac{3}{3} + \frac{29}{12} \cdot \frac{2}{2}$ The LCD is $2 \cdot 2 \cdot 2 \cdot 3$ or $\square$.

$= -\square + \square$ Rename each fraction using the LCD.

$= \frac{\square}{24}$ Add the numerators.

$= \square$ or $-1\square$ Simplify.

Your Turn Find $-5\frac{1}{6} + 3\frac{5}{8}$. Write in simplest form.

EXAMPLE Evaluate Expressions

3 ALGEBRA **Find the value of $p - q$ if $p = \frac{2}{5}$ and $q = -\frac{4}{9}$.**

$p - q = \frac{2}{5} - \left(-\frac{4}{9}\right)$ Replace p with $\frac{2}{5}$ and q with $-\frac{4}{9}$.

$= \square - \left(-\frac{20}{45}\right)$ Rename each fraction using the LCD, $\square$.

$= \frac{\square + (-20)}{45}$ Subtract the numerators.

$= \square$ Simplify.

HOMEWORK ASSIGNMENT

Page(s):

Exercises:

Your Turn Find the value of $c - d$ if $c = \frac{3}{4}$ and $d = -\frac{8}{9}$.

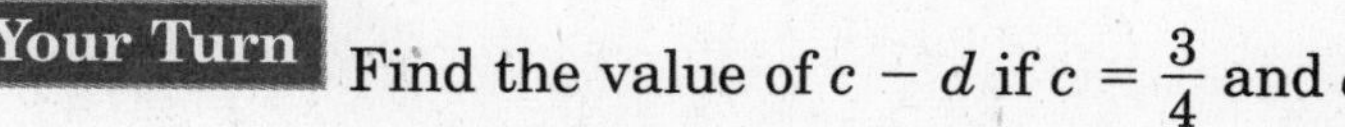

© Glencoe/McGraw-Hill

2–7 Solving Equations with Rational Numbers

EXAMPLES Solve by Using Addition or Subtraction

WHAT YOU'LL LEARN

- Solve equations involving rational numbers.

1 **Solve $g + 2.84 = 3.62$.**

$g +$ ____ $= 3.62$	Write the equation.
$g + 2.84 -$ ____ $= 3.62 -$ ____	Subtract ____ from each side.
$g =$ ____	Simplify.

FOLDABLES

ORGANIZE IT

Under the tab for Lesson 2–7, summarize in your own words what you have learned about solving equations with rational numbers.

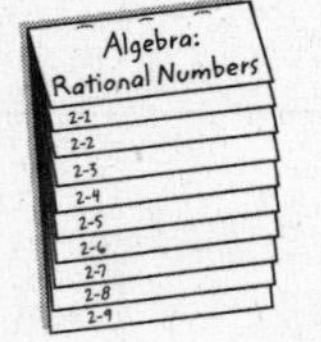

2 **Solve $-\frac{4}{5} = s - \frac{2}{3}$.**

$-\frac{4}{5} = s - \frac{2}{3}$	Write the equation.
$-\frac{4}{5} +$ ____ $= s - \frac{2}{3} +$ ____	Add 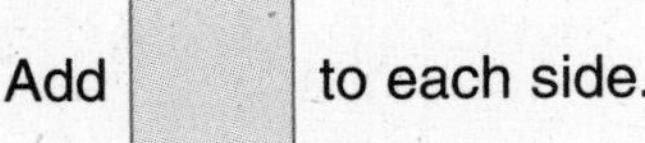 to each side.
$-\frac{4}{5} +$ ____ $= s$	Simplify.
____ $+ \frac{10}{15} = s$	Rename each fraction using the LCD.
____ $= s$	Simplify.

EXAMPLES Solve by Using Multiplication or Division

3 **Solve $\frac{7}{11}c = -21$.**

$\frac{7}{11}c = -21$	Write the equation.
____ $\left(\frac{7}{11}c\right) =$ ____ (-21)	Multiply each side by .
$c =$ ____	Simplify.

© Glencoe/McGraw-Hill

REVIEW IT

What is a mathematical sentence containing equals sign called? (*Lesson 1–8*)

______

______

______

______

4 Solve $9.7t = -67.9$.

$9.7t = -67.9$ Write the equation.

$\frac{9.7t}{\square} = \frac{-67.9}{\square}$ Divide each side by .

$t = \square$ Simplify.

Your Turn **Solve each equation.**

a. $h + 2.65 = 5.73$

b. $-\frac{2}{5} = x - \frac{3}{4}$.

c. $\frac{3}{5}x = -27$

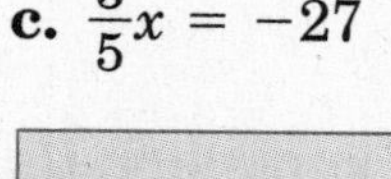

d. $3.4t = -27.2$

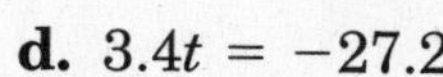

EXAMPLE Write an Equation to Solve a Problem

5 PHYSICS You can determine the rate an object is traveling by dividing the distance it travels by the time it takes to cover the distance $\left(r = \frac{d}{t}\right)$. If an object travels at a rate of 14.3 meters per second for 17 seconds, how far does it travel?

$r = \frac{d}{t}$

$14.3 = \frac{d}{\square}$ Write the equation.

$\square(14.3) = 17\left(\frac{d}{\square}\right)$ Multiply each side by .

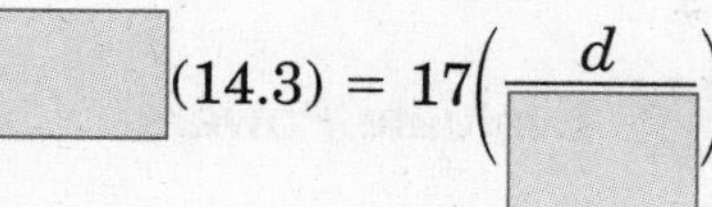

$\square = d$ Simplify.

Your Turn If an object travels at a rate of 73 miles per hour for 5.2 hours, how far does it travel?

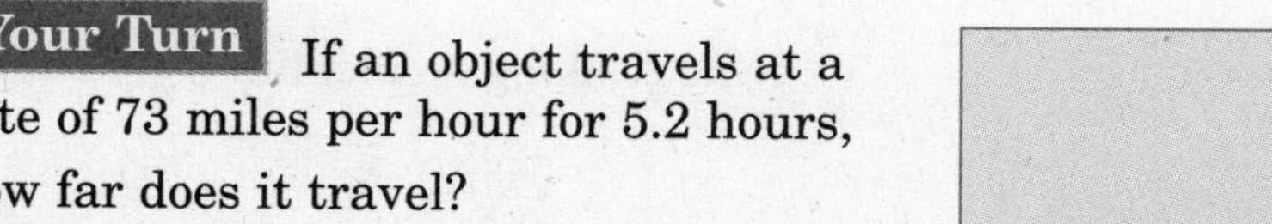

HOMEWORK ASSIGNMENT

Page(s):

Exercises:

© Glencoe/McGraw-Hill

2–8 Powers and Exponents

BUILD YOUR VOCABULARY (pages 32–33)

The **base** is the number that is ______.

The **exponent** tells how many times the ______ is used as a ______.

The number that is expressed using an ______ is called a **power**.

WHAT YOU'LL LEARN

- Use powers and exponents in expressions.

EXAMPLE Write an Expression Using Powers

KEY CONCEPTS

Zero and Negative Exponents Any nonzero number to the zero power is 1. Any nonzero number to the negative *n* power is 1 divided by the number to the *n*th power.

1 **Write $p \cdot p \cdot p \cdot q \cdot p \cdot q \cdot q$ using exponents.**

$p \cdot p \cdot p \cdot q \cdot p \cdot q \cdot q$

$= p \cdot p \cdot p \cdot p \cdot q \cdot q \cdot q$ ______ Property

$= (p \cdot p \cdot p \cdot p) \cdot (q \cdot q \cdot q)$ ______ Property

$=$ ______ $\cdot$ ______ Definition of exponents

Your Turn Write $x \cdot y \cdot x \cdot x \cdot y \cdot y \cdot y$ using exponents.

______

EXAMPLES Evaluate Powers

2 **Evaluate 9^5.**

$9^5 =$ ______ Definition of exponents

$= 59,049$ Simplify.

Check using a calculator.

9 [^] 5 [ENTER =] ______

© Glencoe/McGraw-Hill

FOLDABLES

ORGANIZE IT

On the tab for Lesson 2–8, compare how to evaluate an expression with positive exponents and one with negative exponents.

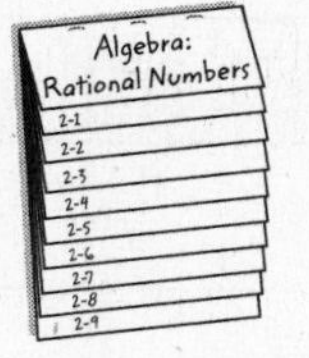

3 Evaluate 3^{-7}.

$3^{-7} = \frac{1}{\square}$ Definition of negative exponents

$= \frac{1}{\square}$ Simplify.

4 ALGEBRA Evaluate $x^3 \cdot y^5$ if $x = 4$ and $y = 2$.

$x^3 \cdot y^5 = \square^3 \cdot \square^5$ Replace x with $\square$ and y with $\square$.

$= (\square) \cdot (\square)$ Definition of exponents

$= 64 \cdot 32$ Simplify.

$= \square$ Simplify.

Your Turn **Evaluate each expression.**

a. 6^5

b. 2^{-5}

c. Evaluate $x^2 \cdot y^4$ if $x = 3$ and $y = 4$.

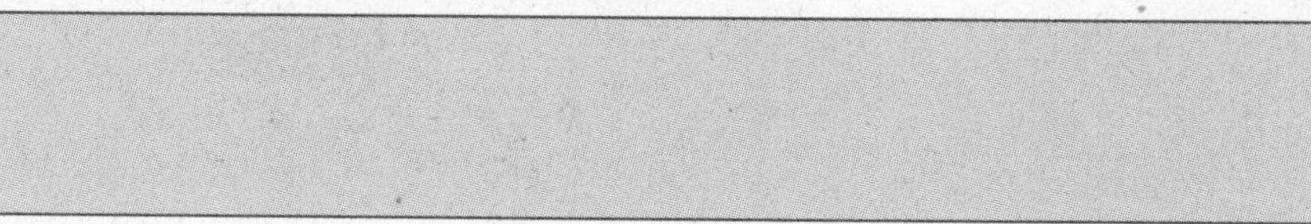

© Glencoe/McGraw-Hill

HOMEWORK ASSIGNMENT

Page(s):

Exercises:

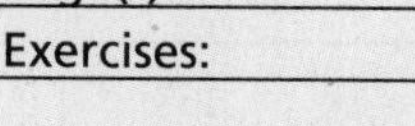

2–9 Scientific Notation

WHAT YOU'LL LEARN

- Express numbers in scientific notation.

BUILD YOUR VOCABULARY (pages 32–33)

A number is expressed in **scientific notation** when it is written as a ______ of a factor and a ______ of 10.

KEY CONCEPT

Scientific Notation A number is expressed in scientific notation when it is written as the product of a factor and a power of 10. The factor must be greater than or equal to 1 and less than 10.

EXAMPLES Express Numbers in Standard Form

1 Write 9.62×10^5 in standard form.

$9.62 \times 10^5 = 9.62 \times$ ______ $\qquad 10^5 = 10 \cdot 10 \cdot 10 \cdot 10 \cdot 10$ or ______

$= 962000$ $\qquad$ Notice that the decimal place moves ______ places to the right.

$=$ ______

2 Write 2.85×10^{-6} in standard form.

$2.85 \times 10^{-6} = 2.85 \times$ ______ $\qquad 10^{-6} =$ ______

$= 2.85 \times 0.000001$

$= 0.00000285$ $\qquad$ Notice that the decimal point moves 6 places to the left.

$=$ 

Your Turn **Write each number in standard form.**

a. 5.32×10^4

b. 3.81×10^{-4}

© Glencoe/McGraw-Hill

FOLDABLES

ORGANIZE IT

Under the tab for Lesson 2–9, collect and record examples of numbers you encounter in your daily life and write them in scientific notation.

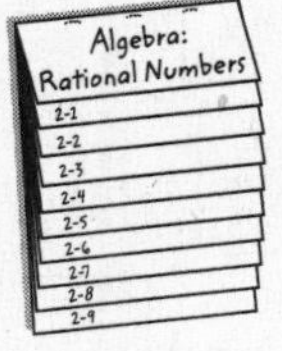

EXAMPLES Write Numbers in Scientific Notation

3 Write 931,500,000 in scientific notation.

$931500000 = 9.315 \times 100{,}000{,}000$ The decimal point moves 8 places.

= ______ The exponent is positive.

4 Write 0.00443 in scientific notation.

$0.00443 =$ ______ $\times 0.001$ The decimal point moves  places.

$= 4.43 \times$ ______ The exponent is .

EXAMPLE Compare Numbers in Scientific Notation

5 PLANETS The following table lists the average radius at the equator for each of the planets in our solar system. Order the planets according to radius from largest to smallest.

Planet	Radius (km)
Earth	6.38×10^3
Jupiter	7.14×10^4
Mars	3.40×10^3
Mercury	2.44×10^3
Neptune	2.43×10^4
Pluto	1.5×10^3
Saturn	6.0×10^4
Uranus	2.54×10^4
Venus	6.05×10^3

Source: *CRC Handbook of Chemistry and Physics*

First order the numbers according to their exponents. Then order the numbers with the same exponents by comparing the factors.

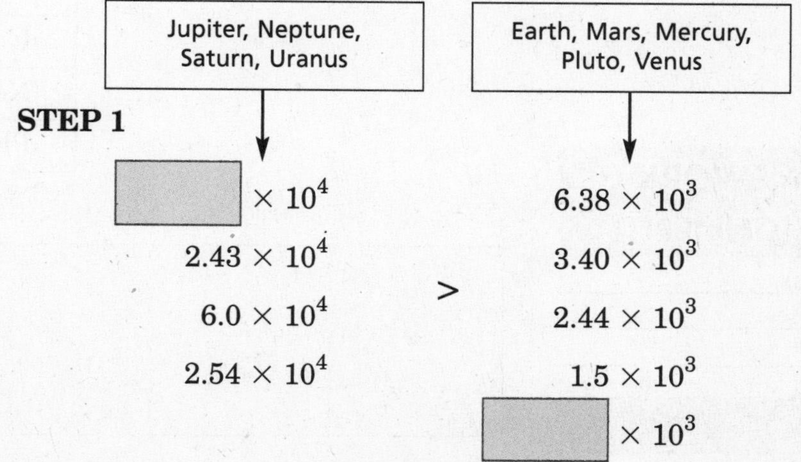

© Glencoe/McGraw-Hill

STEP 2

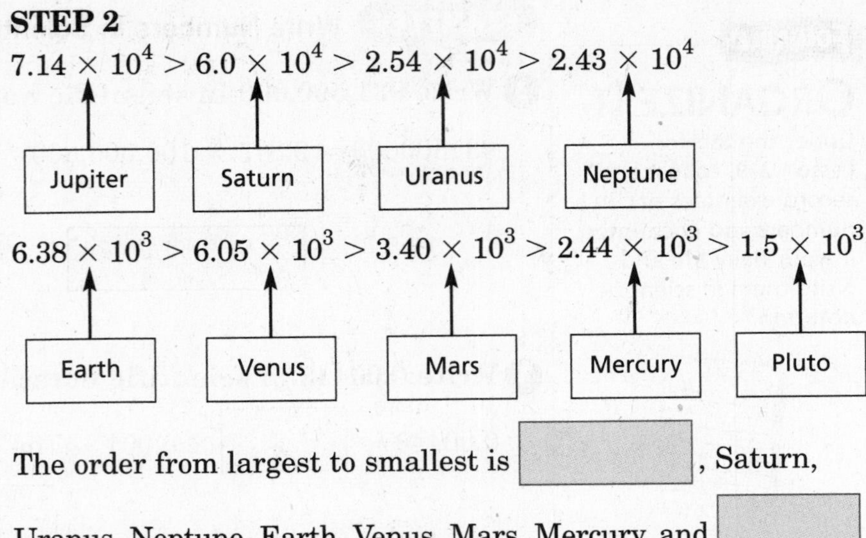

The order from largest to smallest is ______, Saturn, Uranus, Neptune, Earth, Venus, Mars, Mercury, and ______.

Your Turn **Write each number in scientific notation.**

a. 35,600,000

b. 0.000653

c. The following table lists the mass for each of the planets in our solar system. Order the planets according to mass from largest to smallest.

Planet	Mass (in tons)
Mercury	3.64×10^{20}
Venus	5.37×10^{21}
Earth	6.58×10^{21}
Mars	7.08×10^{20}
Jupiter	2.09×10^{24}
Saturn	6.25×10^{23}
Uranus	9.57×10^{22}
Neptune	1.13×10^{23}
Pluto	1.38×10^{19}

Source: nssdc.gsfc.nasa.gov

HOMEWORK ASSIGNMENT

Page(s):

Exercises:

© Glencoe/McGraw-Hill

CHAPTER 2

BRINGING IT ALL TOGETHER

STUDY GUIDE

FOLDABLES™	VOCABULARY PUZZLEMAKER	BUILD YOUR VOCABULARY
Use your **Chapter 2 Foldable** to help you study for your chapter test.	To make a crossword puzzle, word search, or jumble puzzle of the vocabulary words in Chapter 2, go to: www.glencoe.com/sec/math/t_resources/free/index.php	You can use your completed **Vocabulary Builder** *(pages 32–33)* to help you solve the puzzle.

2-1

Fractions and Decimals

Write each fraction or mixed number as a decimal.

1. $-\frac{3}{4}$ **2.**  $3\frac{1}{6}$ **3.** $-7\frac{2}{5}$

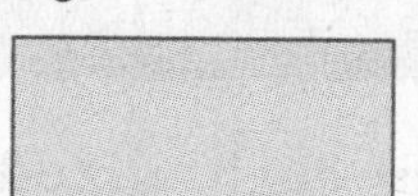

Write each decimal as a fraction or mixed number in simplest form.

4. 9.5 **5.** 0.6 **6.** 8.125

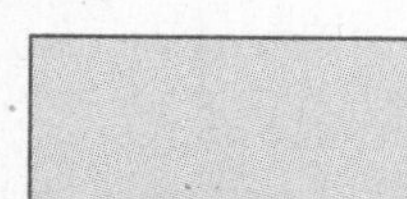

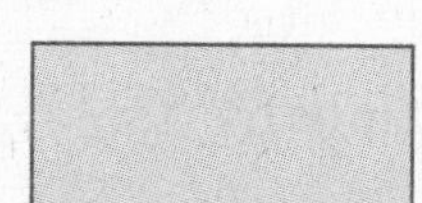

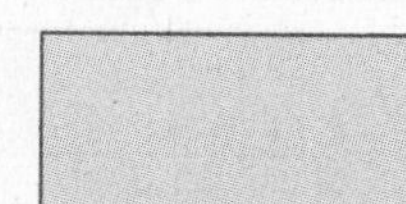

2-2

Comparing and Ordering Rational Numbers

Use <, > or = to make each sentence true.

7. $-\frac{4}{5}$ ☐ $-\frac{2}{3}$ **8.** 4.4 ☐ $4\frac{2}{5}$ **9.** 2.93 ☐ 2.93

Graph each pair of rational numbers on a number line.

10. $\frac{1}{5}, \frac{1}{3}$ **11.** $-\frac{4}{5}, -\frac{9}{10}$

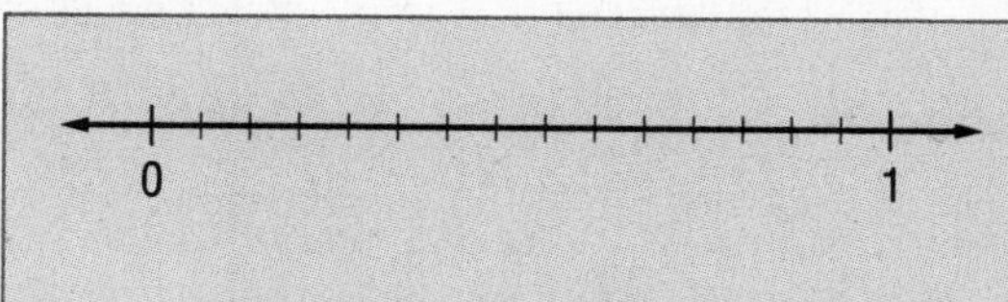

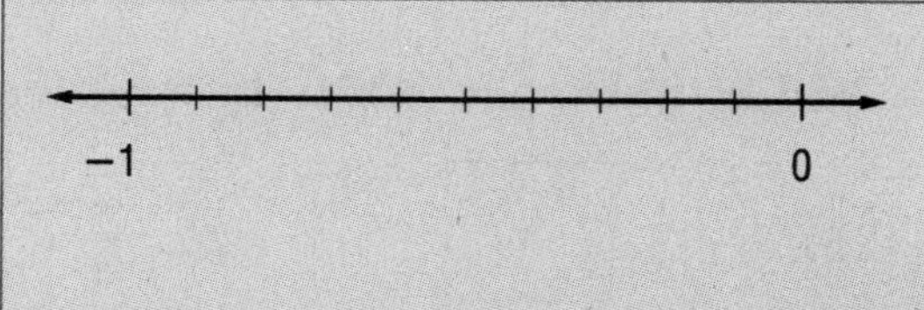

© Glencoe/McGraw-Hill

2-3

Multiplying Rational Numbers

Complete each sentence.

12. The greatest common factor of two numbers is the ______ number that is a ______ of both numbers.

13. Numerators and denominators are ______ by their greatest common factors to ______ the fraction.

Multiply. Write in simplest form.

14. $-\frac{7}{12} \cdot \frac{3}{4}$ ______

15. $4\frac{2}{3} \cdot 5\frac{1}{8}$ ______

Dividing Rational Numbers

Write the multiplicative inverse for each mixed number.

16. $2\frac{1}{5}$ ______

17. $-1\frac{3}{8}$ ______

18. $3\frac{4}{7}$ ______

Complete the sentence.

19. To divide by a ______, multiply by its ______ inverse.

20. To ______ a number by $2\frac{1}{5}$, multiply by $\frac{5}{11}$.

2-5

Adding and Subtracting Like Fractions

Determine whether each pair of fractions are like fractions.

21. $\frac{3}{5}, \frac{3}{7}$ ______

22. $\frac{5}{8}, \frac{7}{8}$ ______

23. $\frac{4}{7}, -\frac{5}{7}$ ______

24. $\frac{5}{9}, -\frac{2}{3}$ ______

Add or subtract. Write in simplest form.

25. $\frac{5}{9} - \frac{2}{9}$ ______

26. $\frac{5}{8} + \frac{7}{8}$ ______

27. $\frac{4}{7} - \frac{5}{7}$ ______

© Glencoe/McGraw-Hill

2-6 Adding and Subtracting Unlike Fractions

Find the LCD.

28. $\frac{3}{5}, \frac{3}{7}$

29. $\frac{5}{8}, \frac{7}{12}$

30. $\frac{5}{9}, \frac{-2}{3}$

Add or subtract. Write in simplest form.

31. $\frac{5}{8} - \frac{7}{12}$

32. $\frac{3}{5} + \frac{3}{7}$

2-7 Solving Equations with Rational Numbers

Match the method of solving with the appropriate equation.

33. $25a = 3.75$

34. $\frac{3}{5}m + \frac{7}{10}$

35. $r - 1.25 = 4.5$

36. $\frac{3}{5} + f = \frac{1}{2}$

a. Subtract $\frac{3}{5}$ from each side

b. Multiply each side by $\frac{5}{3}$

c. Subtract 3.75 from each side

d. Add 1.25 to each side

e. Divide each side by 1.25

2-8 Powers and Exponents

Identify the base, exponent, and power of each expression.

37. 5^4

38. x^8

Evaluate each expression.

38. 5^4

40. 6^3

41. 2^8

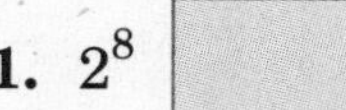

2-9 Scientific Notation

Write each number in scientific notation.

42. 8,790,000

43. 0.0000125

44. 0.00899

45. 402,500,000

© Glencoe/McGraw-Hill

CHAPTER 2

Checklist

ARE YOU READY FOR THE CHAPTER TEST?

Visit **msmath3.net** to access your textbook, more examples, self-check quizzes, and practice tests to help you study the concepts in Chapter 2.

Check the one that applies. Suggestions to help you study are given with each item.

☐ **I completed the review of all or most lessons without using my notes or asking for help.**

- You are probably ready for the Chapter Test.
- You may want take the Chapter 2 Practice Test on page 111 of your textbook as a final check.

☐ **I used my Foldable or Study Notebook to complete the review of all or most lessons.**

- You should complete the Chapter 2 Study Guide and Review on pages 108–110 of your textbook.
- If you are unsure of any concepts or skills, refer back to the specific lesson(s).
- You may also want to take the Chapter 2 Practice Test on page 111 of your text book.

☐ **I asked for help from someone else to complete the review of all or most lessons.**

- You should review the examples and concepts in your Study Notebook and Chapter 2 Foldable.
- Then complete the Chapter 2 Study Guide and Review on pages 108–110 of your textbook.
- If you are unsure of any concepts or skills, refer back to the specific lesson(s).
- You may also want to take the Chapter 2 Practice Test on page 111 of your textbook.

Student Signature

Parent/Guardian Signature

Teacher Signature

© Glencoe/McGraw-Hill

© Glencoe/McGraw-Hill

Algebra: Real Numbers and the Pythagorean Theorem

Use the instructions below to make a Foldable to help you organize your notes as you study the chapter. You will see Foldable reminders in the margin of this Interactive Study Notebook to help you in taking notes.

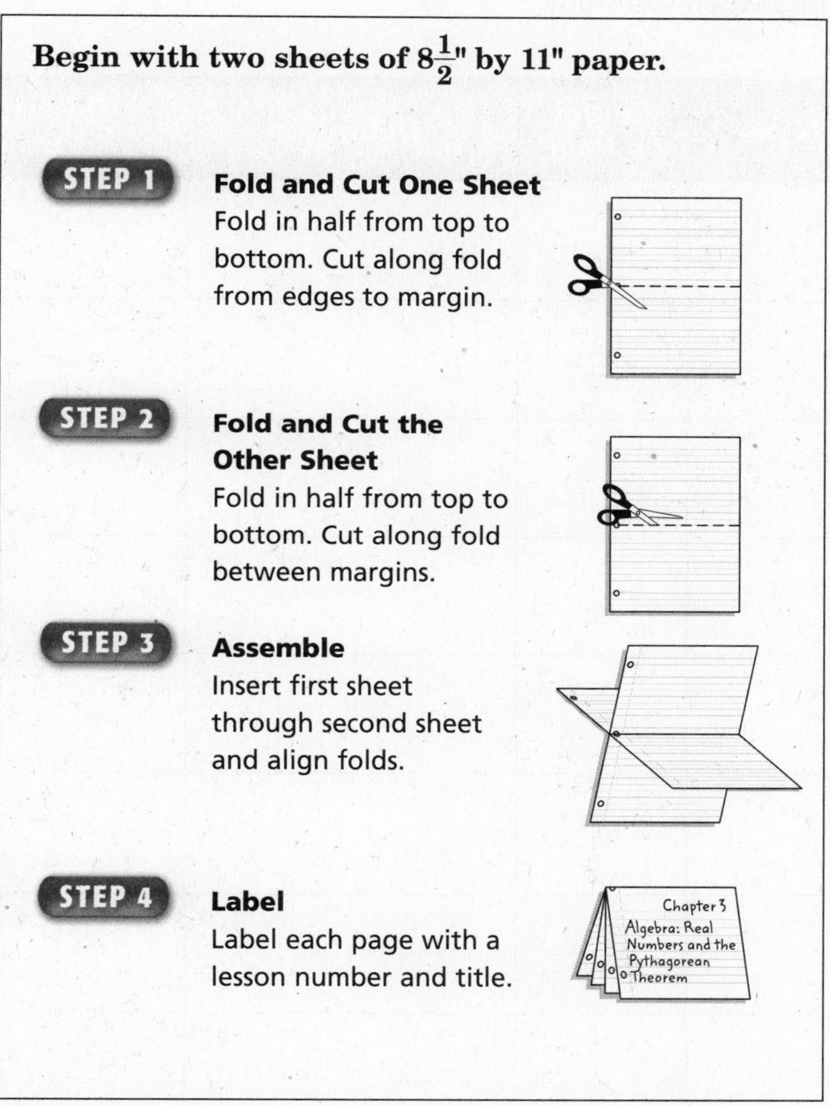

Begin with two sheets of $8\frac{1}{2}$" by 11" paper.
STEP 1
Fold and Cut One Sheet
Fold in half from top to bottom. Cut along fold from edges to margin.
STEP 2
Fold and Cut the Other Sheet
Fold in half from top to bottom. Cut along fold between margins.
STEP 3
Assemble
Insert first sheet through second sheet and align folds.
STEP 4
Label
Label each page with a lesson number and title.
Chapter 3
Algebra: Real Numbers and the Pythagorean Theorem

NOTE-TAKING TIP: When you take notes, clarify terms, record concepts, and write examples for each lesson. You may also want to list ways in which the new concepts can be used in your daily life.

© Glencoe/McGraw-Hill

BUILD YOUR VOCABULARY

This is an alphabetical list of new vocabulary terms you will learn in Chapter 3. As you complete the study notes for the chapter, you will see Build Your Vocabulary reminders to complete each term's definition or description on these pages. Remember to add the textbook page number in the second column for reference when you study.

Vocabulary Term	Found on Page	Definition	Description or Example
abscissa [ab-SIH-suh]			
converse			
coordinate plane			
hypotenuse			
irrational number			
legs			
ordered pair			
ordinate [OR-din-it]			
origin			
perfect square			

© Glencoe/McGraw-Hill

Vocabulary Term	Found on Page	Definition	Description or Example
principal square root			
Pythagorean Theorem			
Pythagorean triple			
quadrants			
radical sign			
real number			
right triangle			
square root			
x-axis			
x-coordinate			
y-axis			
y-coordinate			

© Glencoe/McGraw-Hill

3-1 Square Roots

What You'll Learn

- Find square roots of perfect squares.

BUILD YOUR VOCABULARY (pages 60–61)

Numbers such as 1, 4, 9, and 25 are called **perfect squares** because they are squares of ______ numbers.

The ______ of squaring a number is finding a **square root**.

The symbol $\sqrt{}$ is called a **radical sign** and is used to indicate the positive ______.

A ______ square root is called the **principal square root**.

EXAMPLE Find a Square Root

Key Concept

Square Root A square root of a number is one of its two equal factors.

1 **Find $\sqrt{81}$.**

$\sqrt{81}$ indicates the ______ square root of 81.

Since ______ = 81, $\sqrt{81}$ = ______.

EXAMPLE Find the Negative Square Root

2 **Find $-\sqrt{144}$.**

$-\sqrt{144}$ indicates the ______ square root of 144.

Since (−12) ______ = 144, $-\sqrt{144}$ = ______.

Your Turn **Find each square root.**

a. $\sqrt{64}$ ______

b. $-\sqrt{36}$ ______

© Glencoe/McGraw-Hill

EXAMPLE Use Square Roots to Solve an Equation

3 ALGEBRA Solve $x^2 = \frac{49}{64}$.

$x^2 = \frac{49}{64}$ Write the equation.

$\sqrt{x^2} = \sqrt{\frac{49}{64}}$ or ____ Take the square root of each side.

$x =$ ____ or $-\frac{7}{8}$ Notice that $\frac{7}{8} \cdot \frac{7}{8} =$ ____ and

$\left(-\frac{7}{8}\right)\left(-\frac{7}{8}\right) =$ ____.

The equation has two solutions, ____ and ____.

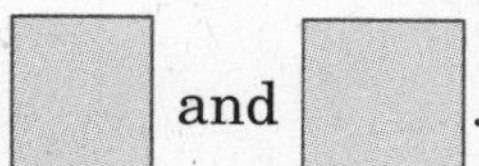

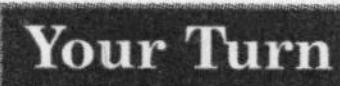

Your Turn Solve $x^2 = \frac{81}{100}$.

FOLDABLES™

ORGANIZE IT

On Lesson 3-1 of your Foldable, explain how to find the square root of a number and give an example.

Chapter 3 Algebra: Real Numbers and the Pythagorean Theorem

EXAMPLE Use an Equation to Solve a Problem

4 MUSIC The art work of the square picture in a compact disc case is approximately 14,161 mm² in area. Find the length of each side of the square.

The area is equal to the square of the length of a side. Let A = the area and let s = the length of the side $A = s^2$

$14{,}161 = s^2$ Write the equation.

____ $= \sqrt{s^2}$ Take the square root of each side.

The length of a side of a compact disc case is about ____ millimeters since distance cannot be negative.

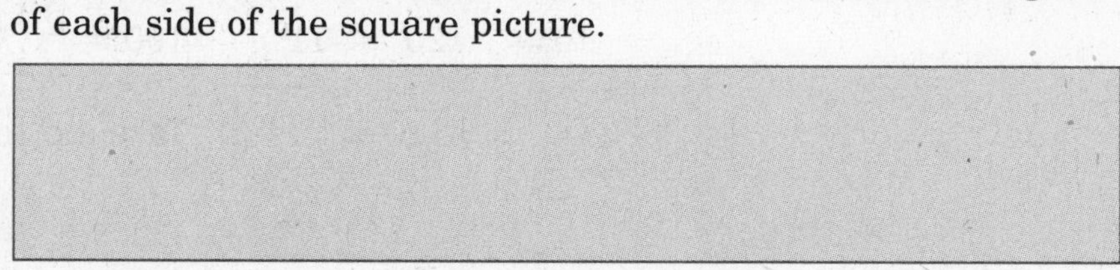

Your Turn A piece of art is a square picture that is approximately 11,025 square inches in area. Find the length of each side of the square picture.

HOMEWORK ASSIGNMENT

Page(s):

Exercises:

© Glencoe/McGraw-Hill

3–2 Estimating Square Roots

EXAMPLE Estimate Square Roots

WHAT YOU'LL LEARN

- Estimate square roots.

1 Estimate $\sqrt{54}$ to the nearest whole number.

The first perfect square less than 54 is ______.

The first perfect square greater than 54 is ______.

$49 < 54 < 64$ — Write an inequality.

______ $< 54 <$ ______ — $49 =$ ______ and $64 =$ ______

$\sqrt{7^2} < \sqrt{54} < \sqrt{8^2}$ — Take the square root of each number.

$7 < \sqrt{54} < 8$ — Simplify.

So, $\sqrt{54}$ is between ______ and ______. Since 54 is closer to 49 than 64, the best whole number estimate for $\sqrt{54}$ is ______.

EXAMPLE Estimate Square Roots

ORGANIZE IT

On Lesson 3-2 of your Foldable, explain how to estimate square roots.

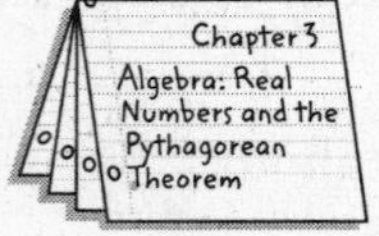

2 FINANCE If you were to invest $100 in a bank account for two years, your money would earn interest daily and be worth more when you withdrew it. If you had $120 after two years, the interest rate, written as a decimal, would be found using the expression $\frac{(\sqrt{120} - 10)}{10}$. Estimate this value.

First estimate the value of $\sqrt{120}$.

$100 < 120 < 121$ — ______ and  ______ are perfect squares.

$10^2 < 120 < 11^2$ — $100 =$ ______ and $121 =$ ______

______ $< \sqrt{120} <$ ______ — Take the square root of each number.

© Glencoe/McGraw-Hill

Since 120 is closer to ☐ than 100, the best whole number estimate for $\sqrt{120}$ is ☐. Use this to evaluate the expression.

$$\frac{(\sqrt{120} - 10)}{10} = \frac{(\square - 10)}{10} \text{ or } \square$$

The approximate interest rate is 0.10 or ☐.

Your Turn

a. Estimate $\sqrt{65}$ to the nearest whole number.

b. If you were to invest $100 in a bank account for two years, your money would earn interest daily and be worth more when you withdrew it. If you had $150 after two years, the interest rate, written as a decimal, would be found using the expression $\frac{(\sqrt{150} - 10)}{10}$.

HOMEWORK ASSIGNMENT

Page(s):

Exercises:

© Glencoe/McGraw-Hill

3–3 The Real Number System

BUILD YOUR VOCABULARY (pages 60–61)

Numbers that are not ______ are called **irrational numbers**.

The set of rational numbers and the set of ______ numbers together make up the set of **real numbers**.

WHAT YOU'LL LEARN

- Identify and classify numbers in the real number system.

KEY CONCEPT

Irrational Number An irrational number is a number that cannot be expressed as $\frac{a}{b}$, where a and b are integers and $b \neq 0$.

EXAMPLES Classify Numbers

Name all sets of numbers to which each real number belongs.

1 $\sqrt{25}$

Since $\sqrt{25}$ = ______, it is a ______ number, an ______, and a rational number.

2 $-\sqrt{12}$

Since the decimal does not repeat or ______, it is an ______ number.

3 **0.090909 . . .**

The decimal ends in a ______ pattern.

It is a ______ number because it is equivalent to ______.

Your Turn **Name all sets of numbers to which each real number belongs.**

a. $\sqrt{64}$ ______

b. 0.1010101010 . . . ______

c. $\sqrt{13}$ ______

© Glencoe/McGraw-Hill

EXAMPLE Graph Real Numbers

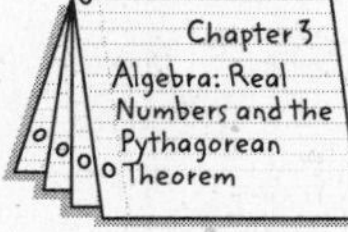

ORGANIZE IT

On Lesson 3-3 of your Foldable, summarize the properties of the real number system.

4 Estimate $\sqrt{8}$ and $-\sqrt{2}$ to the nearest tenth. Then graph $\sqrt{8}$ and $-\sqrt{2}$ on a number line.

Use a calculator to determine the approximate decimal values.

$\sqrt{8} \approx$ ______

$-\sqrt{2} \approx$ ______

Locate these points on a number line.

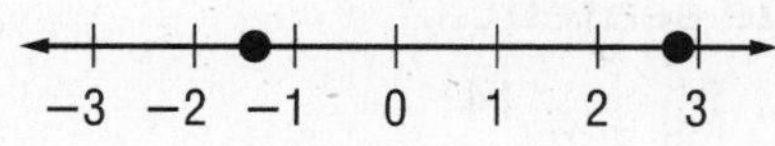

$\sqrt{8} \approx$ ___ and $-\sqrt{2} \approx$ ___.

Your Turn Estimate $\sqrt{3}$ and $-\sqrt{6}$ to the nearest tenth. Then graph $\sqrt{3}$ and $-\sqrt{6}$ on a number line.

EXAMPLES Compare Real Numbers

REMEMBER IT

Always simplify numbers before classifying them.

5 Replace each ● with <, >, or = to make a true sentence.

$3\frac{7}{8}$ ● $\sqrt{15}$

Write each number as a decimal.

$3\frac{7}{8} =$ ______ $\sqrt{15} =$ ______

Since ______ is greater than ______,

$3\frac{7}{8}$ ___ $\sqrt{15}$.

© Glencoe/McGraw-Hill

WRITE IT

Explain why you can determine that $-\sqrt{2}$ is less than 1.2 without computation.

______

______

______

______

______

______

6 $3.\overline{2}$ ● $\sqrt{10.4}$

Write $\sqrt{10.4}$ as a decimal.

$\sqrt{10.4} \approx$ ______

Since $3.\overline{2}$ is ______ than 3.224903099 . . . ,

$3.\overline{2}$ ______ $\sqrt{10.4}$.

Your Turn **Replace each ● with <, >, or = to make a true sentence.**

a. $3\frac{3}{8}$ ● $\sqrt{14}$

b. $1.\overline{5}$ ● $\sqrt{2.25}$

EXAMPLE

7 BASEBALL **The time in seconds that it takes an object to fall *d* feet is $0.25\sqrt{d}$. How many seconds would it take for a baseball that is hit 250 feet straight up in the air to fall from its highest point to the ground?**

Use a calculator to approximate the time it will take for the baseball to fall to the ground.

$0.25\sqrt{d} = 0.25$ ______ Replace *d* with ______.

≈ 3.95 or about ______ Use a calculator.

It will take about ______ for the baseball to fall to the ground.

Your Turn The time in seconds that it takes an object to fall *d* feet is $0.25\sqrt{d}$. How many seconds would it take for a baseball that is hit 450 feet straight up in the air to fall from its highest point to the ground?

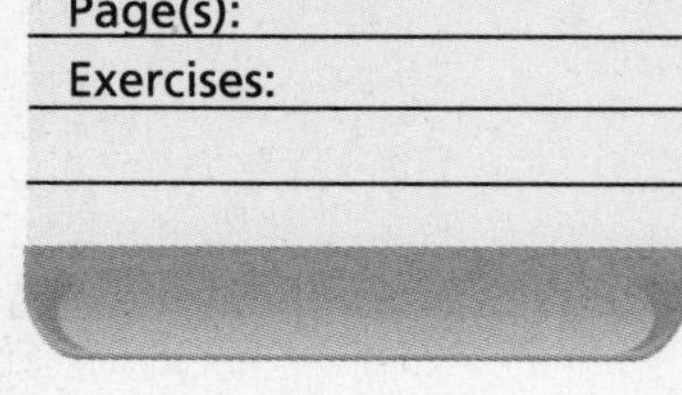

HOMEWORK ASSIGNMENT

Page(s):

Exercises:

© Glencoe/McGraw-Hill

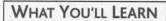

3–4 The Pythagorean Theorem

BUILD YOUR VOCABULARY (pages 60–61)

A **right triangle** is a triangle with one right angle of 90°.

The sides that form the right angle are called **legs**.

The **hypotenuse** is the side opposite the right angle.

The **Pythagorean Theorem** describes the relationship between the lengths of the legs and the hypotenuse for *any* right triangle.

WHAT YOU'LL LEARN

- Use the Pythagorean Theorem.

KEY CONCEPT

Pythagorean Theorem In a right triangle, the square of the length of the hypotenuse is equal to the sum of the squares of the lengths of the legs.

EXAMPLE Find the Length of the Hypotenuse

1 KITES **Find the length of the kite string.**

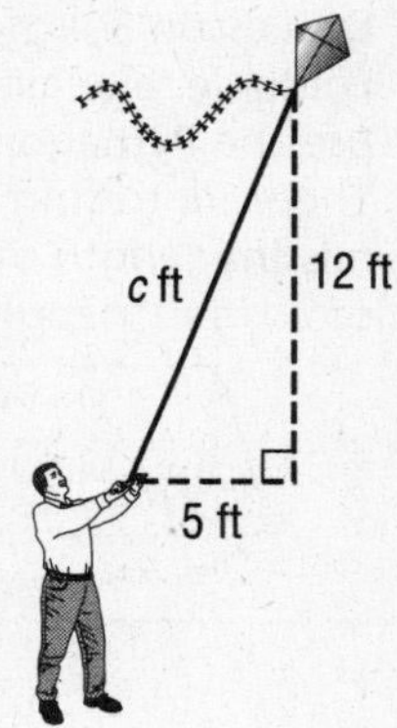

The kite string forms the hypotenuse of a right triangle. The vertical and horizontal distances form the legs.

$c^2 = a^2 + b^2$	Pythagorean Theorem
$c^2 = 5^2 +$ ______	Replace *a* with ______ and *b* with ______.
$c^2 =$ ______ + ______	Evaluate 5^2 and 12^2.
$c^2 =$ ______	Add ______ and 144.
$\sqrt{c^2} = \sqrt{169}$	Take the square root of each side.
$c =$ ______ or ______	Simplify.

The equation has two solutions, ______ and ______.

However, the length of the kite string must be positive.

The kite string is ______ feet long.

© Glencoe/McGraw-Hill

Your Turn Find the length of the kite string.

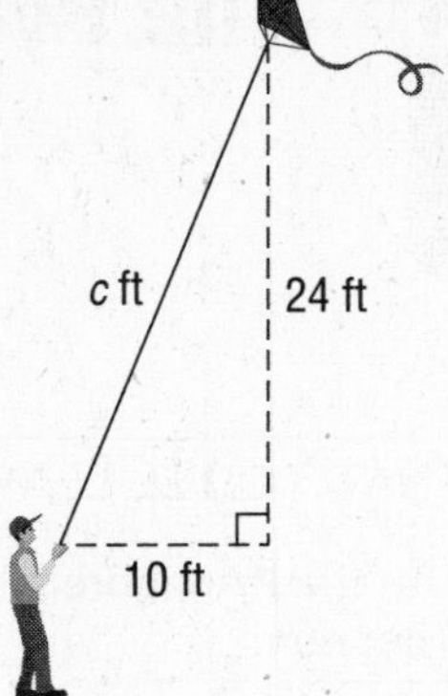

EXAMPLE Find the Length of a Leg

2 **The hypotenuse of a right triangle is 33 centimeters long and one of its legs is 28 centimeters. Find the length of the other leg.**

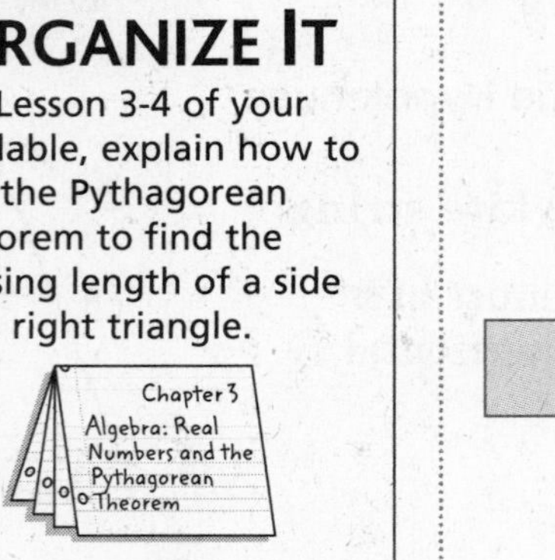

FOLDABLES

ORGANIZE IT

On Lesson 3-4 of your Foldable, explain how to use the Pythagorean Theorem to find the missing length of a side of a right triangle.

$c^2 = a^2 + b^2$	Pythagorean Theorem
$\_\_\_^2 = a^2 + \_\_\_^2$	Replace the variables.
$1{,}089 = a^2 + 784$	Evaluate each power.
$\_\_\_ - \_\_\_ = a^2 + \_\_\_ - \_\_\_$	Subtract.
$\_\_\_ = a^2$	Simplify.
$\sqrt{305} = \sqrt{a^2}$	Take the square root of each side.
$\_\_\_ \approx a$	Use a calculator.

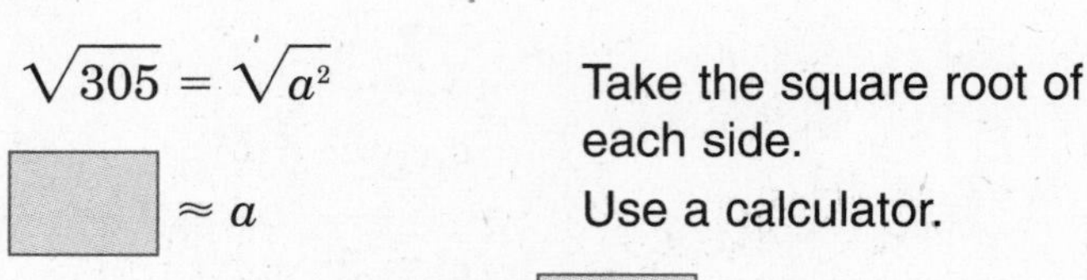

The length of the other leg is about ______ centimeters.

Your Turn The hypotenuse of a right triangle is 26 centimeters long and one of its legs is 17 centimeters. Find the length of the other leg.

REMEMBER IT

The longest side of a right triangle is the hypotenuse. Therefore, *c* represents the length of the longest side.

EXAMPLE Use the Pythagorean Theorem

3 **A 10-foot ramp is extended from the back of a truck to the ground to help movers load furniture onto the truck. If the ramp touches the ground at a point 9 feet behind the truck, how high off the ground is the top of the ramp?**

© Glencoe/McGraw-Hill

$c^2 = a^2 + b^2$	Pythagorean Theorem
$10^2 = a^2 + 9^2$	Replace *c* with 10 and *b* with 9.
$\square = a^2 + \square$	Evaluate each power.
$\square - 81 = a^2 + \square - 81$	Subtract 81 from each side.
$19 = a^2$	Simplify.
$\square = \sqrt{a^2}$	Take the square root of each side.
$\square \approx a$	Simplify.

The top of the ramp is  feet off the ground.

KEY CONCEPT

Converse of the Pythagorean Theorem If the sides of a triangle have lengths *a*, *b*, and *c* units such that $c^2 = a^2 + b^2$, then the triangle is a right triangle.

EXAMPLE **Identify a Right Triangle**

4 **The measures of three sides of a triangle are 24 inches, 7 inches, and 25 inches. Determine whether the triangle is a right triangle.**

$c^2 = a^2 + b^2$	Pythagorean Theorem
$25^2 \stackrel{?}{=} 7^2 + 24^2$	$c = 25, a = 7, b = 24$
$625 \stackrel{?}{=} \square + 576$	Evaluate 25^2, 7^2, and 24^2.
$\square = 625$	Simplify. The triangle is a right triangle.

Your Turn

a. The base of a 12-foot ladder is 5 feet from the wall. How high can the ladder reach?

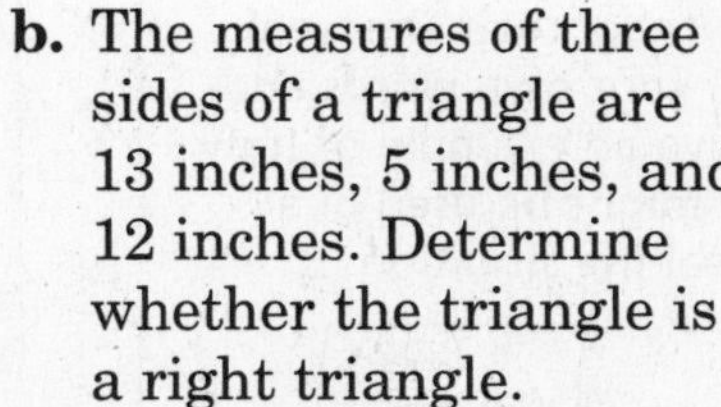

b. The measures of three sides of a triangle are 13 inches, 5 inches, and 12 inches. Determine whether the triangle is a right triangle.

HOMEWORK ASSIGNMENT

Page(s):

Exercises:

© Glencoe/McGraw-Hill

3–5 Using the Pythagorean Theorem

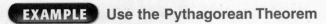

What You'll Learn

- Solve problems using the Pythagorean Theorem.

EXAMPLE Use the Pythagorean Theorem

1 RAMPS **A ramp to the entrance of a newly constructed building must be built according to accessibility guidelines stated in the Americans with Disabilities Act. If the ramp is 24.1 feet long and the top of the ramp is 2 feet off the ground, how far is the end of the ramp from the entrance?**

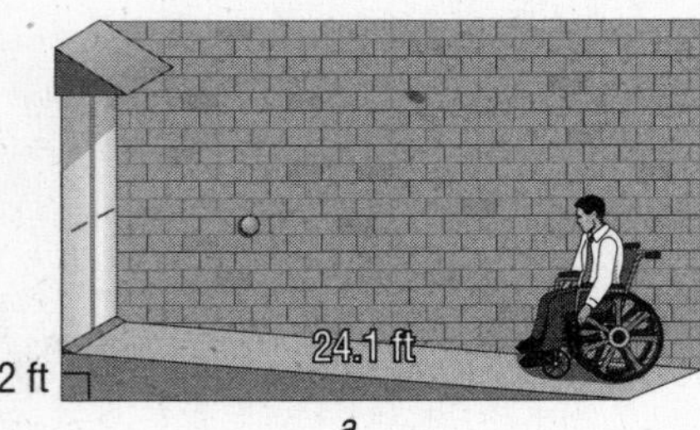

Notice the problem involves a right triangle. Use the Pythagorean Theorem.

$24.1^2 = a^2 + 2^2$	Write the equation.
$\_\_\_ = a^2 + \_\_\_$	Evaluate 24.1^2 and 2^2.
$\_\_\_ - \_\_\_ = a^2 + \_\_\_ - \_\_\_$	Subtract ___ from each side.
$\_\_\_ = a^2$	Simplify.
$\_\_\_ = \sqrt{a^2}$	Take the square root of each side.
$\_\_\_ \approx a$	Simplify.

The end of the ramp is about ______ from the entrance.

Your Turn If a truck ramp is 32 feet long and the top of the ramp is 10 feet off the ground, how far is the end of the ramp from the truck?

FOLDABLES

ORGANIZE IT

On Lesson 3-5 of your Foldable, explain the Pythagorean Theorem in your own words and give an example of how it might be used in a real-life situation.

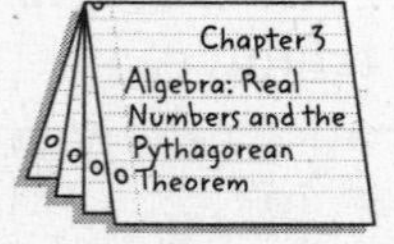

© Glencoe/McGraw-Hill

BUILD YOUR VOCABULARY (pages 60–61)

Whole numbers such as 3, 4, and 5, which satisfy the ____________, are called Pythagorean triples.

EXAMPLE Write Pythagorean Triples

2 **Multiply the triple 5-12-13 by the numbers 2, 3, 4, 5, and 10 to find more Pythagorean triples.**

You can use a table to organize your answers. Multiply each Pythagorean triple entry by the same number.

	a	*b*	*c*
original	5	12	13
× 2	10		26
× 3		36	39
× 4	20		52
× 5		60	
× 10	50		130

Each triple listed in the table is a Pythagorean triple. To verify, you can use the Pythagorean Theorem to show that each triple satisfies the equation.

Your Turn Multiply the triple 8-15-17 by the numbers 2, 3, 4, and 10 to find more Pythagorean triples.

© Glencoe/McGraw-Hill

HOMEWORK ASSIGNMENT

Page(s):

Exercises:

3-6 Distance on the Coordinate Plane

WHAT YOU'LL LEARN

- Find the distance between points on the coordinate plane.

BUILD YOUR VOCABULARY (pages 60–61)

A **coordinate plane** is formed by two number lines that form right angles and intersect at their ______ points.

The point of intersection of the two number lines is the **origin**.

The ______ number line is the ***y*-axis**.

The ______ number line is the ***x*-axis**.

The number lines separate the coordinate plane into ______ sections called **quadrants**.

Any point on the coordinate plane can be graphed by using an **ordered pair** of numbers.

The ______ number in the ordered pair is called the **x-coordinate**.

The ______ number of an ordered pair is the ***y*-coordinate**.

Another name for the ______ is **abscissa**.

Another name for the ______ is **ordinate**.

© Glencoe/McGraw-Hill

FOLDABLES™

ORGANIZE IT

On Lesson 3-6 of your Foldable, explain in writing how to use ordered pairs to find the distance between two points.

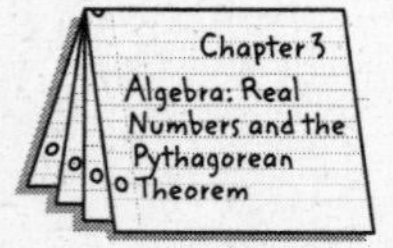

EXAMPLE Find the Distance on the Coordinate Plane

1 **Graph the ordered pairs (0, −6) and (5, −1). Then find the distance between the points.**

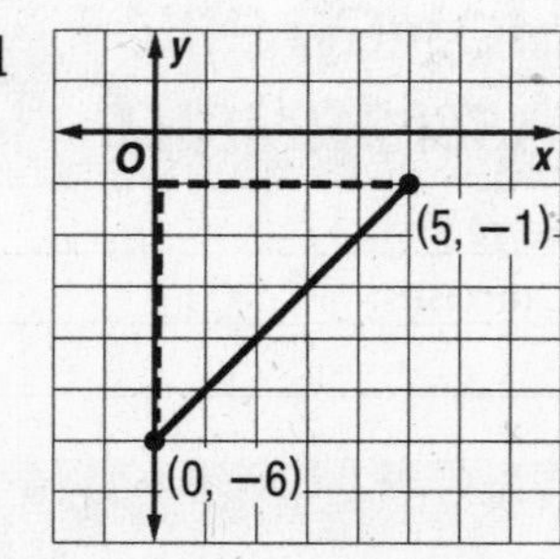

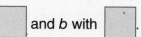

Let c = distance between the two points, $a = 5$, and $b = 5$.

$c^2 = a^2 + b^2$	Pythagorean Theorem
$c^2 = \square + \square$	Replace a with $\square$ and b with $\square$.
$c^2 = \square + \square$	Evaluate $\square$.
$c^2 = \square$	$\square$ 25 and 25.
$\sqrt{c^2} = \square$	Take the $\square$ of each side.
$c = \square$	Simplify.

The points are about $\square$ apart.

REMEMBER IT 

You can use the Pythagorean Theorem to find the distance between two points on a coordinate plane.

Your Turn Graph the ordered pairs (0, −3) and (2, −6). Then find the distance between the points.

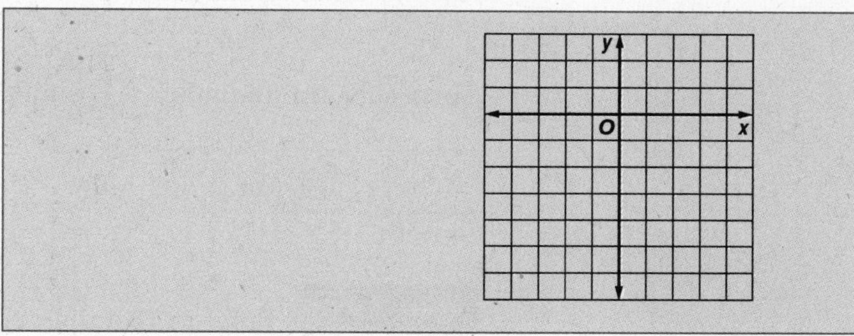

EXAMPLE **Find Distance on a Map**

2 **TRAVEL Melissa lives in Chicago. A unit on the grid of her map shown below is 0.08 mile. Find the distance between McCormickville at (−2, −1) and Lake Shore Park at (2, 2).**

© Glencoe/McGraw-Hill

Let c = the distance between McCormickville and Lake Shore Park. Then a = ____ and b = ____.

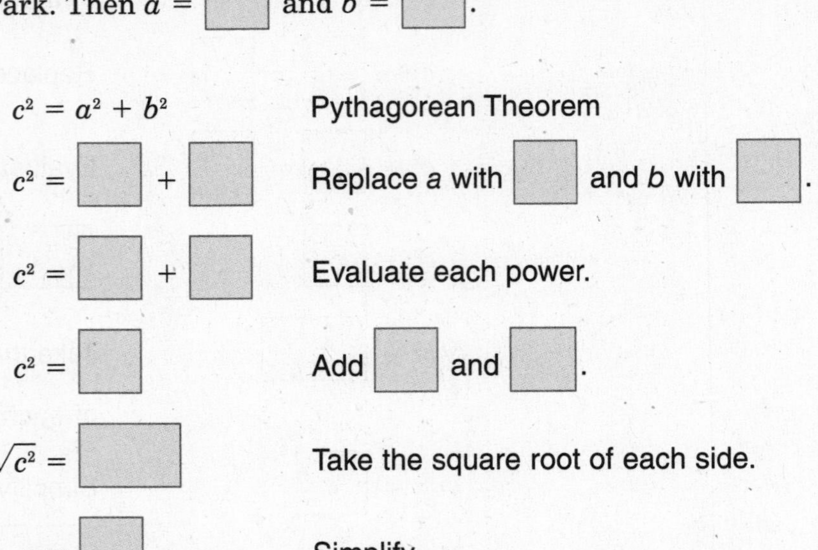

$c^2 = a^2 + b^2$	Pythagorean Theorem
c^2 = ____ + ____	Replace a with ____ and b with ____.
c^2 = ____ + ____	Evaluate each power.
c^2 = ____	Add ____ and ____.
$\sqrt{c^2}$ = ____	Take the square root of each side.
c = ____	Simplify.

The distance between McCormickville and Lake Shore Park is ____ units on the map.

Since each unit equals ____ mile, the distance is

____ × ____ or ____ mile.

Your Turn Refer to Example 2. Find the distance between Shantytown at (2, −1) and the intersection of N. Wabash Ave. and E. Superior St. At (−3, 1).

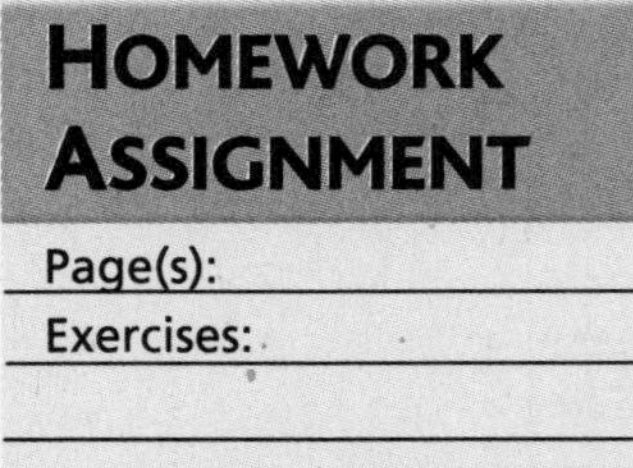

© Glencoe/McGraw-Hill

CHAPTER 3

BRINGING IT ALL TOGETHER

STUDY GUIDE

FOLDABLES	VOCABULARY PUZZLEMAKER	BUILD YOUR VOCABULARY
Use your **Chapter 3 Foldable** to help you study for your chapter test.	To make a crossword puzzle, word search, or jumble puzzle of the vocabulary words in Chapter 3, go to: www.glencoe.com/sec/math/t_resources/free/index.php	You can use your completed **Vocabulary Builder** *(pages 60–61)* to help you solve the puzzle.

3–1 Square Roots

Complete each sentence.

1. The principle square root is the ______ square root of a number.

2. To solve an equation in which one side of the square is a squared term, you can take the ______ of each side of the equation.

Find each square root.

3. $\sqrt{900}$

4. $-\sqrt{\frac{36}{49}}$

5. $-\sqrt{625}$

6. $\sqrt{\frac{25}{121}}$

3–2 Estimating Square Roots

Determine between which two consecutive whole numbers each value is located.

7. $\sqrt{23}$

8. $\sqrt{59}$

9. $\sqrt{27}$

10. $\sqrt{18}$

© Glencoe/McGraw-Hill

3–3 The Real Number System

Match the property of real numbers with the algebraic example.

11. Commutative ______ **a.** $(x + y) + z = x + (y + z)$

12. Associative ______ **b.** $pq = qp$

13. Distributive ______ **c.** $h + 0 = h$

14. Identity ______ **d.** $c + (-c) = 0$

15. Multiplicative Inverse ______ **e.** $x(y + z) = xy + xz$

f. $\frac{a}{b} \cdot \frac{b}{a} = 1$

3–4 The Pythagorean Theorem

Complete each sentence.

16. If you know the lengths of the two legs of a right triangle, you can find the length of the ______ by finding the sum of the squares of their lengths, and then finding the ______ of the sum.

17. If you know the lengths of the hypotenuse and one ______, you can find the length of the other leg by subtracting the square of the length of the known leg from the square of the length of the ______, and then finding the square root of the result.

Use the Pythagorean Theorem to determine whether each of the following measures of the sides of a triangle are the sides of a right triangle.

18. 4, 5, 6 ______ **19.** 9, 12, 15 ______

20. 10, 24, 26 ______ **21.** 5, 7, 9 

© Glencoe/McGraw-Hill

3–5

Using the Pythagorean Theorem

22. The triple 8-15-17 is a Pythagorean Triple. Complete the table to find more Pythagorean triples.

	a	b	c	Check: $c^2 = a^2 + b^2$
original	8	15	17	$289 = 64 + 225$
× 2				
× 3				
× 5				
× 10				

Determine whether each of the following is a Pythagorean triple.

23. 13-84-85

24. 11-60-61

25. 21-23-29

26. 12-25-37

3–6

Distance on the Coordinate Plane

Match each term of the coordinate plane with its description.

27. ordinate

28. y-axis

29. origin

30. abscissa

31. x-axis

a. one of four sections of the coordinate plane

b. x-coordinate

c. y-coordinate

d. vertical number line

e. horizontal number line

f. point where number lines meet

© Glencoe/McGraw-Hill

ARE YOU READY FOR THE CHAPTER TEST?

Math Online

Visit **msmath3.net** to access your textbook, more examples, self-check quizzes, and practice tests to help you study the concepts in Chapter 3.

Check the one that applies. Suggestions to help you study are given with each item.

☐ **I completed the review of all or most lessons without using my notes or asking for help.**

- You are probably ready for the Chapter Test.
- You may want to take the Chapter 3 Practice Test on page 149 of your textbook as a final check.

☐ **I used my Foldable or Study Notebook to complete the review of all or most lessons.**

- You should complete the Chapter 3 Study Guide and Review on pages 146–148 of your textbook.
- If you are unsure of any concepts or skills, refer back to the specific lesson(s).
- You may also want to take the Chapter 3 Practice Test on page 149 of your textbook.

☐ **I asked for help from someone else to complete the review of all or most lessons.**

- You should review the examples and concepts in your Study Notebook and Chapter 3 Foldable.
- Then complete the Chapter 3 Study Guide and Review on pages 146–148 of your textbook.
- If you are unsure of any concepts or skills, refer back to the specific lesson(s).
- You may also want to take the Chapter 3 Practice Test on page 149 of your textbook.

Student Signature

Parent/Guardian Signature

Teacher Signature

© Glencoe/McGraw-Hill

Proportions, Algebra, and Geometry

Use the instructions below to make a Foldable to help you organize your notes as you study the chapter. You will see Foldable reminders in the margin of this Interactive Study Notebook to help you in taking notes.

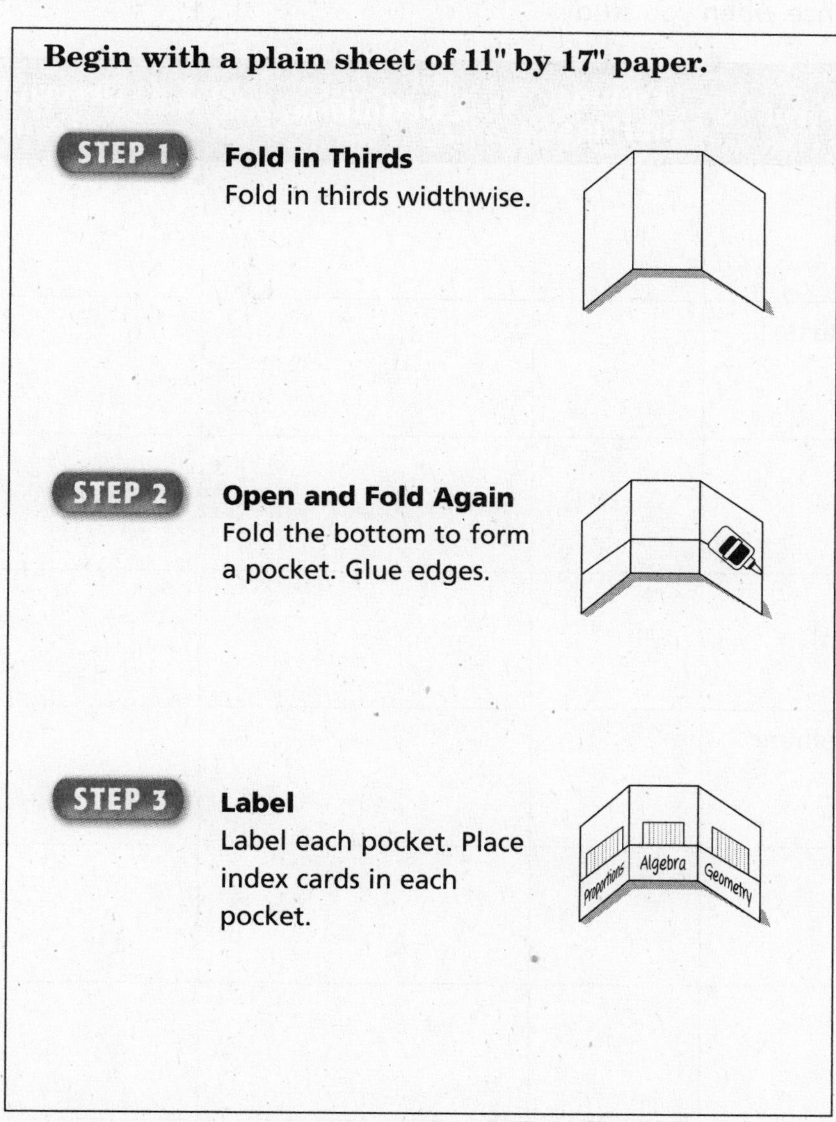

Begin with a plain sheet of 11" by 17" paper.

STEP 1 **Fold in Thirds**
Fold in thirds widthwise.

STEP 2 **Open and Fold Again**
Fold the bottom to form a pocket. Glue edges.

STEP 3 **Label**
Label each pocket. Place index cards in each pocket.

NOTE-TAKING TIP: When you take notes, define new vocabulary words, describe new ideas, and write examples that help you remember the meanings of the words and ideas.

© Glencoe/McGraw-Hill

Build Your Vocabulary

This is an alphabetical list of new vocabulary terms you will learn in Chapter 4. As you complete the study notes for the chapter, you will see Build Your Vocabulary reminders to complete each term's definition or description on these pages. Remember to add the textbook page number in the second column for reference when you study.

Vocabulary Term	Found on Page	Definition	Description or Example
congruent			
corresponding parts			
cross products			
dilation [deye-LAY-shuhn]			
indirect measurement			
polygon			
proportion			
rate			
rate of change			

© Glencoe/McGraw-Hill

Vocabulary Term	Found on Page	Definition	Description or Example
ratio			
rise			
run			
scale			
scale drawing			
scale factor			
scale model			
similar			
slope			
unit rate			

© Glencoe/McGraw-Hill

4-1 Ratios and Rates

Build Your Vocabulary (pages 82–83)

A **ratio** is a comparison of two numbers by ______.

A **rate** is a special kind of ______. It is a comparison of two quantities with different types of units.

When a rate is ______ so it has a denominator of ______, it is called a **unit rate.**

What You'll Learn

- Express ratios as fractions in simplest form and determine unit rates.

EXAMPLE Write Ratios in Simplest Form

1 Express *12 blue marbles out of 18 marbles* in simplest form.

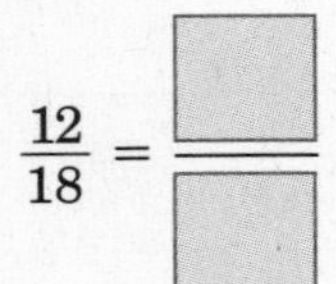

$\frac{12}{18} = \frac{\square}{\square}$ Divide the numerator and denominator by the greatest common factor, ______.

The ratio of blue marbles to marbles is ______ or ______ out of ______.

Review It

What is the greatest common factor of two or more numbers? How can you find it? *(Prerequisite Skill)*

______

______

______

______

______

______

______

EXAMPLE Find a Unit Rate

2 READING **Yi-Mei reads 141 pages in 3 hours. How many pages does she read per hour?**

Write the rate that expresses the comparison of pages to hours. Then find the unit rate.

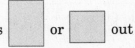

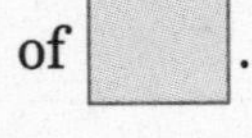

$\frac{141 \text{ pages}}{3 \text{ hours}} = \frac{\square \text{ pages}}{\square \text{ hour}}$ Divide the numerator and denominator by ______ to get a denominator of 1.

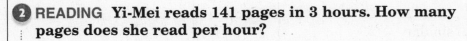

Yi-Mei reads an average of ______ pages per ______.

© Glencoe/McGraw-Hill

Your Turn **Express each ratio in simplest form.**

a. 5 blue marbles out of 20 marbles

b. 14 inches to 2 feet

c. On a trip from Columbus, Ohio, to Myrtle Beach, South Carolina, Lee drove 864 miles in 14 hours. What was Lee's average speed in miles per hour?

EXAMPLE **Compare Unit Rates**

3 SHOPPING **Alex spends $12.50 for 2 pounds of almonds and $23.85 for 5 pounds of jellybeans. Which item costs less per pound? By how much?**

For each item, write a rate that compares the cost to the amount. Then find the unit rates.

Almonds: $\frac{\$12.50}{2 \text{ pounds}} = \frac{\quad}{1 \text{ pound}}$

Jellybeans: $\frac{\$23.85}{5 \text{ pounds}} = \frac{\quad}{1 \text{ pound}}$

The almonds cost ____ per pound and the jellybeans cost

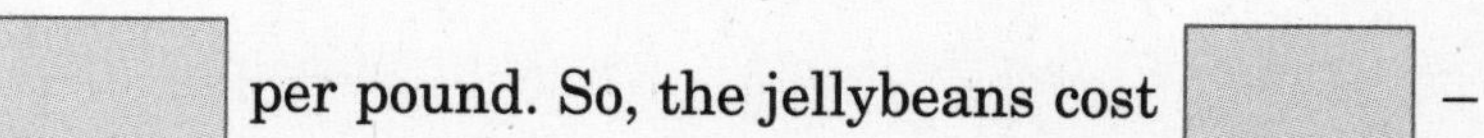

____ per pound. So, the jellybeans cost ____ – ____ or ____ per pound less than the almonds.

Your Turn Cameron spends $22.50 for 2 pounds of macadamia nuts and $31.05 for 3 pounds of cashews. Which item costs less per pound? By how much?

FOLDABLES™

ORGANIZE IT

Write the definitions of *rate* and *unit rate* on an index card. Then on the other side of the card, write examples of how to find and compare unit rates. Include these cards in your Foldable.

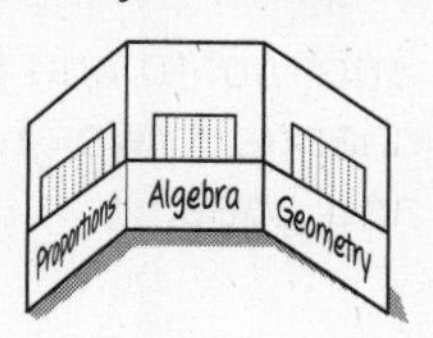

HOMEWORK ASSIGNMENT

Page(s):

Exercises:

© Glencoe/McGraw-Hill

4–2 Rate of Change

BUILD YOUR VOCABULARY (page 82)

A **rate of change** is a rate that describes how one quantity ______ in ______ to another.

WHAT YOU'LL LEARN

- Find rates of change.

EXAMPLE Find a Rate of Change

1 DOGS **The table below shows the weight of a dog in pounds between 4 and 12 months old. Find the rate of change in the dog's weight between 8 and 12 months of age.**

Age (mo)	4	8	12
Weight (lb)	15	28	43

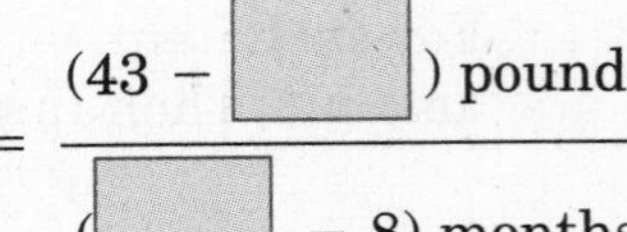

$$\frac{\text{change in weight}}{\text{change in age}} = \frac{(43 - \square)\text{ pounds}}{(\square - 8)\text{ months}}$$

The dog grew from 28 to 43 pounds from ages 8 to 12 months.

$$= \frac{\square\text{ pounds}}{\square\text{ months}}$$

Subtract to find the change in weights and ages.

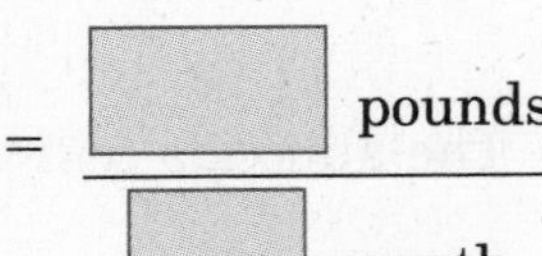

$$= \frac{\square\text{ pounds}}{\square\text{ month}}$$

Express this rate as a ______.

REMEMBER IT

Rate of change is always expressed as a unit rate.

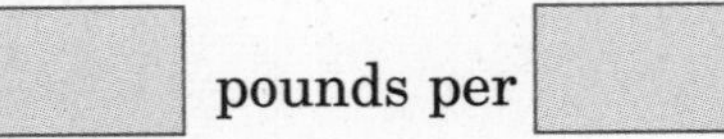

The dog grew an average of ______ pounds per ______.

Your Turn The table below shows Julia's height in inches between the ages of 6 and 11. Find the rate of change in her height between ages 6 and 9.

Age (yr)	6	9	11
Height (in.)	52	58	60

© Glencoe/McGraw-Hill

EXAMPLE Find a Negative Rate of Change

KEY CONCEPT

Rate of Change To find the rate of change, divide the difference in the *y*-coordinate by the difference in the *x*-coordinate.

FOLDABLES™

Record this concept on one side of an index card. Write an example on the other side of the card.

2 SCHOOLS The graph shows the number of students in the 8th grade between 1998 and 2002. Find the rate of change between 2000 and 2002.

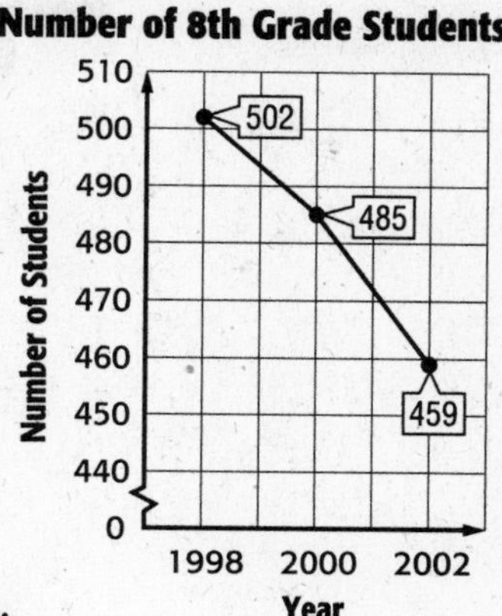

Use the formula for the rate of change.

Let $(x_1, y_1) = (2000, 485)$ and $(x_2, y_2) = (2000, 459)$.

$$\frac{y_2 - y_1}{x_2 - x_1} = \frac{\square - \square}{\square - \square}$$ Write the formula for the rate of change.

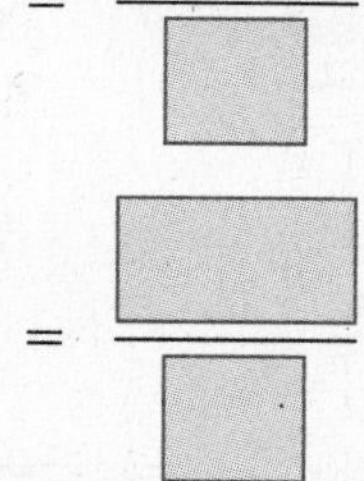

$= \frac{\square}{\square}$ Simplify.

$= \frac{\square}{\square}$ Express this as a unit rate.

The rate of change is ______ students per ______.

REMEMBER IT

Always read graphs from left to right.

Your Turn The graph below shows the number of students in the 6th grade between 1997 and 2003. Find the rate of change between 2001 and 2003.

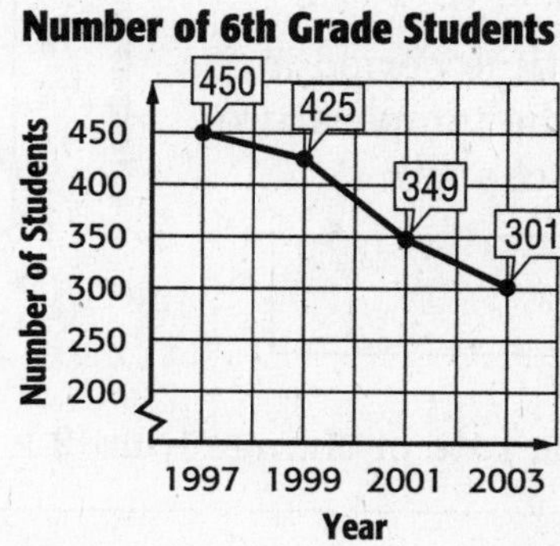

© Glencoe/McGraw-Hill

EXAMPLES Zero Rates of Change

3 TEMPERATURE The graph shows the temperature measured each hour from 10 A.M. to 3 P.M. Find a time period in which the temperature did not change.

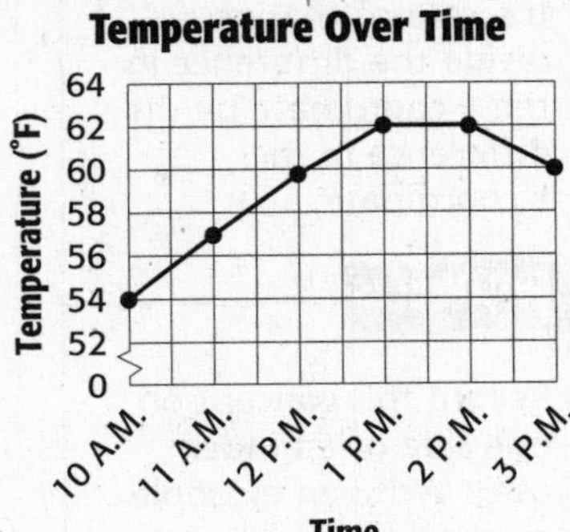

Between ______ and ______, the temperature did not change. This is shown on the graph by a ______ line.

4 TEMPERATURE Refer to Example 3. Find the rate of change from 1 P.M. to 2 P.M.

Let $(x_1, y_1) = (1, 62)$ and $(x_2, y_2) = (2, 62)$.

$\frac{y_2 - y_1}{x_2 - x_1} = \frac{\square}{\square}$ Write the formula for the rate of change.

$= \frac{\square}{\square}$ or $\square$ Simplify.

The rate of change is ______ per hour.

Your Turn The graph shows the temperature measured each hour from 10 A.M. to 4 P.M.

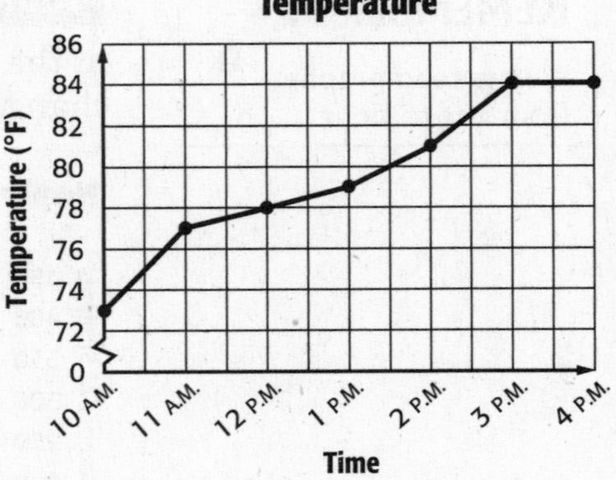

a. Find a time period in which the temperature did not change.

b. Find the rate of change from 2 P.M. to 3 P.M.

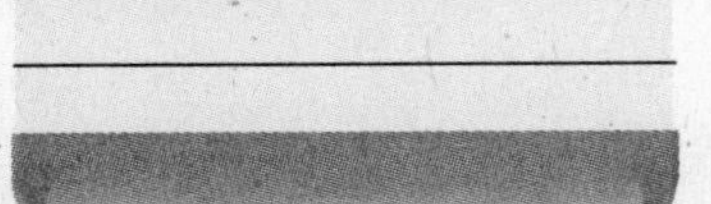

HOMEWORK ASSIGNMENT

Page(s):

Exercises:

© Glencoe/McGraw-Hill

4–3 Slope

WHAT YOU'LL LEARN

- Find the slope of a line.

BUILD YOUR VOCABULARY (page 83)

Slope is the ______ of the **rise,** or ______ change, to the **run,** or ______ change.

EXAMPLE Find Slope Using a Graph

1 **Find the slope of the line.**

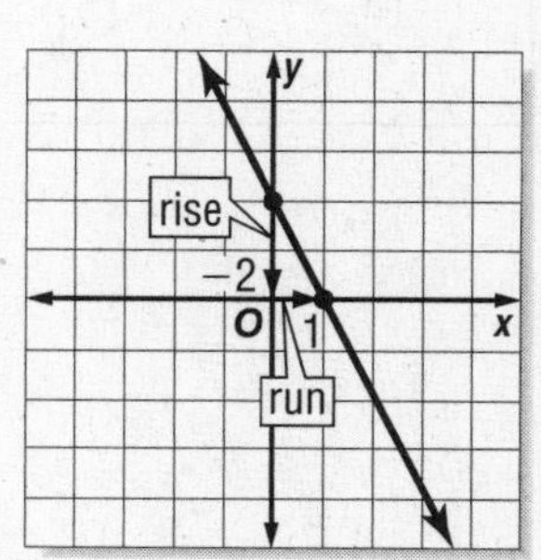

Choose two points on the line. The vertical change is ______ units while the horizontal change is ______ unit.

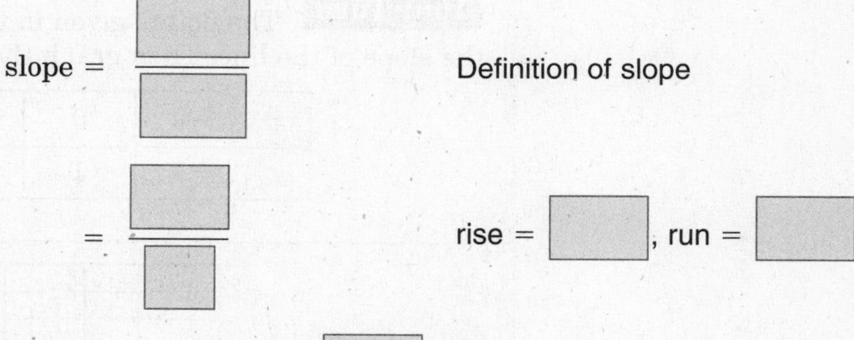

The slope of the line is ______.

REVIEW IT

What is a ratio?
(Lesson 4-1)

______

______

______

______

Your Turn Find the slope of the line graphed.

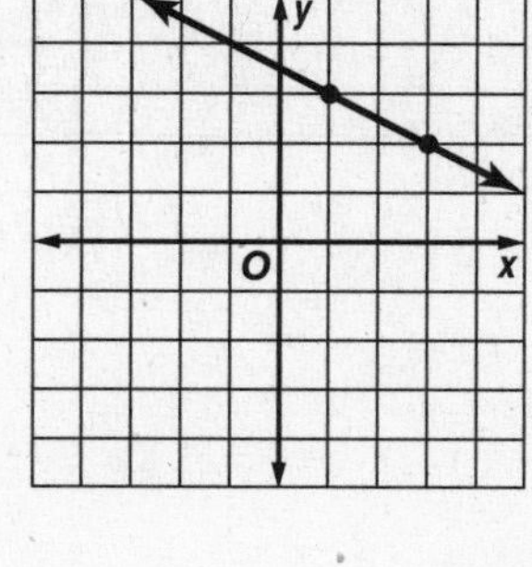

© Glencoe/McGraw-Hill

EXAMPLE Find Slope Using a Table

FOLDABLES™

ORGANIZE IT

On one side of an index card, write the definition of *slope*, including *rise* and *run*. On the other side of the card, draw a line on a coordinate grid and show with a diagram how to find the slope.

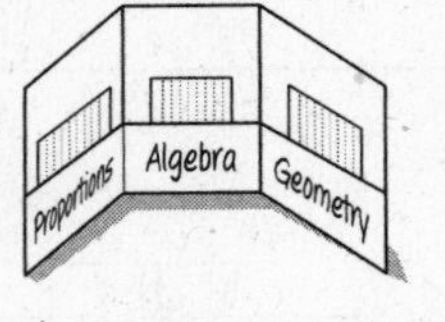

2 The points given in the table lie on a line. Find the slope of the line. Then graph the line.

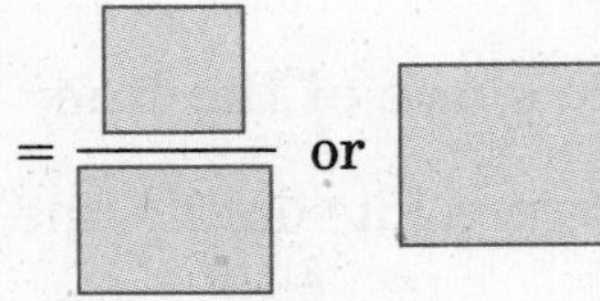

x	5	3	1	−1
y	−1	0	1	2

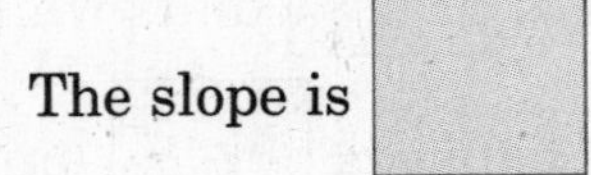

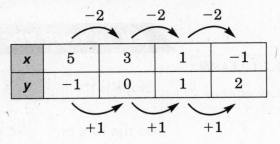

slope $= \dfrac{\text{rise}}{\text{run}}$ ⟵ change in y / ⟵ change in x

= ——— or

The slope is ____.

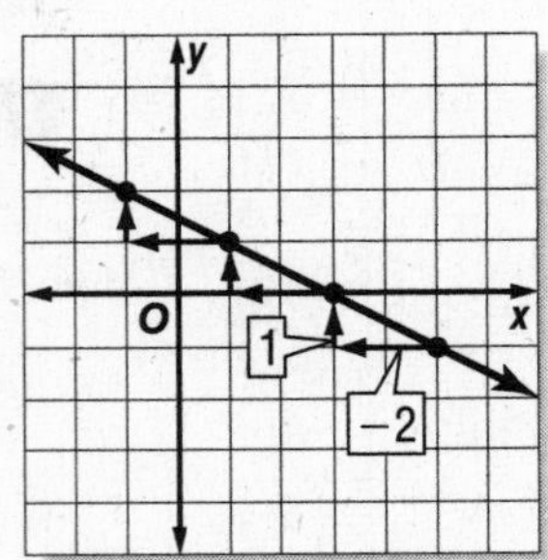

Your Turn The points given in the table lie on a line. Find the slope of the line. Then graph the line.

x	−3	0	3	6
y	−1	1	3	5

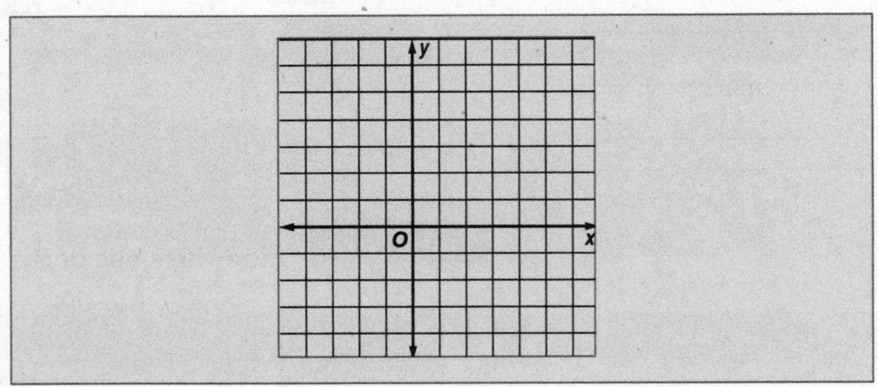

HOMEWORK ASSIGNMENT

Page(s):

Exercises:

© Glencoe/McGraw-Hill

4–4

Solving Proportions

WHAT YOU'LL LEARN

- Use proportions to solve problems.

KEY CONCEPTS

Proportion A proportion is an equation stating that two ratios are equivalent.

Property of Proportions The cross products of a proportion are equal.

FOLDABLES

Be sure to include this definition and property in your Foldable.

BUILD YOUR VOCABULARY (page 82)

In a **proportion,** two ______ are ______.

In a proportion, the **cross products** are ______.

EXAMPLE Identify a Proportion

1 **Determine whether the ratios $\frac{9}{12}$ and $\frac{18}{27}$ form a proportion.**

Find the cross products.

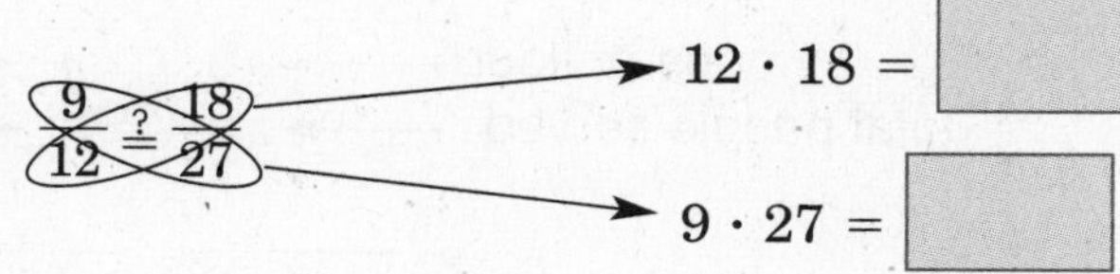

Since the cross products are not ______, the ratios do not form a ______.

EXAMPLE Solve a Proportion

2 **Solve $\frac{x}{4} = \frac{7}{20}$.**

$\frac{x}{4} = \frac{7}{20}$ Write the equation.

$x \cdot$ ______ $= 4 \cdot$ ______ Find the ______.

______ $=$ ______ Multiply.

______ $=$ ______ Divide each side by 20.

$x =$ ______ Simplify.

The solution is ______.

© Glencoe/McGraw-Hill

Your Turn

a. Determine whether the ratios $\frac{7}{21}$ and $\frac{8}{24}$ form a proportion.

b. Solve $\frac{x}{5} = \frac{11}{20}$.

EXAMPLE Use a Proportion to Solve a Problem

3 COOKING **A recipe serves 10 people and calls for 3 cups of flour. If you want to make the recipe for 15 people, how many cups of flour should you use?**

cups of flour → $\frac{3}{10} = \frac{n}{15}$ ← cups of flour
total people served → ← total people served

[ ] = [ ] Find the cross products.

$45 = 10n$ Multiply.

$\frac{45}{[\ ]} = \frac{10n}{[\ ]}$ Divide each side by [ ].

[ ] $= n$ Simplify.

You will need [ ] cups of flour to make the recipe for 15 people.

Your Turn A recipe serves 12 people and calls for 5 cups of sugar. If you want to make the recipe for 18 people, how many cups of sugar should you use?

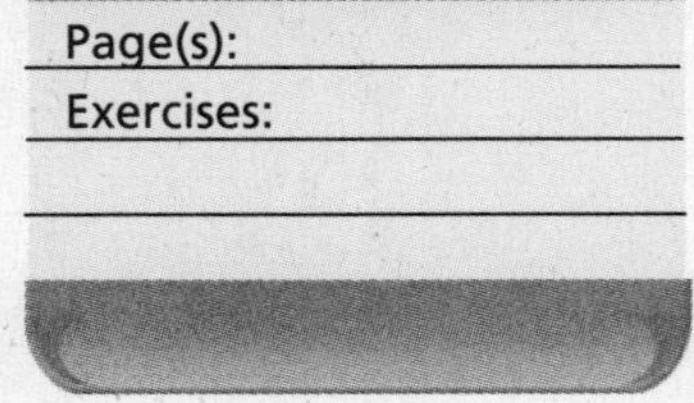

© Glencoe/McGraw-Hill

4–5 Similar Polygons

WHAT YOU'LL LEARN

- Identify similar polygons and find missing measures of similar polygons.

KEY CONCEPT

Similar Polygons If two polygons are similar, then

- their corresponding angles are congruent, or have the same measure, and
- their corresponding sides are proportional.

BUILD YOUR VOCABULARY (pages 82–83)

A **polygon** is a simple closed figure in a plane formed by ______ line segments.

Polygons that have the ______ shape are called **similar polygons.**

The parts of ______ figures that "match" are called **corresponding parts.**

Congruent means to have the ______ measure.

EXAMPLE Identify Similar Polygons

1 Determine whether triangle *DEF* is similar to triangle *HJK*. Explain your reasoning.

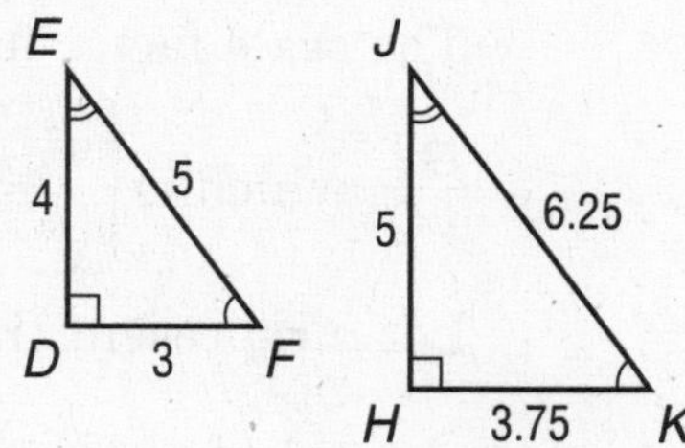

First, check to see if corresponding angles are congruent.

$\angle D \cong \angle H$, $\angle E \cong \angle J$, and $\angle F \cong \angle K$.

Next, check to see if corresponding sides are proportional.

$\frac{DE}{HJ}$ = ______ = 0.8 $\quad$ $\frac{EF}{JK}$ = ______ = 0.8

$\frac{DF}{HK}$ = ______ = 0.8

Since the corresponding angles are congruent and $\frac{4}{5} = \frac{5}{6.25} = \frac{3}{3.75}$, triangle *DEF* is ______ to triangle *HJK*.

© Glencoe/McGraw-Hill

Your Turn Determine whether triangle *ABC* is similar to triangle *TRI*. Explain your reasoning.

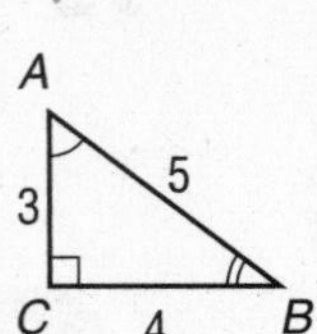

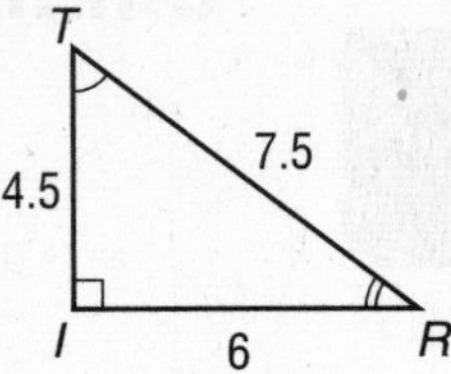

FOLDABLES

ORGANIZE IT

Make vocabulary cards for each term in this lesson. Be sure to place the cards in your Foldable.

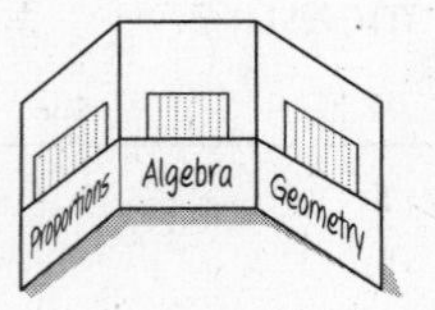

BUILD YOUR VOCABULARY (page 83)

The ______ of the lengths of two ______ sides of two similar polygons is called the **scale factor.**

EXAMPLE Find Missing Measures

2 **Given that rectangle *GHIJ* ~ rectangle *LMNO*, write a proportion to find the measure of $\overline{NO}$. Then solve.**

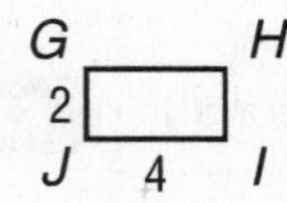

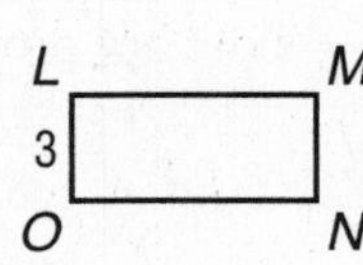

The scale factor from rectangle *GHIJ* to rectangle *LMNO* is

$\frac{GJ}{LO}$, which is 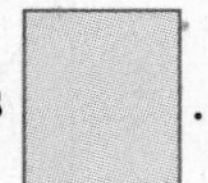. Write a proportion with this scale factor.

Let n represent the measure of $\overline{NO}$.

$\frac{IJ}{NO} = \frac{2}{3}$ — $\overline{IJ}$ corresponds to $\overline{NO}$. The scale factor is $\frac{2}{3}$.

$\frac{\square}{\square} = \frac{2}{3}$ — $IJ = \square$ and $NO = \square$

$\square \cdot 3 = \square \cdot 2$ — Find the cross products.

$\square = \square$ — Multiply.

$\square = \square$ — Divide each side by 2. Simplify.

The measure of $\overline{NO}$ is 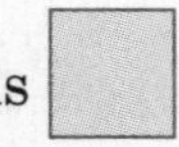.

© Glencoe/McGraw-Hill

Your Turn

Given that rectangle $ABCD \sim$ rectangle $WXYZ$, write a proportion to find the measure of $\overline{ZY}$. Then solve.

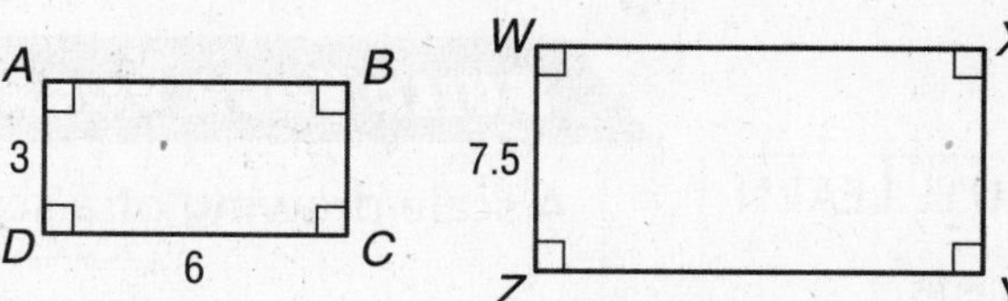

© Glencoe/McGraw-Hill

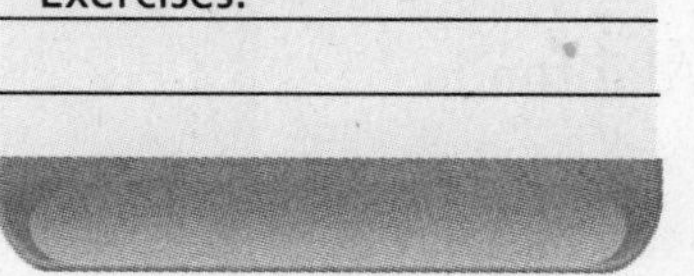

Page(s):

Exercises:

4–6 Scale Drawings and Models

WHAT YOU'LL LEARN

- Solve problems involving scale drawings.

BUILD YOUR VOCABULARY (page 83)

A **scale drawing** or a **scale model** is used to represent an object that is too ______ or too ______ to be drawn or built at actual size.

The **scale** is determined by the ______ of given length on a ______ to the corresponding actual length of the object.

EXAMPLE Find a Missing Measurement

1 MAPS **The distance from Bingston to Alanton is 1.5 inches on the map. Find the actual distance.**

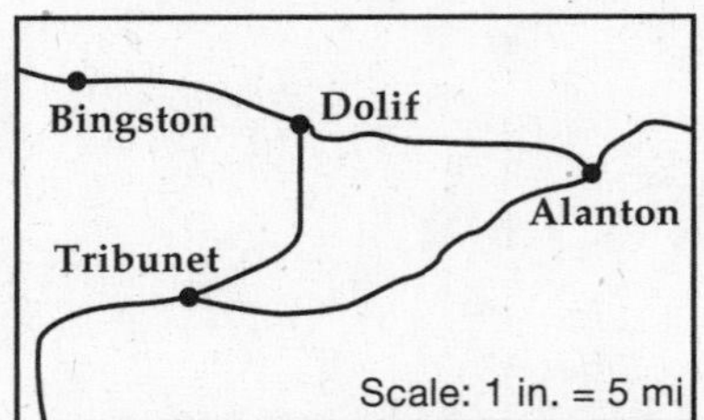

Let x represent the actual distance from Bingston to Alanton. Write and solve a proportion.

REMEMBER IT 

Scales and scale factors are usually written so that the drawing length comes first in the ratio.

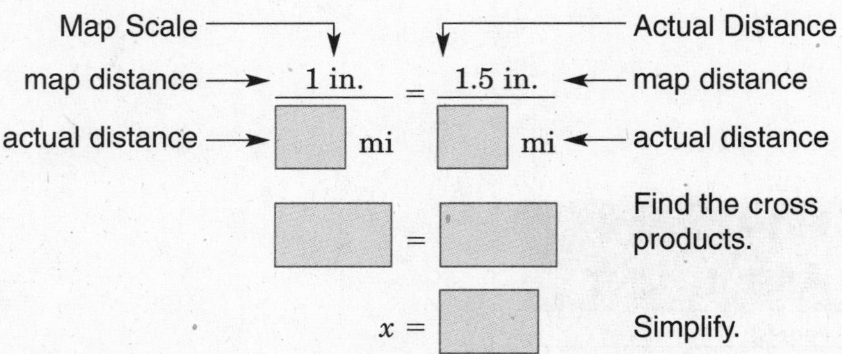

The actual distance from Bingston to Alanton is ______.

© Glencoe/McGraw-Hill

EXAMPLE Find the Scale Factor

2 MAPS Refer to Example 1. Find the scale factor for the map.

$\frac{1 \text{ in.}}{5 \text{ mi}} = \frac{1 \text{ in.}}{316{,}800 \text{ in.}}$ Convert 5 miles to inches.

The scale factor is 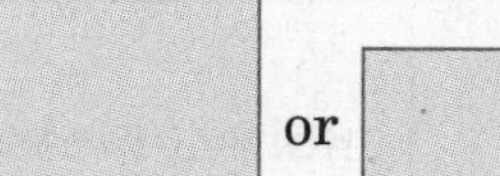 or ______.

Your Turn **The distance from Springfield to Capital City is 1.4 inches on the map.**

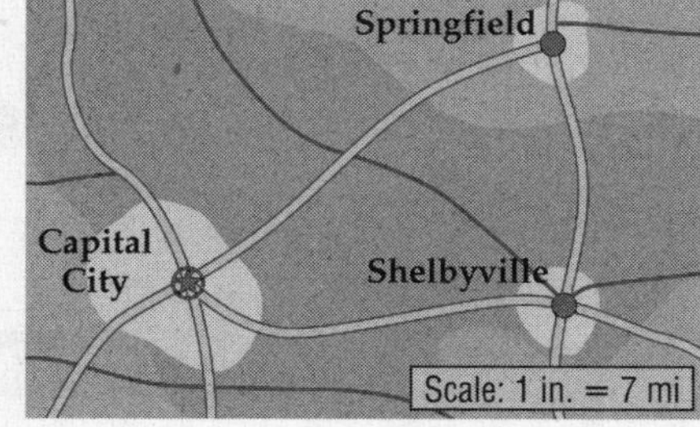

a. Find the actual distance.

b. Find the scale factor for the map.

FOLDABLES

ORGANIZE IT

Write definitions of *scale, scale drawing,* and *scale model* on cards and give your own examples. Be sure to explain how to create a scale for a scale drawing or model.

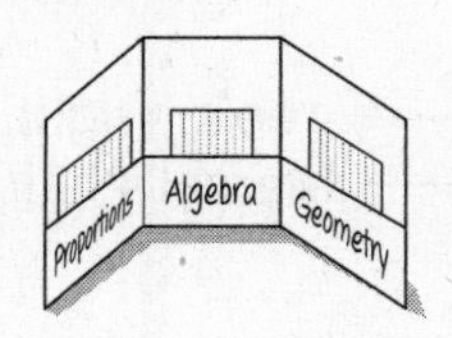

EXAMPLE Find the Scale

3 SCALE DRAWINGS A wall in a room is 15 feet long. On a scale drawing it is shown as 6 inches. What is the scale of the drawing?

Write and solve a proportion to find the scale of the drawing.

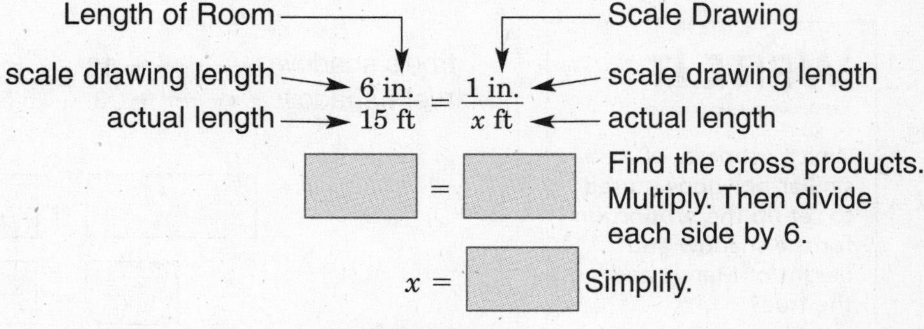

$$\frac{6 \text{ in.}}{15 \text{ ft}} = \frac{1 \text{ in.}}{x \text{ ft}}$$

______ = ______ Find the cross products. Multiply. Then divide each side by 6.

$x =$ ______ Simplify.

So, the scale is 1 inch = ______.

Your Turn The length of a garage is 24 feet. On a scale drawing the length of the garage is 10 inches. What is the scale of the drawing?

HOMEWORK ASSIGNMENT

Page(s):

Exercises:

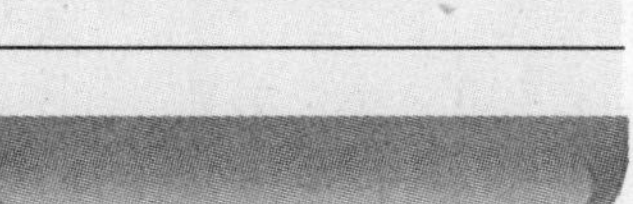

© Glencoe/McGraw-Hill

4–7 Indirect Measurement

What You'll Learn

- Solve problems involving similar triangles.

Build Your Vocabulary (page 82)

Indirect measurement uses the properties of ______ polygons and ______ to measure distance of lengths that are too ______ to measure directly.

EXAMPLE Use Shadow Reckoning

1 **TREES A tree in front of Marcel's house has a shadow 12 feet long. At the same time, Marcel has a shadow 3 feet long. If Marcel is 5.5 feet tall, how tall is the tree?**

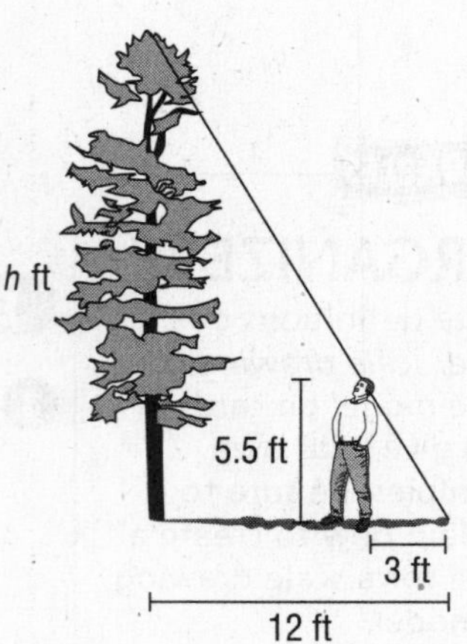

tree's shadow → $\frac{12}{3} = \frac{h}{5.5}$ ← tree's height
Marcel's shadow → ← Marcel's height

______ = ______ Find the cross products.

______ = ______ Multiply.

$\frac{\_\_\_\_}{\_\_\_\_} = \frac{\_\_\_\_}{\_\_\_\_}$ Divide each side by ______.

______ $= h$ Simplify.

The tree is ______ feet tall.

Write It

Which property of similar polygons is used to set up the proportion for the shadow and height of Marcel and the tree?

______

______

______

______

© Glencoe/McGraw-Hill

Your Turn Jayson casts a shadow that is 10 feet. At the same time, a flagpole casts a shadow that is 40 feet. If the flagpole is 20 feet tall, how tall is Jayson?

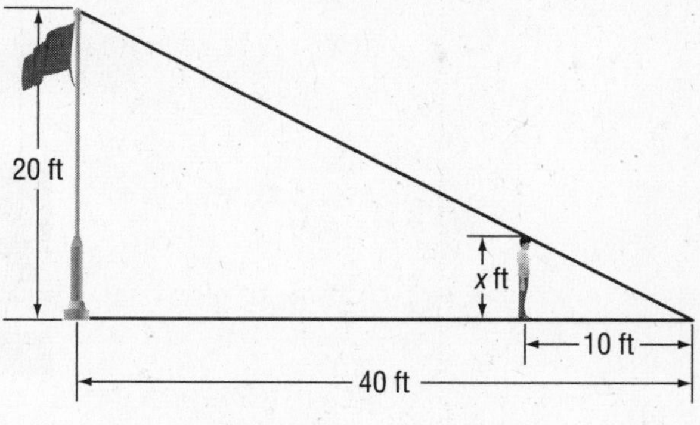

EXAMPLE Use Indirect Measurement

2 SURVEYING The two triangles shown in the figure are similar. Find the distance *d* across the stream.

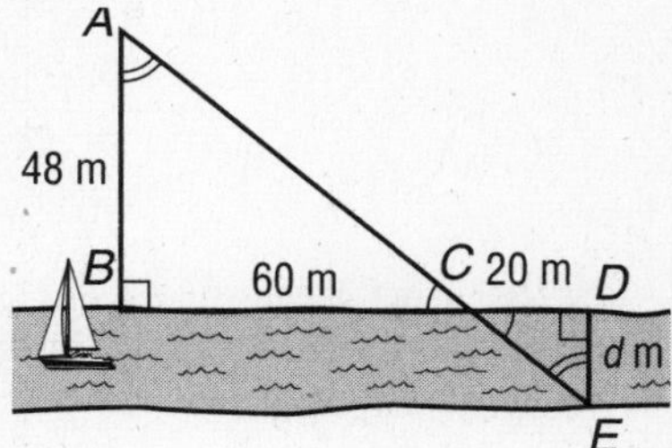

In this figure $\triangle ABC \sim \triangle EDC$.

So, $\overline{AB}$ corresponds to $\overline{ED}$, and $\overline{BC}$ corresponds to ______.

$\frac{AB}{ED} = \frac{BC}{DC}$ Write a ______.

______ = ______ $AB = 48$, $ED = d$, $BC = 60$, and $DC = 20$

______ = ______ Find the cross products.

______ = ______ Multiply. Then divide each side by ______.

______ $= d$ Simplify.

The distance across the stream is ______.

FOLDABLES

ORGANIZE IT

Include a definition of *indirect measurement*. Also include an explanation of how to use indirect measurement with your own words or sketch.

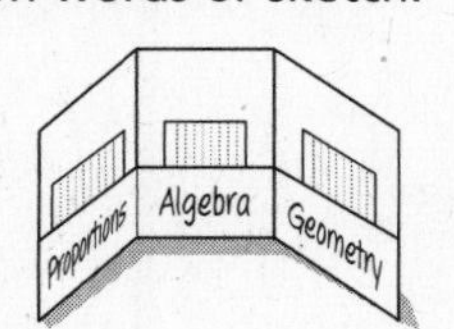

© Glencoe/McGraw-Hill

Your Turn The two triangles shown in the figure are similar. Find the distance d across the river.

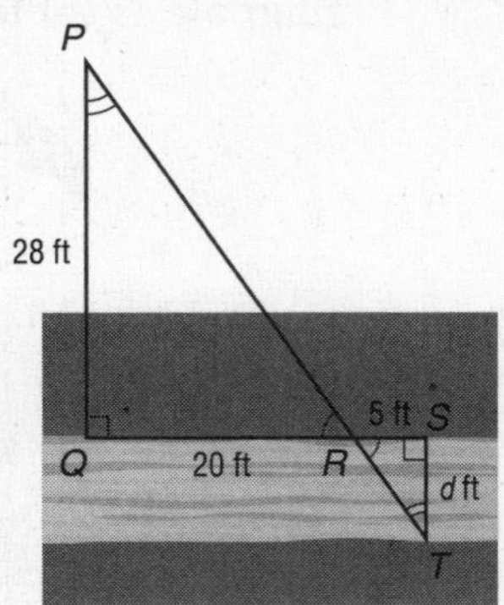

HOMEWORK ASSIGNMENT

Page(s):

Exercises:

© Glencoe/McGraw-Hill

4-8 Dilations

BUILD YOUR VOCABULARY (page 82)

The image produced by ______ or reducing a ______ is called a **dilation.**

WHAT YOU'LL LEARN

- Graph dilations on a coordinate plane.

REVIEW IT

What is a scale factor of similar polygons? *(Lesson 4-5)*

______

______

______

______

______

EXAMPLE Graph a Dilation

1 Graph $\triangle MNO$ with vertices $M(3, -1)$, $N(2, -2)$, and $O(0, 4)$. Then graph its image $\triangle M'N'O'$ after a dilation with a scale factor of $\frac{3}{2}$.

To find the vertices of the dilation, multiply each coordinate in the ordered pairs by $\frac{3}{2}$. Then graph both images on the same axes.

$M(3, -1) \rightarrow$ ______ $\rightarrow M'\left(\frac{9}{2}, -\frac{3}{2}\right)$

$N(2, -2) \rightarrow \left(2 \cdot \frac{3}{2}, -2 \cdot \frac{3}{2}\right) \rightarrow N'$ ______

$O(0, 4) \rightarrow$ ______ $\rightarrow O'$ ______

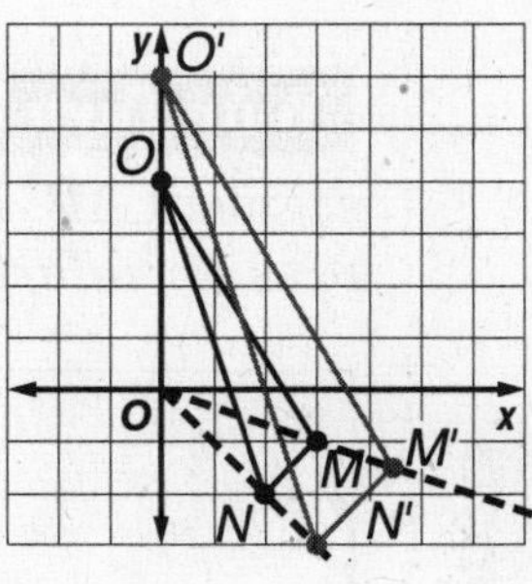

CHECK Draw lines through the origin and each of the vertices of the original figure. The vertices of the dilation should lie on those same lines.

© Glencoe/McGraw-Hill

Your Turn Graph ΔJKL with vertices $J(2, 4)$, $K(4, -6)$, and $L(0, -4)$. Then graph its image $\Delta J'K'L'$ after a dilation with a scale factor of $\frac{1}{2}$.

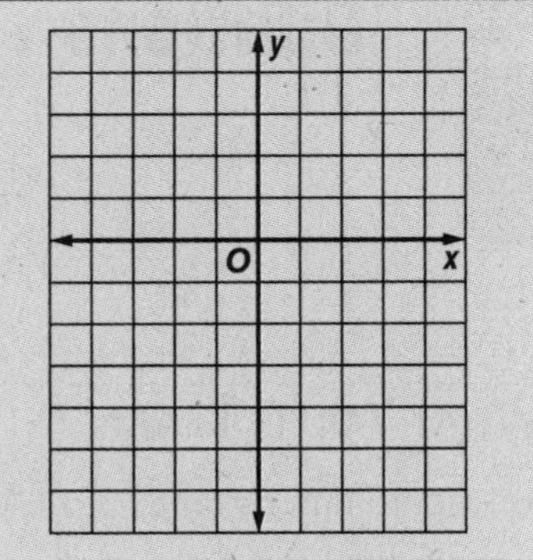

EXAMPLE Find and Classify a Scale Factor

REMEMBER IT

- If the scale factor is between 0 and 1, the dilation is a reduction.
- If the scale factor is greater than 1, the dilation is an enlargement.
- If the scale factor is equal to 1, the dilation is the same size as the original figure.

2 In the figure, segment $X'Y'$ is a dilation of segment XY. Find the scale factor of the dilation, and classify it as an enlargement or as a reduction.

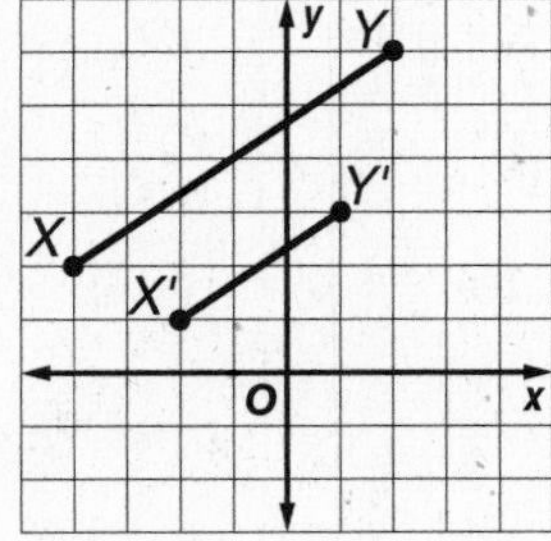

Write a ratio of the x- or y-coordinate of one vertex of the dilation to the x- or y-coordinate of the corresponding vertex of the original figure. Use the y-coordinates of $X(-4, 2)$ and $X'(-2, 1)$.

$$\frac{y\text{-coordinate of } X'}{y\text{-coordinate of } X} = \square$$

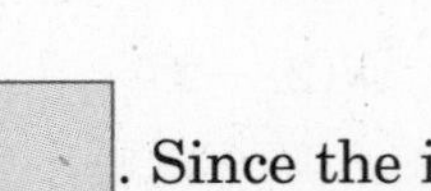

The scale factor is ▭. Since the image is smaller than the original figure, the dilation is a ▭.

Your Turn In the figure, segment $A'B'$ is a dilation of segment AB. Find the scale factor of the dilation, and classify it as an *enlargement* or as a *reduction*.

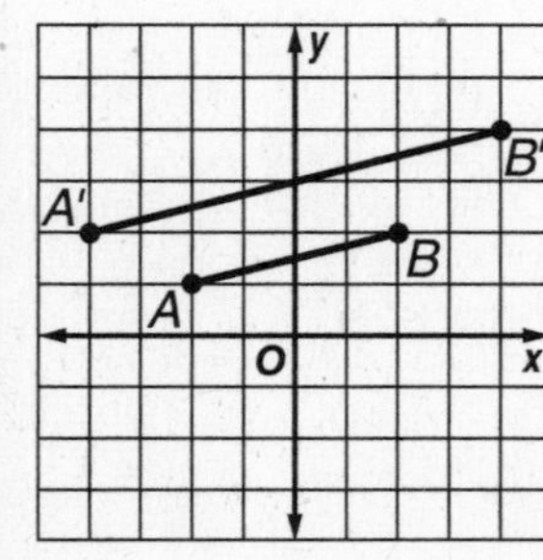

HOMEWORK ASSIGNMENT

Page(s):

Exercises:

© Glencoe/McGraw-Hill

CHAPTER 4

BRINGING IT ALL TOGETHER

STUDY GUIDE

FOLDABLES™	VOCABULARY PUZZLEMAKER	BUILD YOUR VOCABULARY
Use your **Chapter 4 Foldable** to help you study for your chapter test.	To make a crossword puzzle, word search, or jumble puzzle of the vocabulary words in Chapter 4, go to: www.glencoe.com/sec/math/t_resources/free/index.php	You can use your completed **Vocabulary Builder** *(pages 82–83)* to help you solve the puzzle.

4-1

Ratios and Rates

Match each phrase with the term they describe.

1. a comparison of two numbers

2. a comparison of two quantities with different types of units

3. a rate that is simplified so it has a denominator of 1

a. unit rate
b. numerator
c. ratio
d. rate

4. Express 12 wins to 14 losses in simplest form.

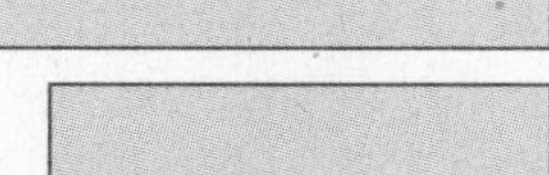

5. Express 6 inches of rain in 4 hours as a unit rate.

4-2

Rate of Change

Use the table shown to answer each question.

Week	Bicycles Sold
2	2
4	14
6	14
8	12

6. Find the rate of change in the number of bicycles sold between weeks 2 and 4.

7. Between which weeks is the rate of change negative?

© Glencoe/McGraw-Hill

4-3

Slope

8. Complete the statement.

Slope is the ratio of the vertical ________ between two points to the ________ change between the points.

9. The points given in the table lie on a line. Find the slope of the line.

x	−4	−2	2	4
y	2	1	−1	−2

4-4

Solving Proportions

10. Do the ratios $\frac{a}{b}$ and $\frac{c}{d}$ always form a proportion? Why or why not?

Solve each proportion.

11. $\frac{7}{b} = \frac{35}{5}$

12. $\frac{a}{16} = \frac{3}{8}$

13. $\frac{4}{13} = \frac{3}{c}$

4-5

Similar Polygons

14. If two polygons have corresponding angles that are congruent, does that mean that the polygons are similar? Why or why not?

15. Rectangle *ABCD* has side lengths of 30 and 5. Rectangle *EFGH* has side lengths of 15 and 3. Determine whether the rectangles are similar.

© Glencoe/McGraw-Hill

4-6 Scale Drawings and Models

16. The scale on a map is 1 inch = 20 miles.

Find the actual distance for the map distance of $\frac{5}{8}$ inch.

17. What is the scale factor for a model if part of the model that is 4 inches corresponds to a real-life object that is 16 inches.

4-7 Indirect Measurement

18. Complete the following sentence.

When you solve a problem using shadow reckoning, the objects being compared and their shadows form two sides of ______ triangles.

19. **STATUE** If a statue casts a 6-foot shadow and a 5-foot mailbox casts a 4-foot shadow, how tall is the statue?

4-8 Dilations

20. If you are given the coordinates of a figure and the scale factor of a dilation of that figure, how can you find the coordinates of the new figure?

21. Complete the table.

If the scale factor is	Then the dilation is
between 0 and 1	
greater than 1	
equal to 1	

© Glencoe/McGraw-Hill

ARE YOU READY FOR THE CHAPTER TEST?

Visit **msmath3.net** to access your textbook, more examples, self-check quizzes, and practice tests to help you study the concepts in Chapter 4.

Check the one that applies. Suggestions to help you study are given with each item.

☐ **I completed the review of all or most lessons without using my notes or asking for help.**

- You are probably ready for the Chapter Test.
- You may want to take the Chapter 4 Practice Test on page 201 of your textbook as a final check.

☐ **I used my Foldable or Study Notebook to complete the review of all or most lessons.**

- You should complete the Chapter 4 Study Guide and Review on pages 198–200 of your textbook.
- If you are unsure of any concepts or skills, refer to the specific lesson(s).
- You may also want to take the Chapter 4 Practice Test on page 201.

☐ **I asked for help from someone else to complete the review of all or most lessons.**

- You should review the examples and concepts in your Study Notebook and Chapter 4 Foldable.
- Then complete the Chapter 4 Study Guide and Review on pages 198–200 of your textbook.
- If you are unsure of any concepts or skills, refer to the specific lesson(s).
- You may also want to take the Chapter 4 Practice Test on page 201.

Student Signature

Parent/Guardian Signature

Teacher Signature

© Glencoe/McGraw-Hill

Percent

Use the instructions below to make a Foldable to help you organize your notes as you study the chapter. You will see Foldable reminders in the margin of this Interactive Study Notebook to help you in taking notes.

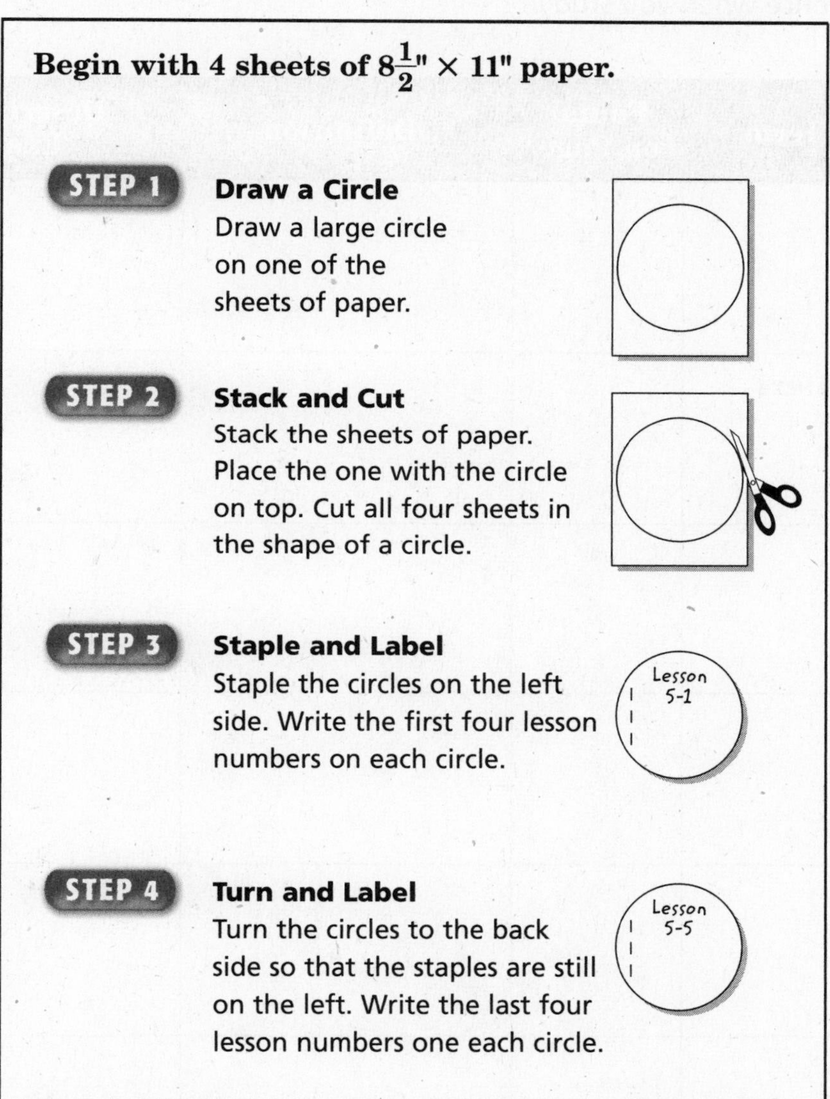

Begin with 4 sheets of $8\frac{1}{2}$" × 11" paper.

STEP 1 **Draw a Circle**
Draw a large circle on one of the sheets of paper.

STEP 2 **Stack and Cut**
Stack the sheets of paper. Place the one with the circle on top. Cut all four sheets in the shape of a circle.

STEP 3 **Staple and Label**
Staple the circles on the left side. Write the first four lesson numbers on each circle.

STEP 4 **Turn and Label**
Turn the circles to the back side so that the staples are still on the left. Write the last four lesson numbers one each circle.

NOTE-TAKING TIP: When you take notes, it may help to create a visual representation, such as a drawing or a chart, to organize the information you learn. When you use a visual, be sure to clearly label it.

© Glencoe/McGraw-Hill

BUILD YOUR VOCABULARY

This is an alphabetical list of new vocabulary terms you will learn in Chapter 5. As you complete the study notes for the chapter, you will see Build Your Vocabulary reminders to complete each term's definition or description on these pages. Remember to add the textbook page number in the second column for reference when you study.

Vocabulary Term	Found on Page	Definition	Description or Example
base			
compatible numbers			
discount			
interest			
markup			
part			
percent			

© Glencoe/McGraw-Hill

Vocabulary Term	Found on Page	Definition	Description or Example
percent equation			
percent of change			
percent of decrease			
percent of increase			
percent proportion			
principal			
selling price			

© Glencoe/McGraw-Hill

5-1 Ratios and Percents

BUILD YOUR VOCABULARY (page 108)

[ ] such as 27 out of 100 or 8 out of 25 can be written as **percents**.

WHAT YOU'LL LEARN

- Write ratios as percents and vice versa.

KEY CONCEPT

Percent A percent is a ratio that compares a number to 100.

EXAMPLES Write Ratios as Percents

1 POPULATION **According to the 2000 U.S. Census Bureau, 13 out of every 100 people living in Delaware were 65 or older. Write this ratio as a percent.**

13 out of every [ ] = 13%

2 BASEBALL **In 2001, Manny Ramirez got on base 40.5 times for every 100 times he was at bat. Write this ratio as a percent.**

40.5 out of [ ] = 40.5%

Your Turn **Write each ratio as a percent.**

a. 59 out of 100

b. 68 out of 100

EXAMPLES Write Ratios and Fractions as Percents

3 TRANSPORTATION **About 4 out of 5 commuters in the United States drive or carpool to work. Write this ratio as a percent.**

$\times$ [ ]

$\frac{4}{5} = \frac{80}{100}$

$\times$ [ ]

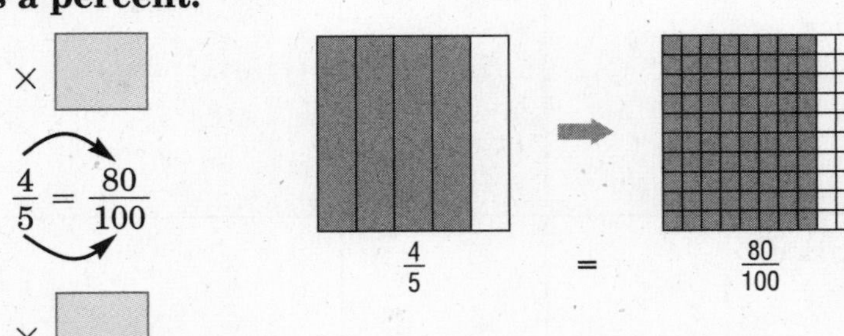

So, [ ] out of [ ] equals [ ].

© Glencoe/McGraw-Hill

FOLDABLES

ORGANIZE IT

Write in words and symbols what you've learned about expressing ratios as percents.

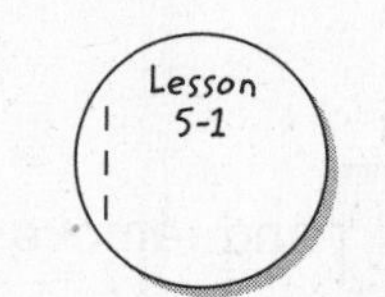

4 INTERNET In 2000, about $\frac{3}{200}$ of the population in Peru used the Internet. Write this fraction as a percent.

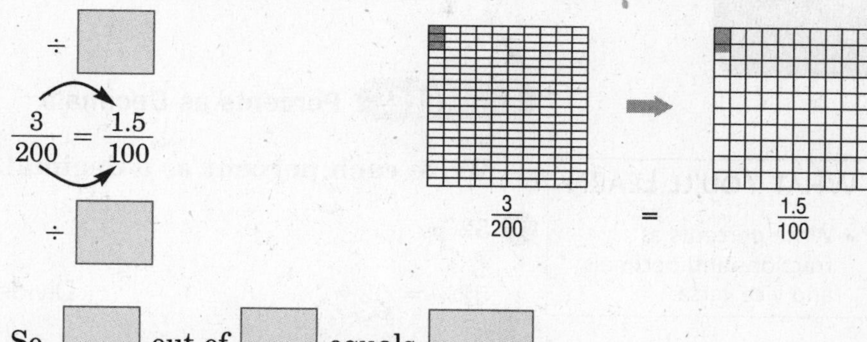

So, ___ out of ___ equals ___.

Your Turn **Write each ratio or fraction as a percent.**

a. 3 out of 5

b. $\frac{122}{200}$ of teens

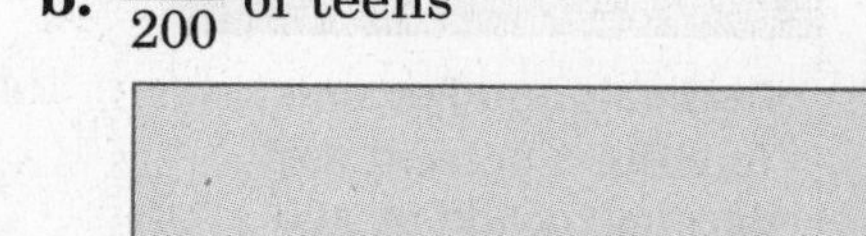

EXAMPLE **Write Percents as Fractions**

5 SCHEDULE The circle graph shows an estimate of the percent of his day that Peter spends on each activity. Write the percents for eating and sleeping as fractions in simplest form.

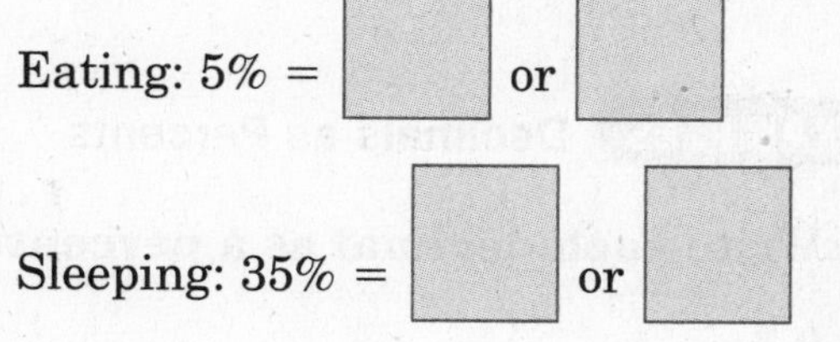

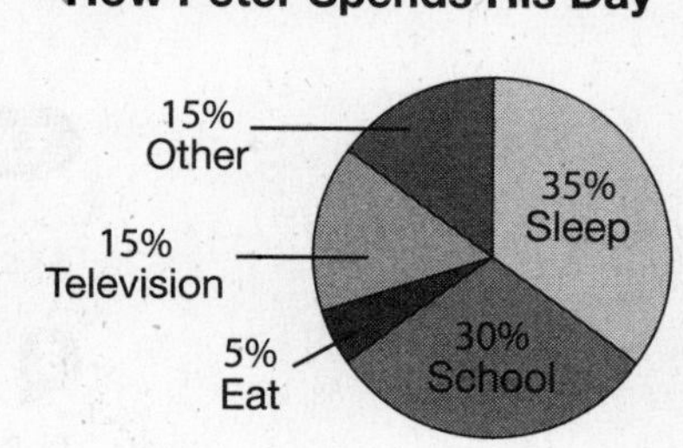

Your Turn The circle graph shows an estimate of the percent of his day that Leon spends on each activity. Write the percents for school and television as fractions in simplest form.

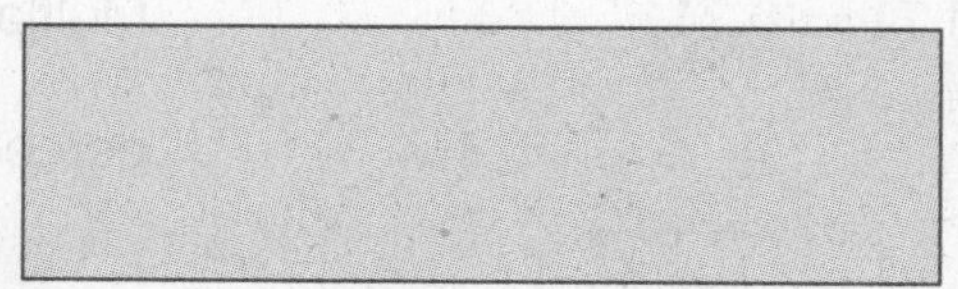

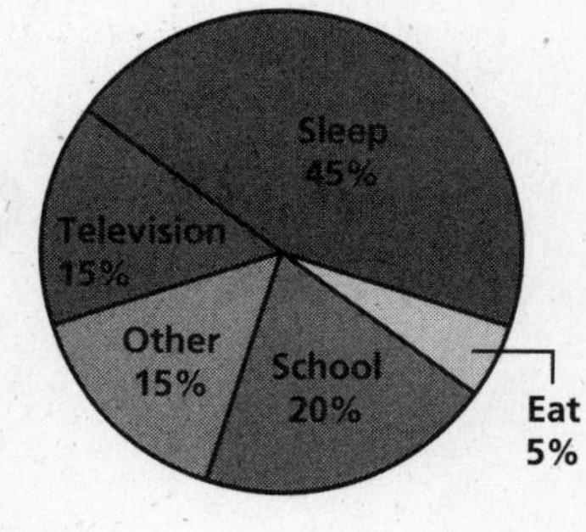

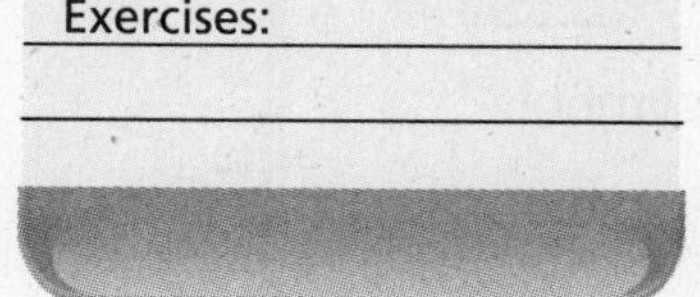

© Glencoe/McGraw-Hill

5–2 Fractions, Decimals, and Percents

WHAT YOU'LL LEARN

- Write percents as fractions and decimals and vice versa.

KEY CONCEPTS

Decimals and Percents
To write a percent as a decimal, divide by 100 and remove the percent symbol.

To write a decimal as a percent, multiply by 100 and add the percent symbol.

EXAMPLES Percents as Decimals

Write each percent as a decimal.

1 52%

52% = 52%   Divide by ____ and remove the percent symbol.

= ____

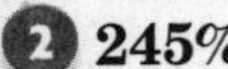

2 245%

245% = 245%   Divide by ____ and remove the percent symbol.

= ____

Your Turn **Write each percent as a decimal.**

a. 28%

b. 135%

EXAMPLES Decimals as Percents

Write each decimal as a percent.

3 0.3

0.3 = 0.30   Multiply by ____ and add the percent symbol.

= ____

4 0.71

0.71 = 0.71   Multiply by ____ and add the percent symbol.

= ____

© Glencoe/McGraw-Hill

Your Turn **Write each decimal as a percent.**

a. 0.91

b. 1.65

EXAMPLES Fractions as Percents

5 Write $\frac{3}{4}$ as a percent.

Method 1

Use a proportion.

$\frac{3}{4} = \frac{x}{100}$

$3 \cdot 100 =$ _____

$300 =$ _____

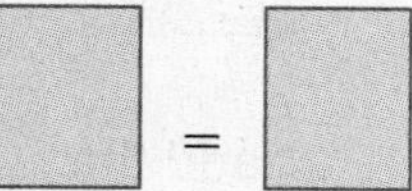

_____ $= x$

Method 2

Write as a decimal.

$\frac{3}{4} = 0.75$

= _____

$$\begin{array}{r} 0.75 \\ 4\overline{)3.00} \\ \underline{28} \\ 20 \\ \underline{20} \\ 0 \end{array}$$

So, $\frac{3}{4}$ can be written as _____.

6 Write $\frac{1}{6}$ as a percent.

Method 1

Use a proportion.

$\frac{1}{6} = \frac{x}{100}$

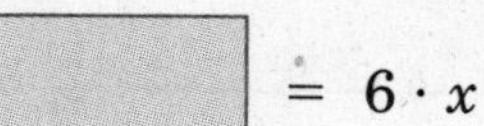

_____ $= 6 \cdot x$

_____ $= 6x$

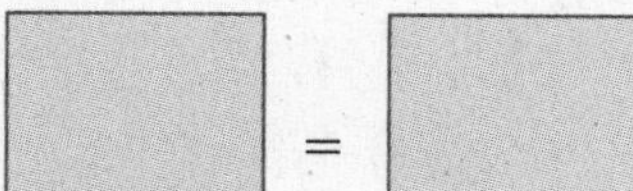

_____ $= x$

Method 2

Write as a decimal.

$\frac{1}{6} = 0.16\overline{6}$

= _____

$$\begin{array}{r} 0.116... \\ 6\overline{)1.0000} \\ \underline{6} \\ 40 \\ \underline{36} \\ 40 \\ \underline{36} \\ 4 \end{array}$$

So, $\frac{1}{6}$ can be written as _____.

REVIEW IT

Show an example of how to write fractions as decimals. *(Lesson 2–1)*

____

____

____

____

____

____

FOLDABLES™

ORGANIZE IT

Write in words and symbols what you have learned about the relationship between percents, decimals, and fractions.

Your Turn **Write each fraction as a percent.**

a. $\frac{1}{4}$ **b.** $\frac{1}{9}$

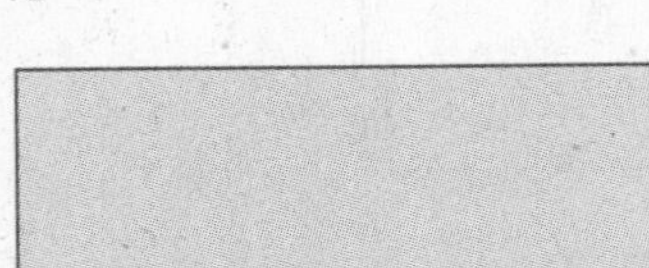

EXAMPLE **Compare Numbers**

7 **POLITICS In Sun City, $\frac{9}{20}$ of voters are Democrats. In Moon Town, 48% of voters are Democrats. In which town is there a greater proportion of Democrats?**

Write $\frac{9}{20}$ as a percent.

$\frac{9}{20} = 0.45$ $9 \div 20 = 0.45$

=  ______ and

add the ______ symbol.

Since  is less than 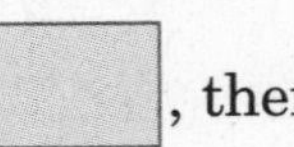, there are 

Democrats in Moon Town.

Your Turn In Star City, $\frac{3}{20}$ of voters are Republicans. In Meteorville, 13% of voters are Republicans. In which town is there a greater proportion of Republicans?

HOMEWORK ASSIGNMENT

Page(s):

Exercises:

© Glencoe/McGraw-Hill

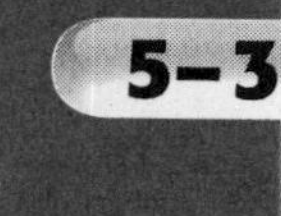

5–3 The Percent Proportion

WHAT YOU'LL LEARN

- Solve problems using the percent proportion.

KEY CONCEPT

Percent Proportion

$\frac{\text{part}}{\text{base}} = \frac{\text{percent}}{100}$

BUILD YOUR VOCABULARY (pages 108–109)

In a **percent proportion,** ______ one of the numbers, called the **part**, is being compared to the ______ quantity, called the **base**. The other ratio is the percent, written as a fraction, whose base is ______.

EXAMPLE Find the Percent

1 **34 is what percent of 136?**

Since 34 is being compared to 136, ______ is part and ______ is the base. You need to find the percent.

$\frac{a}{b} = \frac{p}{100} \rightarrow \frac{34}{136} = \frac{p}{100}$ $a =$ ______ , $b =$ ______

______ · ______ = ______ $\cdot\, p$ Find the cross products.

______ = ______   Multiply.

______ = ______   Divide each side by ______.

______ = ______   Simplify.

So, 34 is ______ of 136.

Your Turn 63 is what percent of 210?

______

© Glencoe/McGraw-Hill

FOLDABLES

ORGANIZE IT

Be sure to explain how to find the percent, the part, and the base of a percent proportion. You also may want to show the ideas in a chart like the Concept Summary in your text.

EXAMPLE Find the Part

2 What number is 70% of 600?

The percent is 70, and the base is 600. You need to find the part.

$\frac{a}{b} = \frac{p}{100} \rightarrow \frac{a}{600} = \frac{70}{100}$ — $b = 600, p = 70$

$a \cdot 100 = 600 \cdot 70$ — Find the cross products.

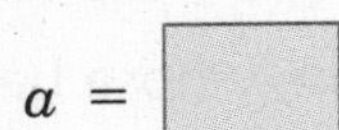

$100a = \square$ — Multiply.

$\frac{100a}{\square} = \frac{42{,}000}{\square}$ — Divide each side by $\square$.

$a = \square$ — Simplify.

So, $\square$ is 70% of 600.

Your Turn What number is 40% of 400?

EXAMPLE Find the Base

3 BASEBALL From 1999 to 2001, Derek Jeter had 11 hits with bases loaded. This was about 30% of his at bats with bases loaded. How many times was he at bat with bases loaded?

The percent is 30 and the part is 11. Find the base.

$\frac{a}{b} = \frac{p}{100} \rightarrow \frac{11}{b} = \frac{30}{100}$ — $a = 11, p = 30$

$11 \cdot 100 = b \cdot 30$ — Find the cross products.

$\square = \square$ — Multiply.

$\frac{1{,}100}{\square} = \frac{30b}{\square}$ — Divide each side by $\square$.

$\square \approx b$ — Simplify.

Derek Jeter was at bat about $\square$ times with bases loaded.

Your Turn In the 2002 season, Barry Bonds had 149 hits. This was about 37% of his at bats. How many times was he at bat?

HOMEWORK ASSIGNMENT

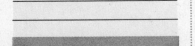

Page(s):

Exercises:

© Glencoe/McGraw-Hill

5–4 Finding Percents Mentally

EXAMPLES Use Fractions to Compute Mentally

WHAT YOU'LL LEARN

- Compute mentally with percents.

Compute mentally.

1 **40% of 80**

40% of 80 = ____ of 80 or ____ Use the fraction form of 40%, which is ____.

2 **$66\frac{2}{3}$% of 75**

$66\frac{2}{3}$% of 75 = ____ of 75 or ____. Use the fraction form of $66\frac{2}{3}$%, which is ____.

EXAMPLES Use Decimals to Compute Mentally

Compute mentally.

3 **10% of 65**

10% of 65 = ____ of 65 or ____

4 **1% of 304**

1% of 304 = ____ of 304 or ____

WRITE IT

Explain how you can move the decimal point to mentally multiply 0.1 by 1.1.

______

______

______

______

Your Turn **Compute mentally.**

a. 20% of 60

b. $66\frac{2}{3}$% of 300

c. 10% of 13

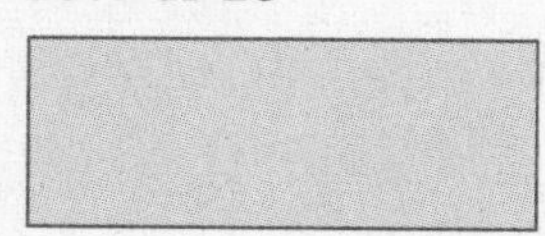

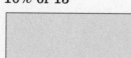

d. 1% of 244

© Glencoe/McGraw-Hill

EXAMPLE Use Mental Math to Solve a Problem

5 TECHNOLOGY **A company produces 2,500 of a particular printer. They later discover that 25% of the printers have defects. How many printers from this group have defects?**

Method 1 Use a fraction.

25% of 2,500 = ☐ of 2,500

THINK $\frac{1}{4}$ of 2,000 is ☐ and $\frac{1}{4}$ of 500 is ☐.

So, ☐ of 2,500 is ☐ + ☐ or ☐.

Method 2 Use a decimal.

25% of 2,500 = ☐ of 2,500

THINK 0.5 of 2,500 is ☐.

So, 0.25 of 2,500 is ☐ · ☐ or ☐.

There were ☐ printers that had defects.

Your Turn A company produces 1,400 of a particular monitor. They later discover that 20% of the monitors have defects. How many monitors from this group have defects?

FOLDABLES

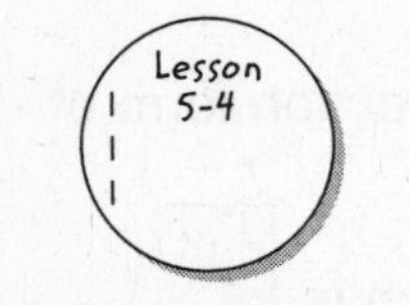

ORGANIZE IT

In your Foldable, be sure to include examples that show how to estimate percents of numbers.

HOMEWORK ASSIGNMENT

Page(s):

Exercises:

© Glencoe/McGraw-Hill

5–5 Percent and Estimation

BUILD YOUR VOCABULARY (page 108)

Compatible numbers are two numbers that are easy to add, subtract, multiply, or divide mentally.

WHAT YOU'LL LEARN

- Estimate by using equivalent fractions, decimals, and percents.

EXAMPLES **Estimate Percents of Numbers**

Estimate.

1 **48% of 70**

48% is about ______ or ______. ______ and 70 are compatible numbers.

______ of 70 is ______.

So, 48% of 70 is about ______.

2 **75% of 98**

75% is $\frac{3}{4}$, and 98 is about ______. $\frac{3}{4}$ and ______ are compatible numbers.

$\frac{3}{4}$ of ______ is ______.

So, 75% of 98 is about ______.

3 **12% of 81**

12% is about 12.5% or ______, and 81 is about ______.

______ and ______ are compatible numbers.

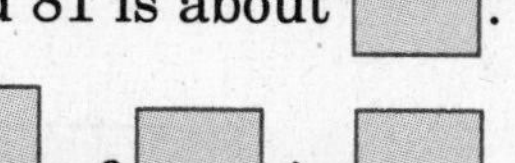

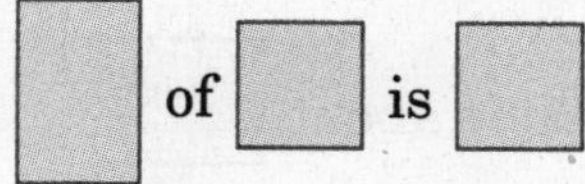

______ of ______ is ______.

So, 12% of 81 is about ______.

© Glencoe/McGraw-Hill

Your Turn **Estimate.**

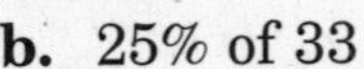

a. 51% of 60 **b.** 25% of 33 **c.** 37% of 17

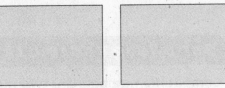

EXAMPLES **Estimate Percents**

FOLDABLES

ORGANIZE IT

Include the meaning of the symbol "≈." You may wish to include an example of estimating a percent in which the symbol ≈ is used.

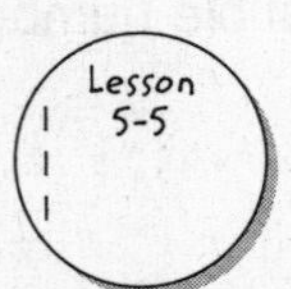

Estimate each percent.

4 **12 out of 47**

$\frac{12}{47} \approx$ ____ or $\frac{1}{4}$ 47 is about ____.

$\frac{1}{4} =$ ____ %

So, 12 out of 47 is about ____.

5 **19 out of 31**

$\frac{19}{31} \approx$ ____ or $\frac{2}{3}$ 19 is about ____, and

31 is about ____.

$\frac{2}{3} =$ ____ %

So, 19 out of 31 is about ____.

6 **41 out of 200**

$\frac{41}{200} \approx$ ____ or $\frac{1}{5}$ 41 is about ____.

$\frac{1}{5} =$ ____

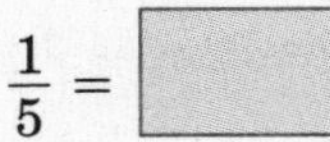

So, 41 out of 200 is about ____.

© Glencoe/McGraw-Hill

Your Turn **Estimate each percent.**

a. 15 out of 76 **b.** 14 out of 47 **c.** 11 out of 100

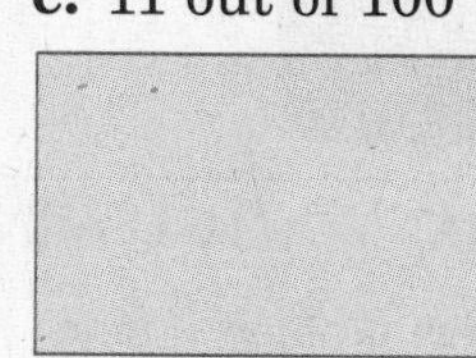

EXAMPLE Estimate Percent of an Area

7 ART **An artist creates a design on a grid. Estimate the percent of the grid that has been painted so far.**

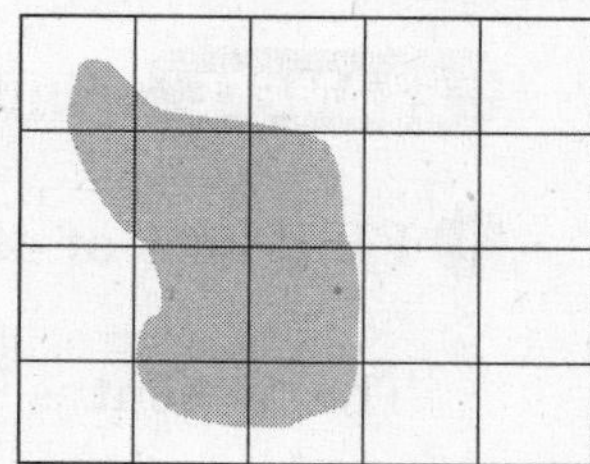

About 6 squares out of 20 squares are covered with paint.

$\frac{6}{20} = \frac{3}{10} \longrightarrow \frac{3}{10} =$ ______

So, about ______ of the grid has been painted.

Your Turn An artist created a design on a grid. Estimate the percent of the grid has been painted so far.

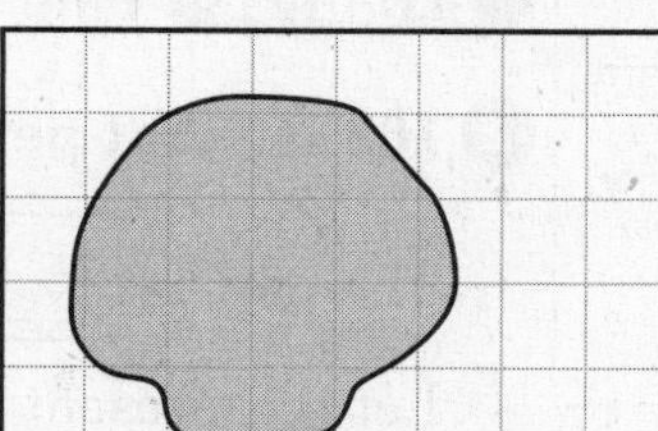

HOMEWORK ASSIGNMENT

Page(s):

Exercises:

© Glencoe/McGraw-Hill

5–6 The Percent Equation

WHAT YOU'LL LEARN

- Solve problems using the percent equation.

BUILD YOUR VOCABULARY (page 109)

The **percent equation** is an equivalent form of the percent proportion in which the ______ is written as a ______.

REVIEW IT

Explain how to write a decimal as a percent. *(Lesson 5-2)*

______

______

______

______

______

______

EXAMPLE Find the Part

1 **Find 30% of 450.**

The percent is ______%, and the base is ______.

Let n represent the part.

$n =$ ______ · ______ Write 30% as the decimal ______.

$n =$ ______ Simplify.

So, 30% of 450 is ______.

EXAMPLE Find the Percent

2 **102 is what percent of 150?**

The part is ______, and the base is ______.

Let n represent the percent.

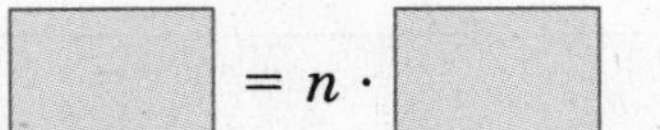

______ $= n \cdot$ ______ Write the equation.

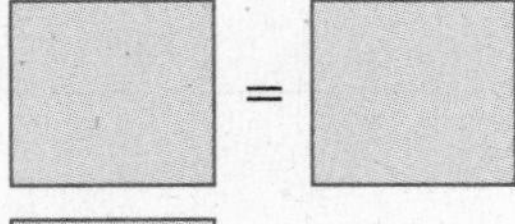

______ = ______ Divide each side by 150.

______ $= n$ Simplify.

In the percent equation, the percent is written as a ______. Since ______ = 68%, 102 is 68% of 150.

© Glencoe/McGraw-Hill

FOLDABLES

ORGANIZE IT

Write the percent equation in words and symbols. Explain why the rate in a percent equation is usually written as a decimal.

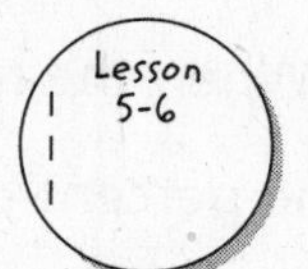

EXAMPLE Find the Base

3 **144 is 45% of what number?**

The part is ______, and the percent is ______%.

Let n represent the ______.

$144 = \_\_\_\_\_\_ n$ Write 45% as a decimal.

$\frac{144}{\_\_\_\_} = \frac{\_\_\_\_ n}{\_\_\_\_}$ Divide each side by 0.45.

$\_\_\_\_ = n$ Simplify.

Your Turn **Find the part, percent, or base.**

a. Find 20% of 315.

b. 135 is what percent of 250?

c. 186 is 30% of what number?

EXAMPLE Solve a Real-Life Problem

4 SALES TAX **The price of a sweater is \$75. The sales tax is $5\frac{3}{4}$ percent. What is the total price of the sweater?**

You need to find what amount is $5\frac{3}{4}$% of \$75.

Let t = the amount of tax.

$t = \_\_\_\_ \cdot \_\_\_\_$ Write the equation.

$t = \_\_\_\_$ Simplify.

The amount of tax is ______. The total cost of the sweater is \$75 + ______ or ______.

Your Turn The price of a pair of tennis shoes is \$60. The sales tax is 5 percent. What is the total price of the shoes?

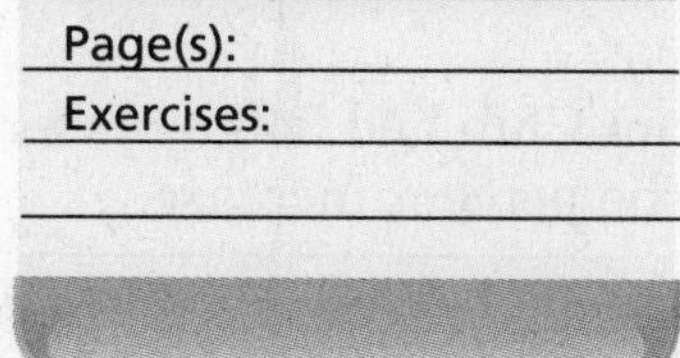

© Glencoe/McGraw-Hill

5–7 Percent of Change

WHAT YOU'LL LEARN

- Find and use the percent of increase or decrease.

BUILD YOUR VOCABULARY (page 109)

A **percent of change** is a ______ that compares the change in quantity to the original amount. When the new amount is ______ than the original, the percent of change is called a **percent of increase**.

When the new amount is ______ than the original, the percent of change is called a **percent of decrease**.

KEY CONCEPT

Percent of Change A percent of change is a ratio that compares the change in quantity to the original amount.

EXAMPLE Find the Percent of Increase

1 HOMES **The Nietos bought a house several years ago for $120,000. This year, they sold it for $150,000. Find the percent of increase.**

Step 1 Subtract to find the amount of change.

$150{,}000 - 120{,}000 =$ ______

Step 2 Write a ratio that compares the amount of ______ to the ______ amount they paid for the house. Express the ratio as a percent.

$\text{Percent of change} = \dfrac{\text{amount of change}}{\text{original amount}}$ Definition of percent of change

$= \dfrac{\_\_\_\_}{\_\_\_\_}$

$= 0.25$ or ______%

The percent of increase is ______%.

Your Turn Last year Cedar Park Swim Club had 340 members. This year they have 391 members. Find the percent increase.

______

© Glencoe/McGraw-Hill

EXAMPLE Find the Percent of Change

FOLDABLES™

ORGANIZE IT

Be sure to include an explanation and examples showing the difference between percent of increase and percent of decrease.

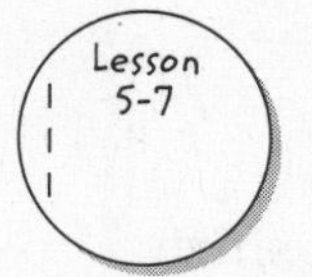

2 SCHOOLS Johnson Middle School had 240 students last year. This year, there are 192 students. Find the percent of change. State whether the percent of change is an increase or a decrease.

Step 1 Subtract to find the amount of change.

$240 - 192 = \square$

Step 2 Write a ratio that compares the amount of change to the number of students last year. Express the ratio as a percent.

$$\text{Percent of change} = \frac{\text{amount of change}}{\text{original amount}}$$

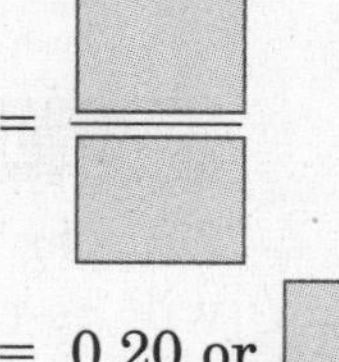

$$= \frac{\square}{\square}$$

$$= 0.20 \text{ or } \square$$

The percent of change is ______. Since the new amount is ______ than the original, it is a percent of ______.

Your Turn Meagan bought a new car several years ago for \$14,000. This year she sold the car for \$9,100. Find the percent of change. State whether the percent of change is an *increase* or a *decrease*.

______

BUILD YOUR VOCABULARY (pages 108–109)

The **markup** is the amount the price of an item is ______ above the price the store ______ for the item.

The **selling price** is the amount the ______ pays.

The amount by which a ______ is ______ is called the **discount**.

© Glencoe/McGraw-Hill

EXAMPLE Find the Selling Price

3 MARKUP **Shirts bought by a sporting goods store cost them $20 per shirt. The want to mark them up 40%. What will be the selling price?**

REMEMBER IT

There may be more than one way to solve a problem. See page 238 of your textbook for other methods you can use to solve Examples 3 and 4.

Find the amount of the markup. That is, find 40% of $20. Let m represent the markup.

$m = \square \cdot 20$ Write 40% as a decimal.

$m = \square$ Multiply.

Add the markup to the price they paid for the shirts.

$\$20 + \square = \square$

Your Turn Silk flowers bought by a craft store cost them $10 per yard. They want to mark them up 35 percent. What will be the selling price?

EXAMPLE Find the Sale Price

4 SHOPPING **A computer usually sells for $1,200. This week it is on sale for 30% off. What is the sale price?**

Find 30% of $1,200. Let d represent the discount.

$d = \square \cdot \square$

$d = \square$

Subtract the amount of the discount from the original price.

$\$1{,}200 - \square = \square$

Your Turn A DVD sells for $28. This week it is on sale for 20% off. What is the sale price?

HOMEWORK ASSIGNMENT

Page(s):

Exercises:

© Glencoe/McGraw-Hill

5–8 Simple Interest

What You'll Learn

- Solve problems involving simple interest.

Build Your Vocabulary (pages 108–109)

Interest is the amount of money paid or ______ for the use of money.

Principal is the amount of money ______ or borrowed.

EXAMPLE Find Simple Interest

1 Find the simple interest for $2,000 invested at 5.5% for 4 years.

$I = prt$ Write the simple interest formula.

$I =$ ______ · ______ · ______ Replace *p* with ______, *r* with ______, and *t* with ______.

$I =$ ______ The simple interest is ______.

EXAMPLE Find the Total Amount

Remember It

The *t* in the simple interest formula represents time in years. If time is given in months, weeks, or days, the time must be changed to time in years.

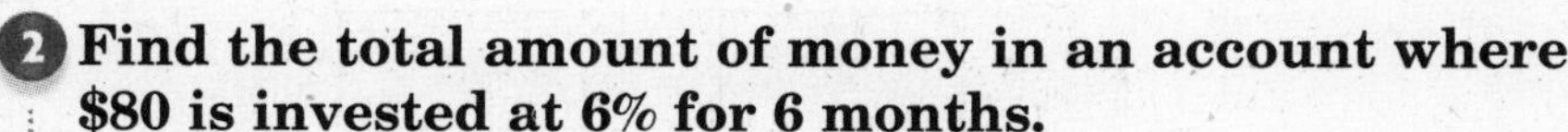

2 Find the total amount of money in an account where $80 is invested at 6% for 6 months.

You need to find the total amount in an account. The time is given in months. Six months is $\frac{6}{12}$ or ______ year.

$I = prt$

$I =$ ______ · ______ · ______

$I =$ ______

The amount in the account is $80 + ______ or ______.

© Glencoe/McGraw-Hill

ORGANIZE IT

Explain what you have learned about computing simple interest. Be sure to include the simple interest formula.

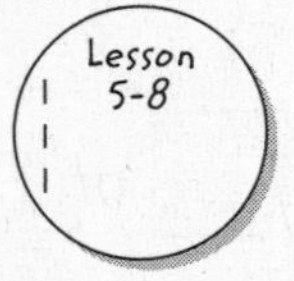

Your Turn

a. Find the simple interest for $1,500 invested at 5% for 3 years.

b. Find the total amount of money in an account where $60 is invested at 8% for 3 months.

EXAMPLE Find the Interest Rate

3 LOANS **Gerardo borrowed $4,500 from his bank for home improvements. He will repay the loan by paying $120 a month for the next four years. Find the simple interest rate of the loan.**

First, find the total amount of money Gerardo will pay.

$120 · 48 = ______.

He will pay ______ − $4,500 or ______ in interest.

The loan will be for 48 months or 4 years. Use the simple interest formula to find the interest rate.

$I = p \cdot r \cdot t$

______ = ______ $\cdot r \cdot$ ______

______ = ______ Simplify.

______ = ______ Divide each side by 18,000.

______ $= r$ Simplify.

The simple interest rate is ______.

Your Turn Jocelyn borrowed $3,600 from her bank for home improvements. She will repay the loan by paying $90 a month for the next 5 years. Find the simple interest rate of the loan.

HOMEWORK ASSIGNMENT

Page(s):

Exercises:

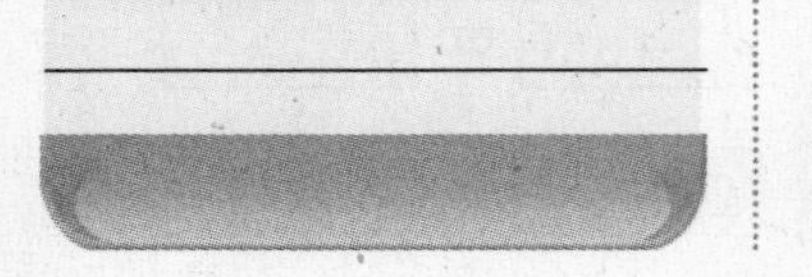

© Glencoe/McGraw-Hill

CHAPTER 5

BRINGING IT ALL TOGETHER

STUDY GUIDE

FOLDABLES™	VOCABULARY PUZZLEMAKER	BUILD YOUR VOCABULARY
Use your **Chapter 5 Foldable** to help you study for your chapter test.	To make a crossword puzzle, word search, or jumble puzzle of the vocabulary words in Chapter 5, go to: www.glencoe.com/sec/math/t_resources/free/index.php	You can use your completed **Vocabulary Builder** *(pages 108–109)* to help you solve the puzzle.

5-1 Ratios and Percents

Write each ratio or fraction as a percent.

1. 21 out of 100

2. 4:10

3. $\frac{9}{25}$

Write each percent as a fraction in simplest form.

4. 27%

5. 50%

6. 80%

5-2 Fractions, Decimals, and Percents

Write each percent as a decimal.

7. 29%

8. 376%

9. 5%

Write each decimal or fraction as a percent.

10. 3.9

11. $\frac{7}{8}$

12. $\frac{1}{3}$

© Glencoe/McGraw-Hill

5-3 The Percent Proportion

13. In a percent proportion, the ______ is being compared to the whole quantity, called the ______.

Solve.

14. What percent of 48 is 6?

15. 14 is 20% of what number?

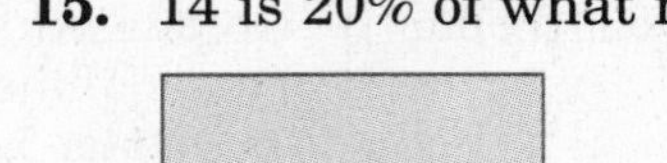

5-4 Finding Percents Mentally

Complete each statement.

16. 40% of 25 = ______ of 25 or ______

17. ______ of 36 = $\frac{1}{4}$ of 36 or ______

18. $66\frac{2}{3}$% of 48 = ______ of 48 or ______

19. ______ of 89 = 0.1 of 89 or ______

Compute mentally.

20. $33\frac{1}{3}$ % of 300

21. 1% of 268

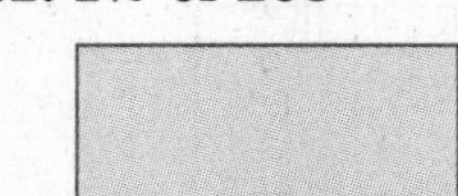

22. 25% of 120

5-5 Percent and Estimation

23. Are $\frac{1}{8}$ and 56 compatible numbers? Explain.

24. Describe how to estimate 65% of 64 using compatible numbers.

© Glencoe/McGraw-Hill

5-6
The Percent Equation

Write each percent proportion as a percent equation.

25. $\frac{16}{64} = \frac{25}{100}$

26. $\frac{a}{14} = \frac{2}{100}$

27. $\frac{96}{b} = \frac{48}{100}$

28. $\frac{13}{100} = \frac{p}{675}$

5-7
Percent of Change

Find the percent of change. Round to the nearest tenth if necessary. State whether the change is an *increase* or *decrease*.

29. Original: 29
New: 64

30. Original: 51
New: 42

31. Find the selling price for the sweater.

Cost to store: $15
Mark up: 35%

5-8
Simple Interest

Write *interest* or *principal* to complete each sentence.

32. ______ is the amount of money paid or earned for the use of money.

33. ______ equals ______ times rate times time.

34. Find the total amount in the account where $560 is invested at 5.6% for 6 months.

First, find the ______ earned.

Then, add the ______ earned and the ______ to find the total amount in the account.

What is the total amount for $560 at 5.6% for 6 months?

© Glencoe/McGraw-Hill

ARE YOU READY FOR THE CHAPTER TEST?

Visit **msmath3.net** to access your text book, more examples, self-check quizzes, and practice tests to help you study the concepts in Chapter 5.

Check the one that applies. Suggestions to help you study are given with each item.

- [ ] **I completed the review of all or most lessons without using my notes or asking for help.**
  - You are probably ready for the Chapter Test.
  - You may want to take the Chapter 5 Practice Test on page 249 of your textbook as a final check.

- [ ] **I used my Foldable or Study Notebook to complete the review of all or most lessons.**
  - You should complete the Chapter 5 Study Guide and Review on pages 246–248 of your textbook.
  - If you are unsure of any concepts or skills, refer back to the specific lesson(s).
  - You may want to take the Chapter 5 Practice Test on page 249.

- [ ] **I asked for help from someone else to complete the review of all or most lessons.**
  - You should review the examples and concepts in your Study Notebook and Chapter 5 Foldable.
  - Then complete the Chapter 5 Study Guide and Review on pages 246–248 of your textbook.
  - If you are unsure of any concepts or skills, refer back to the specific lesson(s).
  - You may want to take the Chapter 5 Practice Test on page 249.

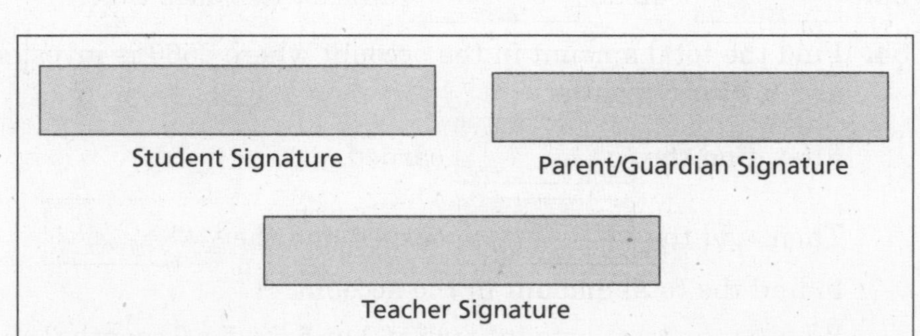

© Glencoe/McGraw-Hill

Geometry

FOLDABLES™ Use the instructions below to make a Foldable to help you organize your notes as you study the chapter. You will see Foldable reminders in the margin of this Interactive Study Notebook to help you in taking notes.

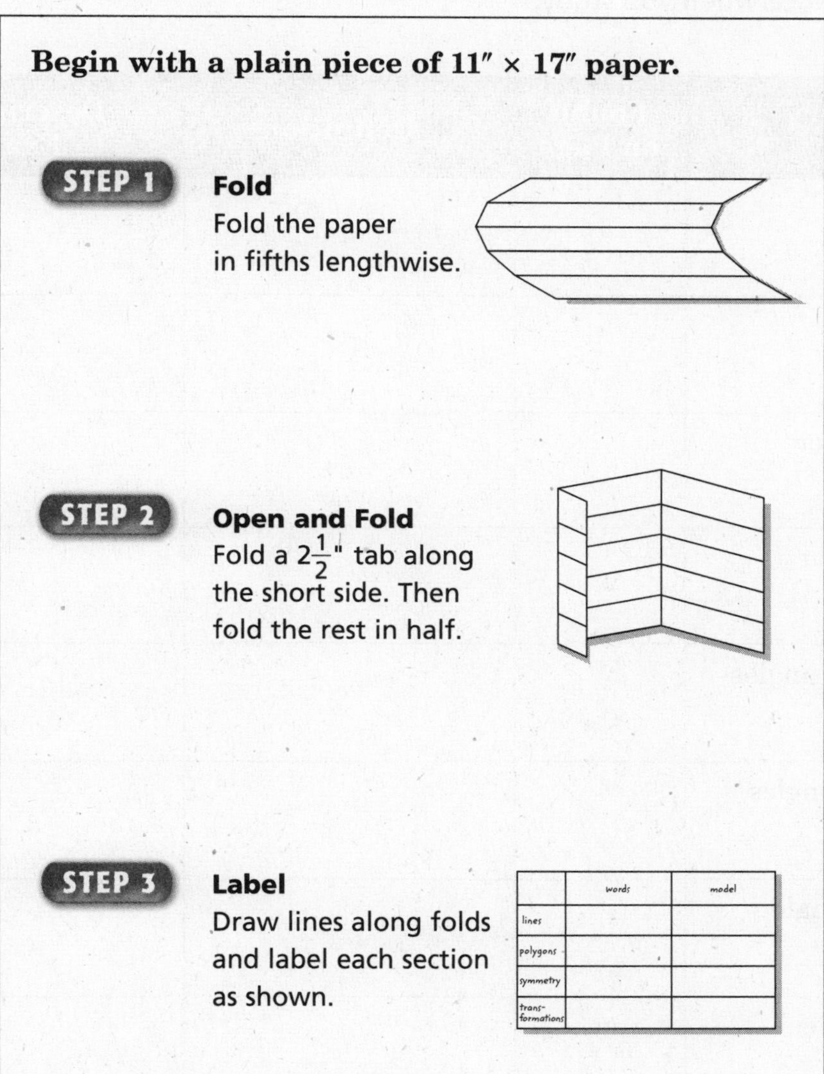

Begin with a plain piece of 11″ × 17″ paper.

STEP 1 **Fold**
Fold the paper in fifths lengthwise.

STEP 2 **Open and Fold**
Fold a $2\frac{1}{2}$" tab along the short side. Then fold the rest in half.

STEP 3 **Label**
Draw lines along folds and label each section as shown.

NOTE-TAKING TIP: When you read and learn new concepts, help yourself remember these concepts by taking notes, writing definitions and explanations, and draw models as needed.

Chapter 6

© Glencoe/McGraw-Hill

BUILD YOUR VOCABULARY

This is an alphabetical list of new vocabulary terms you will learn in Chapter 6. As you complete the study notes for the chapter, you will see Build Your Vocabulary reminders to complete each term's definition or description on these pages. Remember to add the textbook page number in the second column for reference when you study.

Vocabulary Term	Found on Page	Definition	Description or Example
acute triangle			
adjacent angles			
alternate exterior angles			
alternate interior angles			
complementary angles			
corresponding angles			
equilateral triangle			
isosceles triangle			
line symmetry			
obtuse triangle			

© Glencoe/McGraw-Hill

Vocabulary Term	Found on Page	Definition	Description or Example
parallelogram			
quadrilateral			
reflection			
rhombus			
right triangle			
rotation			
scalene triangle			
supplementary angles			
translation			
transversal			
trapezoid			
vertical angles			

© Glencoe/McGraw-Hill

6–1 Line and Angle Relationships

What You'll Learn

- Identify special pairs of angles and relationships of angles formed by two parallel lines cut by a transversal.

Key Concepts

Special Pairs of Angles

Vertical angles are opposite angles formed by intersecting lines. Vertical angles are congruent.

Adjacent angles have the same vertex, share a common side, and do not overlap.

The sum of the measures of **supplementary angles** is 180°.

The sum of the measures of **complementary angles** is 90°.

Build Your Vocabulary (pages 134–135)

Acute angles have measures less than ____.

Right angles have measures ____ to 90°.

Obtuse angles have measures between ____ and ____.

Straight angles have measures equal to ____.

EXAMPLES Classify Angles and Angle Pairs

Classify each angle or angle pair using all names that apply.

1

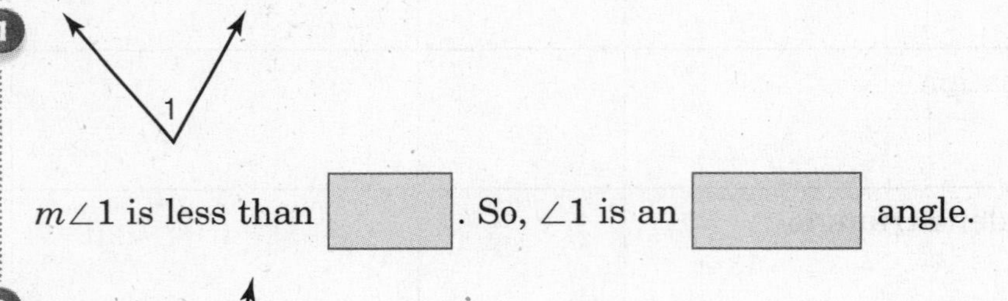

$m\angle 1$ is less than ____. So, $\angle 1$ is an ____ angle.

2

1 2

$\angle 1$ and $\angle 2$ are ____ angles since they have the same vertex, share a common side, and do not overlap. Together they form a straight angle measuring ____. So, $\angle 1$ and $\angle 2$ are ____ angles and ____ angles.

© Glencoe/McGraw-Hill

REMEMBER IT 

Supplementary and complementary angles may or may not be adjacent angles.

Your Turn **Classify each angle or angle pair using all names that apply.**

a.

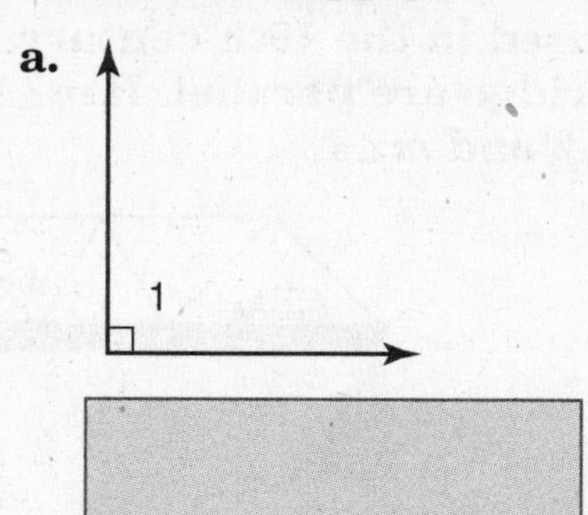

b.

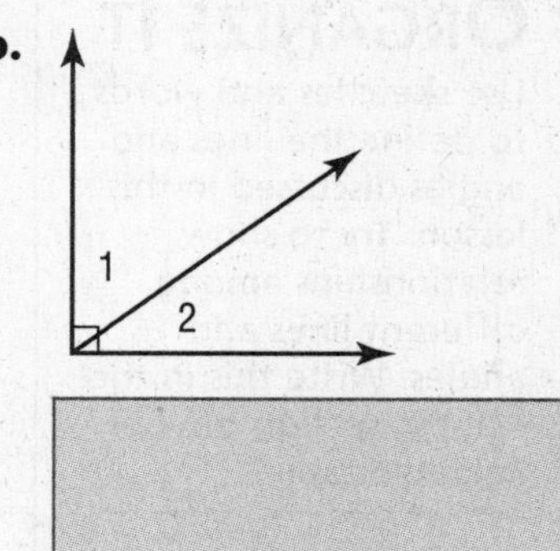

EXAMPLE Finding a Missing Angle Measure

3 **The two angles below are supplementary. Find the value of x.**

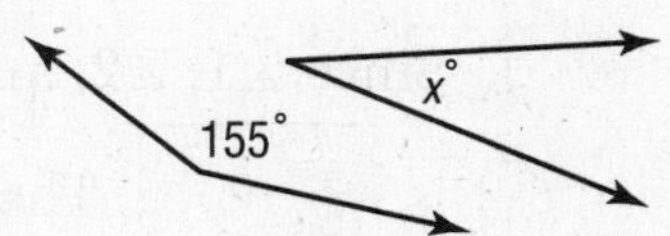

$155 + x$	$=$	180	Definition of supplementary angles
______	$=$	______	Subtract ______ from each side.
x	$=$	25	Simplify.

KEY CONCEPT

If two parallel lines are cut by a transversal, then the following statements are true.

Alternate interior angles, those on opposite sides of the transversal and inside the other two lines, are congruent.

Alternate exterior angles, those on opposite sides of the transversal and outside the other two lines, are congruent.

Corresponding angles, those in the same position on the two lines in relation to the transversal, are congruent.

Your Turn The two angles shown are complementary. Find the value of x.

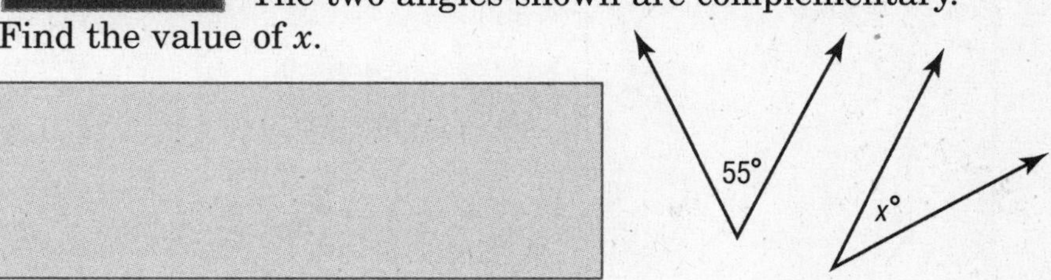

BUILD YOUR VOCABULARY (page 135)

Lines that intersect at ______ angles are called **perpendicular lines.**

Two lines in a plane that never ______ or cross are called **parallel lines.**

A **transversal** is a line that ______ two or more lines.

© Glencoe/McGraw-Hill

EXAMPLE Find an Angle Measure

FOLDABLES™

ORGANIZE IT

Use sketches and words to define the lines and angles discussed in this lesson. Try to show relationships among different lines and angles. Write this in the "Lines" section of your Foldable table.

	words	model
lines		
polygons		
symmetry		
trans-formations		

4 BRIDGES **The sketch below shows a simple bridge design used in the 19th century. The top beam and floor of the bridge are parallel. If $m\angle 1 = 60°$ and $m\angle 3 = 45°$, find $m\angle 2$ and $m\angle 4$.**

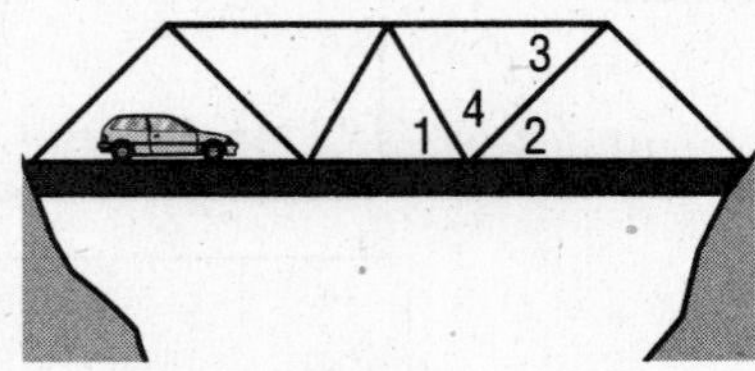

Since $\angle 2$ and $\angle 3$ are alternate interior angles, they are ______.

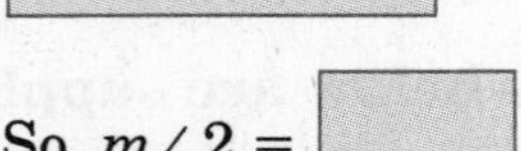

So, $m\angle 2 =$ ______.

Since $\angle 1$, $\angle 2$, and $\angle 4$ form a line, the sum of their measures is ______. Therefore, $m\angle 4 = 180° -$ ______ $- 60°$ or ______.

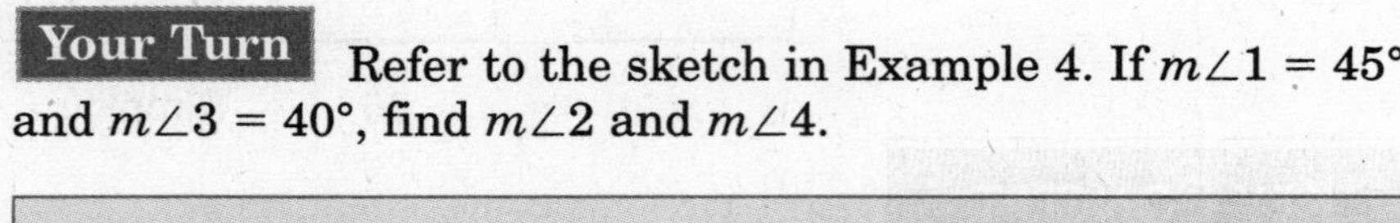

Your Turn Refer to the sketch in Example 4. If $m\angle 1 = 45°$ and $m\angle 3 = 40°$, find $m\angle 2$ and $m\angle 4$.

HOMEWORK ASSIGNMENT

Page(s):

Exercises:

© Glencoe/McGraw-Hill

6–2 Triangles and Angles

What You'll Learn

- Find the missing angle measures in triangles and classify triangles by their angles and sides.

Key Concept

Angles of a Triangle The sum of the measures of the angles of a triangle is 180°.

Build Your Vocabulary (pages 134–135)

An **acute triangle** has ______ acute angles.

An **obtuse triangle** has one ______ angle.

A **right triangle** has one ______ angle.

A **scalene triangle** has ______ sides.

An **isosceles triangle** has at least ______ congruent sides.

An **equilateral triangle** has ______ congruent sides.

EXAMPLE Find a Missing Angle Measure

1 **Find the value of x in $\triangle PQR$.**

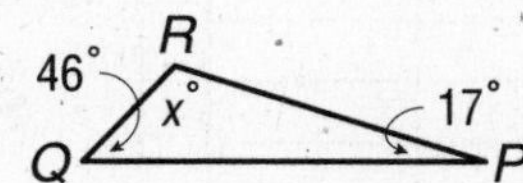

$m\angle P + m\angle Q + m\angle R = 180$ — The sum of the measures is 180.

______ + ______ $+ x = 180$ — Replace $m\angle P$ with 17, $m\angle Q$ with ______, and $m\angle R$ with x.

______ $+ x = 180$ — Simplify.

$-$ ______ $\quad -$ ______ — Subtract ______ from each side.

$x =$ ______

Your Turn Find the value of x in $\triangle TRI$.

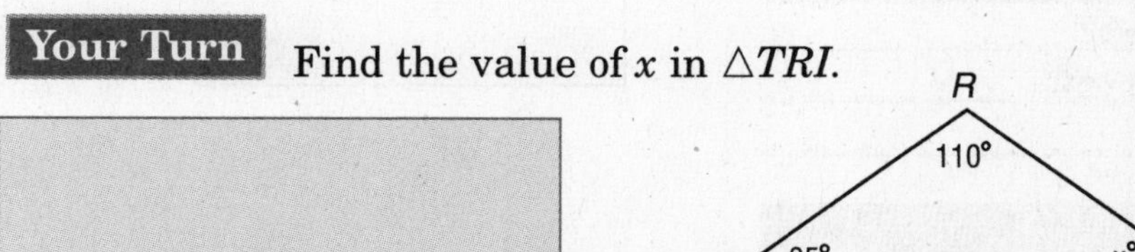

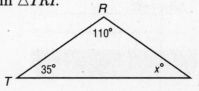

© Glencoe/McGraw-Hill

EXAMPLES Classify Triangles

Classify each triangle by its angles and its sides.

2

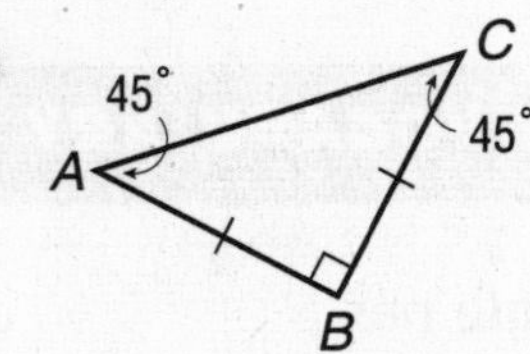

Angles $\triangle ABC$ has ______ right angle.

Sides $\triangle ABC$ has ______ congruent sides.

So, $\triangle ABC$ is a right ______ triangle.

3

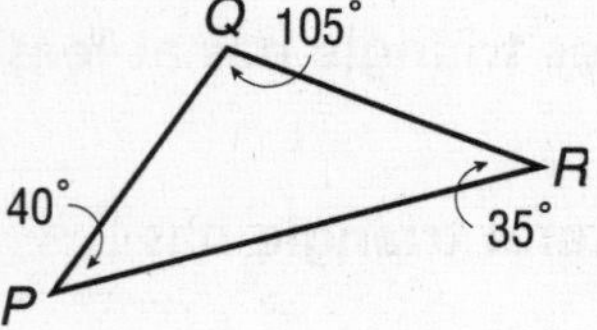

Angles $\triangle PQR$ has ______ obtuse angle.

Sides $\triangle PQR$ has ______ congruent sides.

So, $\triangle PQR$ is an obtuse ______ triangle.

FOLDABLES

ORGANIZE IT

Using sketches, words, and symbols show what you learned about triangles and their angles. Write this in the "Polygon" section of your Foldable table.

	words	model
lines		
polygons		
symmetry		
trans-formations		

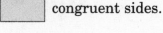

Your Turn **Classify each triangle by its angles and its sides.**

a.

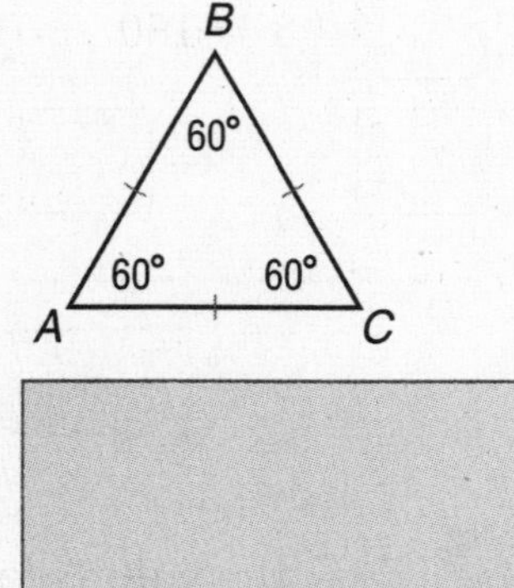

b.

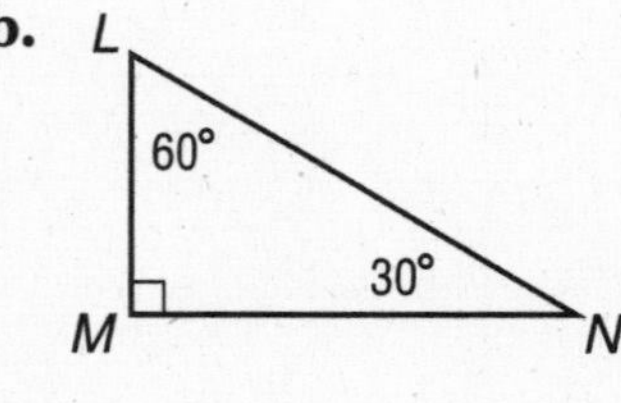

HOMEWORK ASSIGNMENT

Page(s):

Exercises:

© Glencoe/McGraw-Hill

6–3 Special Right Triangles

EXAMPLE Find Lengths of a 30°-60° Right Triangle

WHAT YOU'LL LEARN

- Find the missing measures in 30°-60° right triangles and 45°-45° right triangles.

1 **Find the missing lengths. Round to the nearest tenth if necessary.**

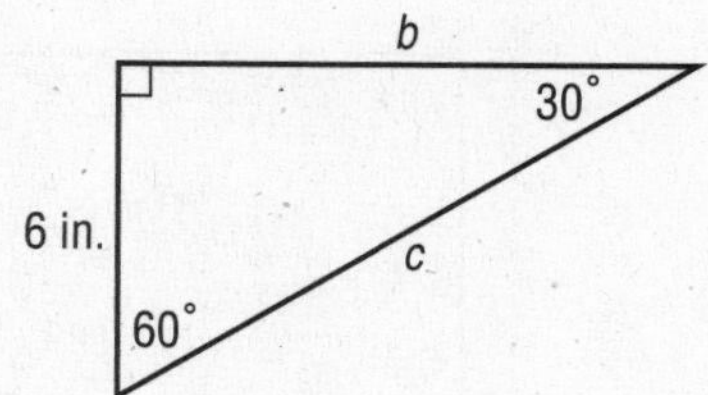

Step 1 Find c.

$a = \frac{1}{2}c$ — Write the equation.

$\_\_\_\_ = \frac{1}{2}c$ — Replace a with ____.

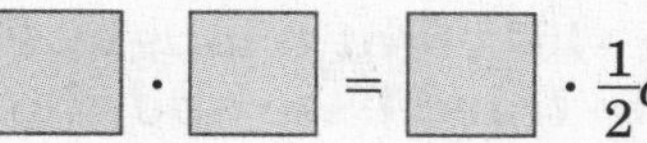

$\_\_\_\_ \cdot \_\_\_\_ = \_\_\_\_ \cdot \frac{1}{2}c$ — Multiply each side by ____.

$\_\_\_\_ = c$ — Simplify.

Step 2 Find b.

$c^2 = a^2 + b^2$ — Pythagorean Theorem

$\_\_\_\_ = \_\_\_\_ + b^2$ — $c =$ ____, $a =$ ____.

$\_\_\_\_ = \_\_\_\_ + b^2$ — Evaluate ____ and ____.

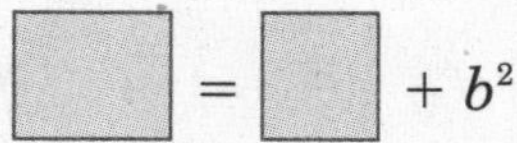

$\_\_\_\_ - \_\_\_\_ = \_\_\_\_ + b^2 - \_\_\_\_$ — Subtract.

$\_\_\_\_ = b^2$ — Simplify.

$\_\_\_\_ = \_\_\_\_$ — Take the square root of each side.

$\_\_\_\_ \approx b$ — Use a calculator.

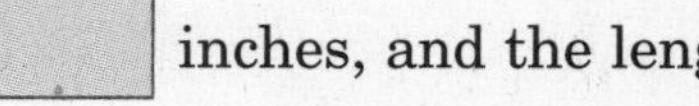

The length of b is about ____ inches, and the length of c is ____ inches.

WRITE IT

What does the Pythagorean Theorem say about the relationship between the hypotenuse and the legs of a right triangle?

____

____

____

____

© Glencoe/McGraw-Hill

Your Turn Find each missing length. Round to the nearest tenth if necessary.

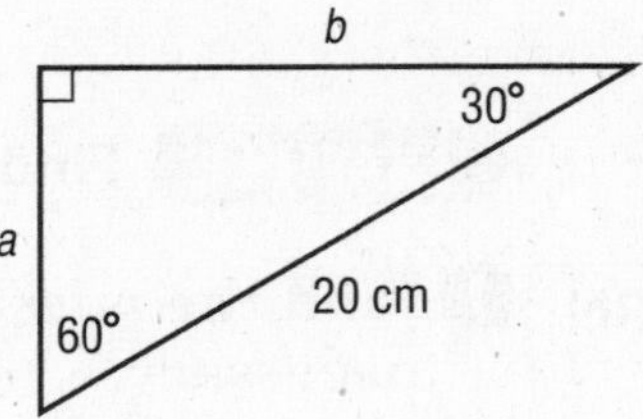

EXAMPLE Find Lengths of a 45°-45° Right Triangle

FOLDABLES

ORGANIZE IT

Explain the relationships among the sides and angles in 30°-60° right triangles and in 45°-45° right triangles. Add this to the "Polygon" section of your Foldable table.

	words	model
lines		
polygons		
symmetry		
trans-formations		

2 **BASEBALL The figure shows a baseball diamond. The distance between home plate and first base is 90 feet. The area between first base, third base, and home plate forms a 45°-45° right triangle. Find the distance from first base to third base and the distance from third base to home plate.**

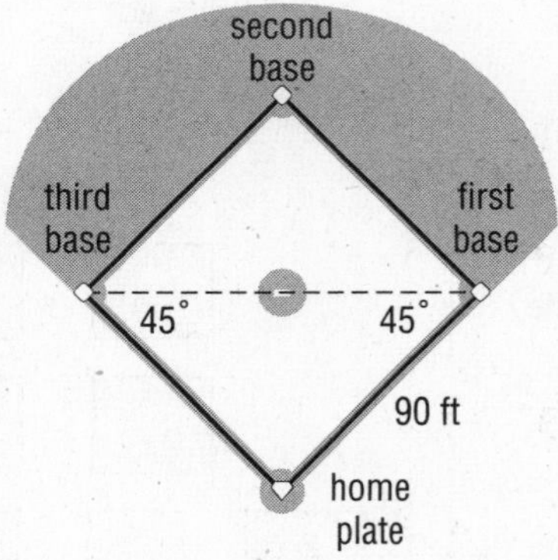

Let a equal the distance from home plate to third base.

Let b equal the distance from home plate to first base.

Let c equal the distance from first base to third base.

Step 1 Find a.

Sides a and b are the ______ length.

Since $b =$ ____ feet, $a =$ ____ feet.

© Glencoe/McGraw-Hill

Step 2 Find c.

$c^2 = a^2 + b^2$

$c^2 = \square^2 + \square^2$ Replace a with $\square$ and b with $\square$.

$c^2 = \square + \square$ Evaluate $\square$.

$c^2 = \square$ Simplify.

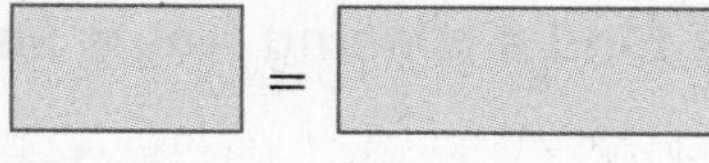

$\square = \square$ Take the square root of each side.

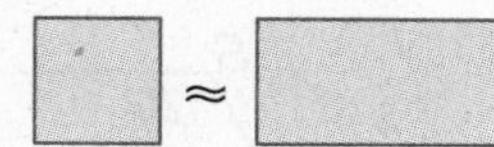

$\square \approx \square$ Use a calculator.

The distance from first base to third base is about  feet,

and the distance from third base to home plate is  feet.

Your Turn The sail of a sailboat is in the shape of a 45°-45° right triangle. The height of the sail is 12 feet. Find each missing length.

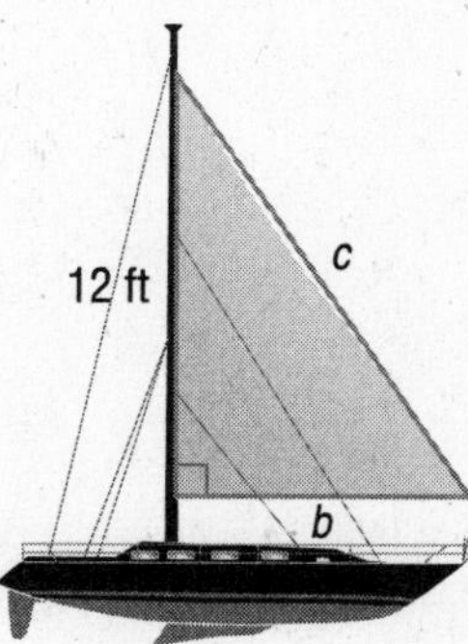

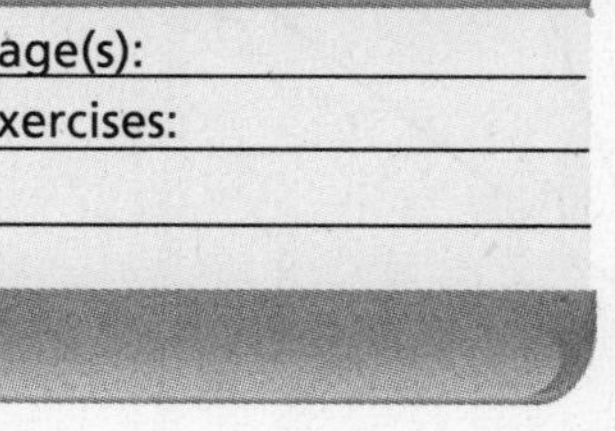
HOMEWORK ASSIGNMENT

Page(s):

Exercises:

© Glencoe/McGraw-Hill

6–4 Classifying Quadrilaterals

BUILD YOUR VOCABULARY (pages 134–135)

A **quadrilateral** is a polygon that has ______ sides and ______ angles.

WHAT YOU'LL LEARN

- Find missing angle measures in quadrilaterals and classify quadrilaterals.

KEY CONCEPT

Angles of a Quadrilateral The sum of the measures of the angles of a quadrilateral is 360°.

EXAMPLE Find a Missing Angle Measure

1 **Find the value of q in quadrilateral *PQRS*.**

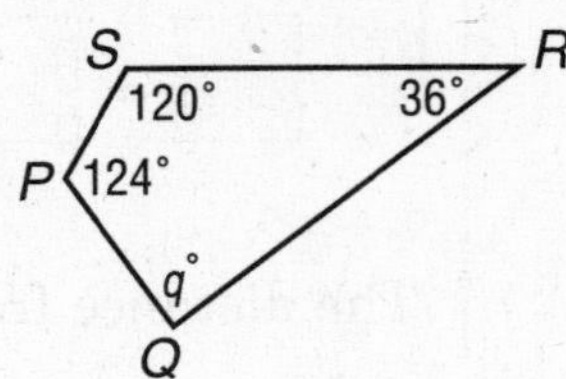

$m\angle P + m\angle Q + m\angle R + m\angle S =$ ______ The sum of the measures is ______.

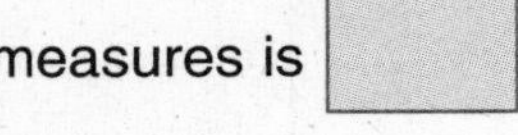

______ $+ q +$ ______ $+$ ______ $=$ ______ Replace the known angle measure and $m\angle Q$ with q.

______ $+ q =$ ______ Simplify.

$-$ ______ $-$ ______ Subtract ______ from each side.

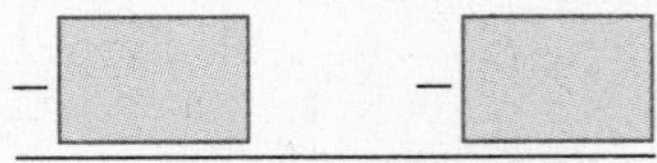

$q =$ ______ Simplify.

Your Turn Find the value of q in quadrilateral *QUAD*.

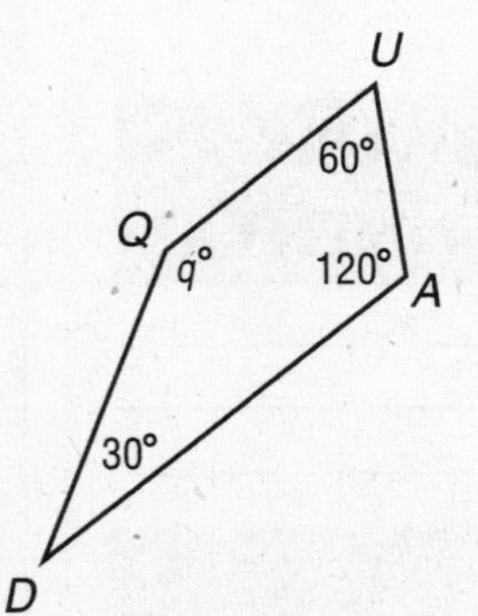

© Glencoe/McGraw-Hill

BUILD YOUR VOCABULARY (page 135)

A **trapezoid** is a quadrilateral with one pair of ______ opposite sides.

A **parallelogram** is a quadrilateral with both pairs of opposite sides ______ and ______.

A **rhombus** is a quadrilateral with ______ congruent sides.

A **rectangle** is a parallelogram with 4 ______ angles.

A **square** is a parallelogram with 4 ______ sides and 4 ______ angles.

EXAMPLES Classify Quadrilaterals

Classify each quadrilateral using the name that *best* describes it.

2 The quadrilateral has ______ congruent sides and ______ special angles.

It is a ______.

3 The quadrilateral has ______ sides congruent and ______ right angles.

It is a ______.

FOLDABLES

ORGANIZE IT

Using sketches, symbols, and words, describe how to classify quadrilaterals. Add this to the "Polygons" section of your Foldable table.

	words	model
lines		
polygons		
symmetry		
trans-formations		

Your Turn **Classify each quadrilateral using the name that *best* describes it.**

a. ______

b. ______

HOMEWORK ASSIGNMENT

Page(s):

Exercises:

© Glencoe/McGraw-Hill

6–5 Congruent Polygons

WHAT YOU'LL LEARN

- Identify congruent polygons.

Congruent Polygons
If two polygons are congruent, their corresponding sides are congruent and their corresponding angles are congruent.

EXAMPLE Identify Congruent Polygons

1 **Determine whether the trapezoids shown are congruent. If so, name the corresponding parts and write a congruence statement.**

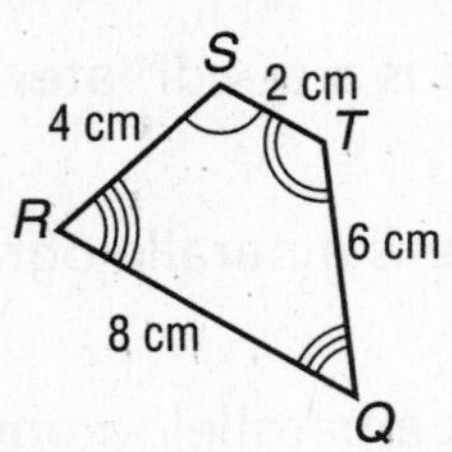

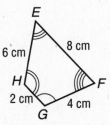

Angles The arcs indicate that $\angle S \cong \angle G$, $\angle T \cong \angle H$, $\angle Q \cong \angle E$, and ________.

Sides The side measures indicate that $\overline{ST} \cong \overline{GH}$, $\overline{TQ} \cong \overline{HE}$, $\overline{QR} \cong \overline{EF}$, and ________.

Since ________ pairs of corresponding angles and sides are ________, the two trapezoids are ________. One congruence statement is trapezoid $EFGH \cong$ trapezoid ________.

Your Turn Determine whether the triangles shown are congruent. If so, name the corresponding parts and write a congruence statement.

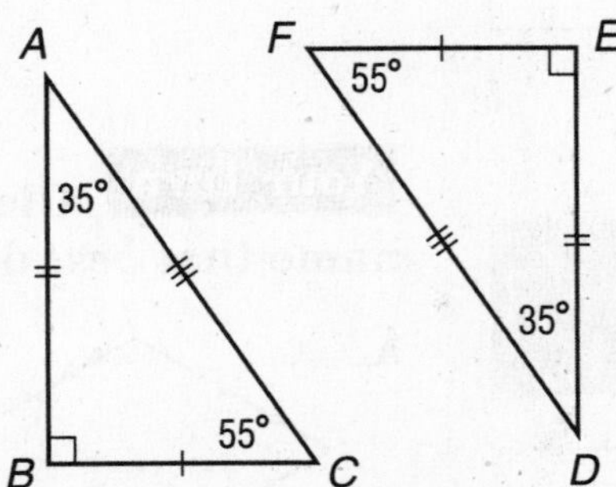

____

© Glencoe/McGraw-Hill

EXAMPLES Find Missing Measures

In the figure, $\triangle FGH$ $\triangle QRS$.

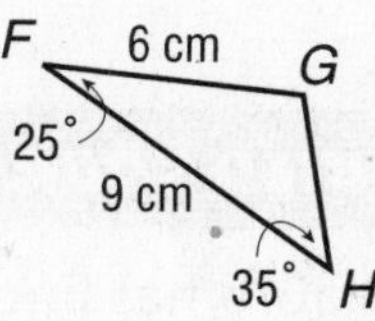

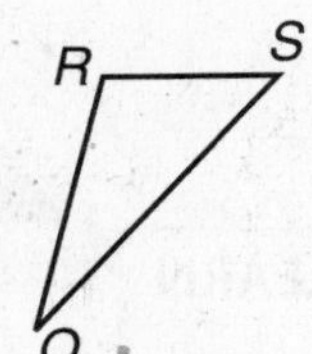

2 Find $m\angle S$.

According to the congruence statement, $\angle H$ and $\angle S$ are corresponding angles. So, ____ ____.

Since $m\angle H =$ ____, $m\angle S =$ ____.

3 Find QR.

$\overline{FG}$ corresponds to ____. So, ____ $\cong$ ____.

Since $FG =$ ____ centimeters, $QR =$ ____ centimeters.

Your Turn **In the figure, $\triangle ABC \cong \triangle LMN$.**

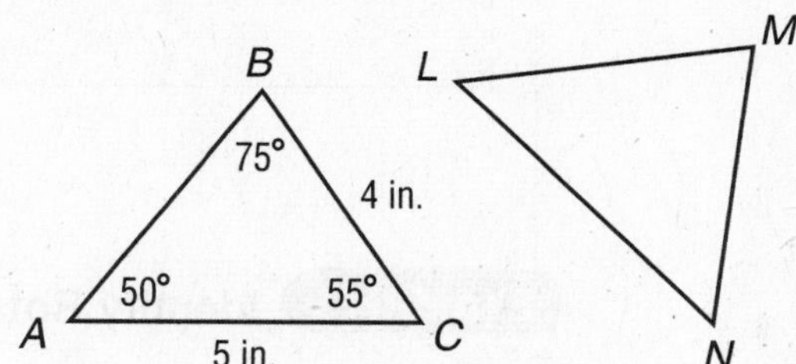

a. Find $m\angle N$.

b. Find LN.

© Glencoe/McGraw-Hill

HOMEWORK ASSIGNMENT

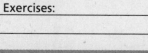

Page(s):

Exercises:

6–6 Symmetry

WHAT YOU'LL LEARN

- Identify line symmetry and rotational symmetry.

BUILD YOUR VOCABULARY (page 134)

A figure has **line symmetry** if it can be folded over a line so that one half of the figure ______ the other half.

EXAMPLE Identify Line Symmetry

1 TRILOBITES **The trilobite is an animal that lived millions of years ago. Determine whether the figure has line symmetry. If it does, draw all lines of symmetry. If not, write *none*.**

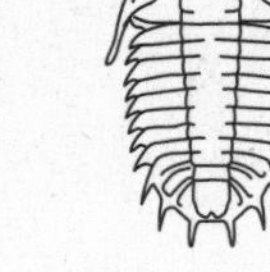

This figure has ______ line of symmetry.

Your Turn Determine whether the leaf has line symmetry. If it does, draw all lines of symmetry. If not, write *none*.

______

EXAMPLES Identify Rotational Symmetry

FLOWERS **Determine whether each flower design has rotational symmetry. Write *yes* or *no*. If *yes*, name its angle(s) of rotation.**

WRITE IT

How many degrees does one complete turn of a figure measure? Why is it this number of degrees?

______

______

______

2

Yes, this figure has  symmetry.

It will match itself after being rotated 90°, 180°, and ______.

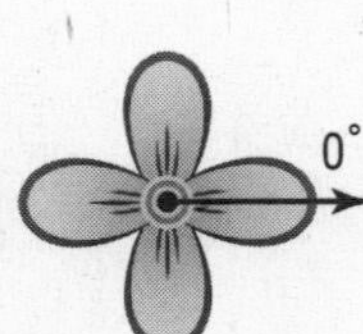

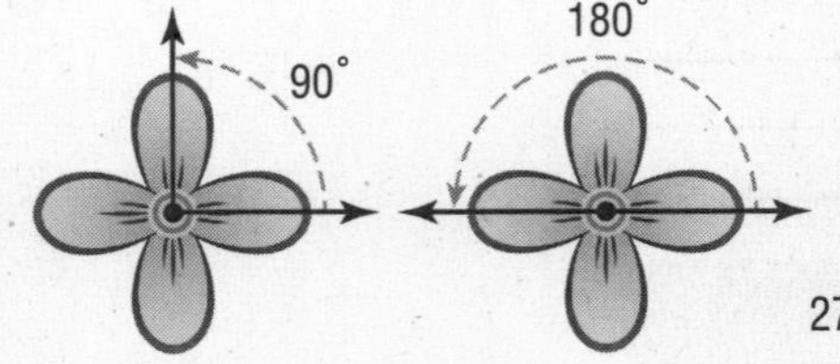

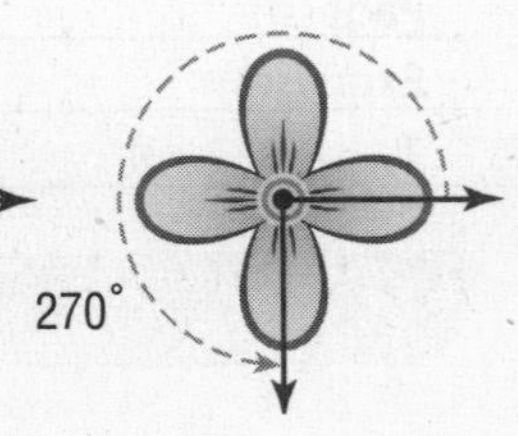

© Glencoe/McGraw-Hill

FOLDABLES™

ORGANIZE IT

Use sketches and words to show lines of symmetry and line symmetry. Write this in the "Symmetry" section of your Foldable table.

	words	model
lines		
polygons		
symmetry		
trans-formations		

3

Yes, this figure has  symmetry.

It will match itself after being rotated ,

120°, 180°, 240°, and 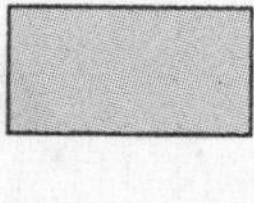.

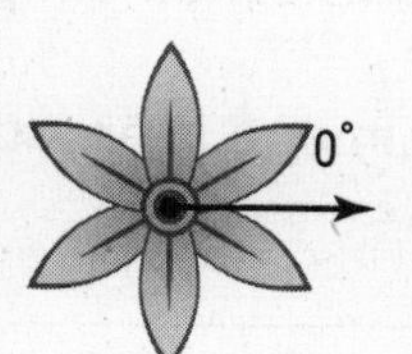

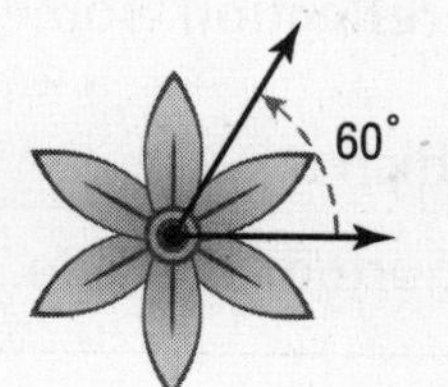

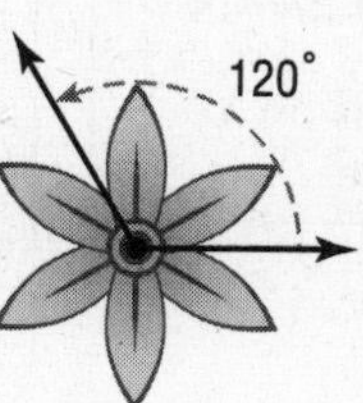

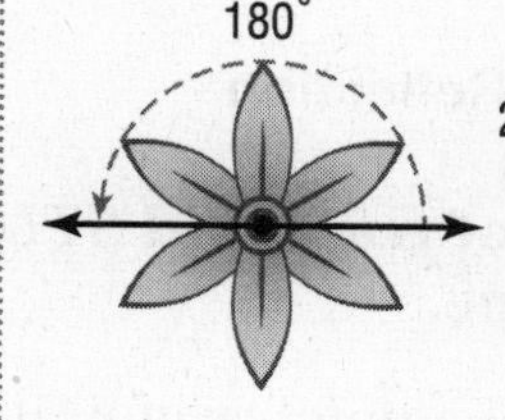

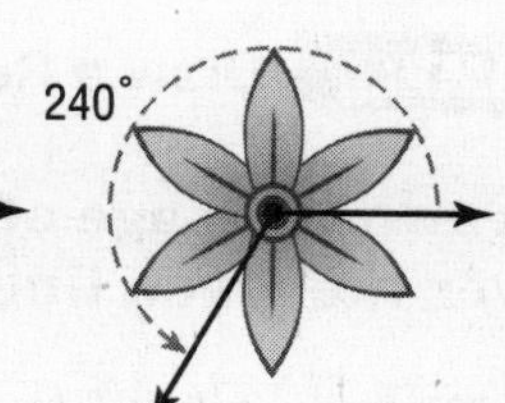

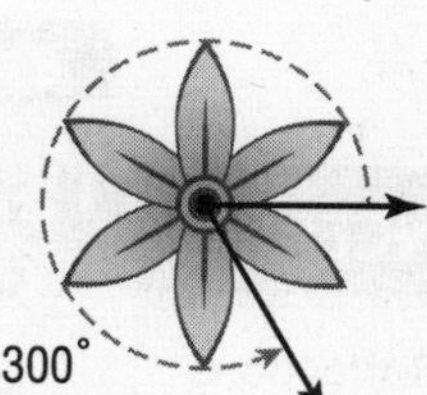

Your Turn **Determine whether the flower design has rotational symmetry. Write *yes* or *no*. If *yes*, name its angle(s) of rotation.**

a.

b.

HOMEWORK ASSIGNMENT

Page(s):

Exercises:

© Glencoe/McGraw-Hill

6–7 Reflections

BUILD YOUR VOCABULARY (page 135)

A **reflection** (sometimes called a *flip*) is a transformation in which a ______ image is produced by ______ a figure over a line.

WHAT YOU'LL LEARN

- Graph reflections on a coordinate plane.

EXAMPLE Draw a Reflection

KEY CONCEPT

Properties of Reflections

1. Every point on a reflection is the same distance from the line of reflection as the corresponding point on the original figure.
2. The image is congruent to the original figure, but the orientation of the image is *different* from that of the original figure.

1 Draw the image of trapezoid *STUV* after a reflection over the given line.

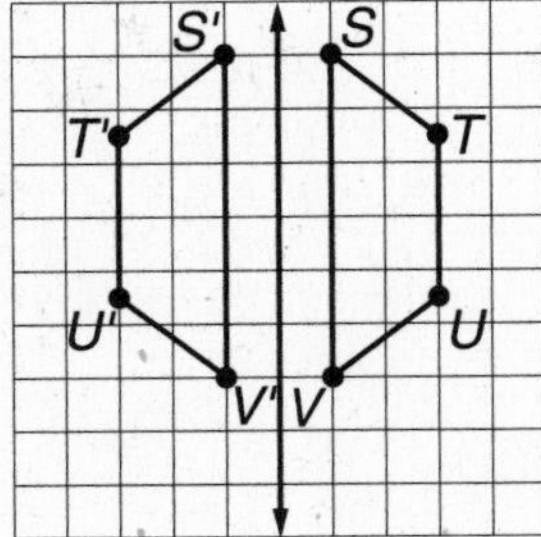

Step 1 Count the number of units between each vertex and the line of ______.

Step 2 Plot a point for each vertex the ______ distance away from the line on the other side.

Step 3 Connect the new ______ to form the image of trapezoid *STUV*, trapezoid *S′T′U′V′*.

Your Turn Draw the image of trapezoid *TRAP* after a reflection over the given line.

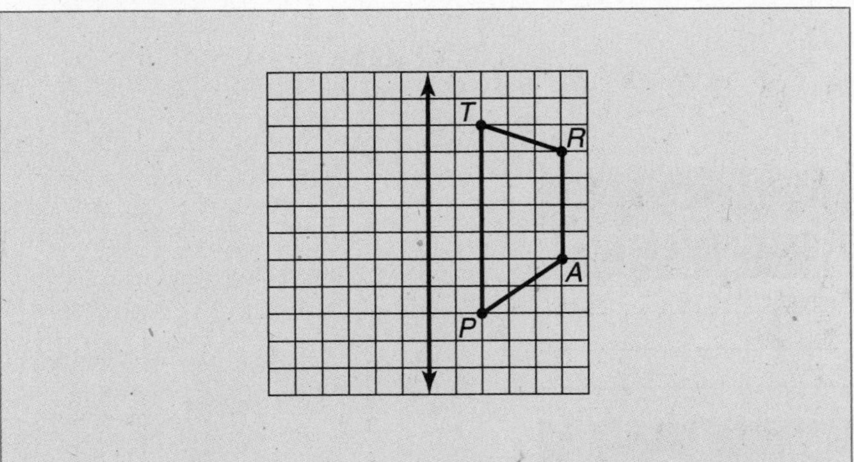

© Glencoe/McGraw-Hill

FOLDABLES™

ORGANIZE IT

Draw a triangle or simple quadrilateral on graph paper. Reflect your figure over the *x*-axis. Add your work to the "Symmetry" section of your Foldable table.

	words	model
lines		
polygons		
symmetry		
trans-formations		

EXAMPLE Reflect a Figure over the *x*-axis

2 **Graph quadrilateral *EFGH* with verticles *E*(−4, 4), *F*(3, 3), *G*(4, 2), and *H*(−2, 1). Then graph the image of *EFGH* after a reflection over the *x*-axis and write the coordinates of its vertices.**

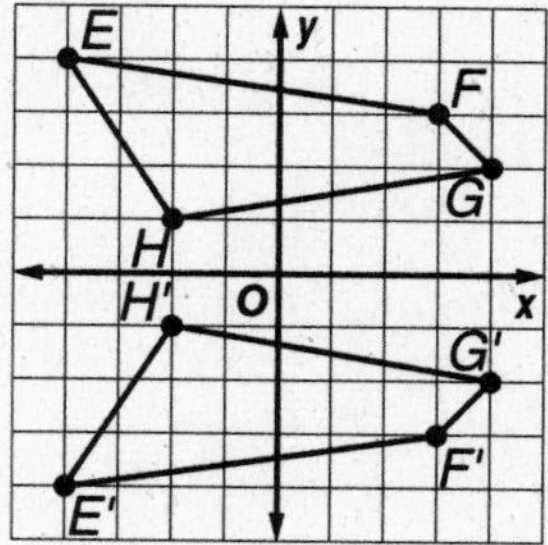

The coordinates of the verticles of the image are *E'* ______,

F' ______, *G'* ______ and *H'* ______.

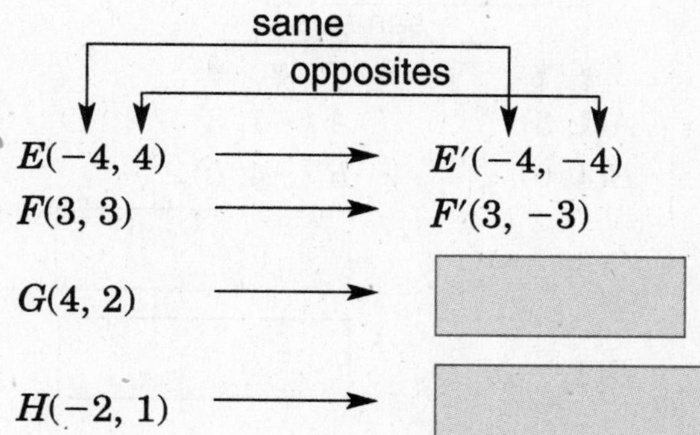

Notice that the *y*-coordinate of a point reflected over the *x*-axis

is the ______ of the *y*-coordinate of the original point.

Your Turn Graph quadrilateral *QUAD* with vertices *Q*(2, 4), *U*(4, 1), *A*(−1, 1), and *D*(−3, 3). Then graph the image of *QUAD* after a reflection over the *x*-axis, and write the coordinates of its vertices.

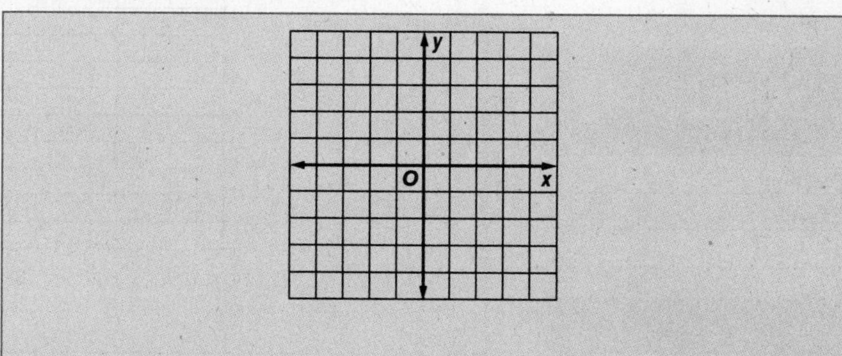

© Glencoe/McGraw-Hill

EXAMPLE Reflect a Figure over the *y*-axis

3 Graph trapezoid *ABCD* with vertices *A*(1, 3), *B*(4, 0), *C*(3, −4), and *D*(1, −2). Then graph the image of *ABCD* after a reflection over the *y*-axis, and write the coordinates of its vertices.

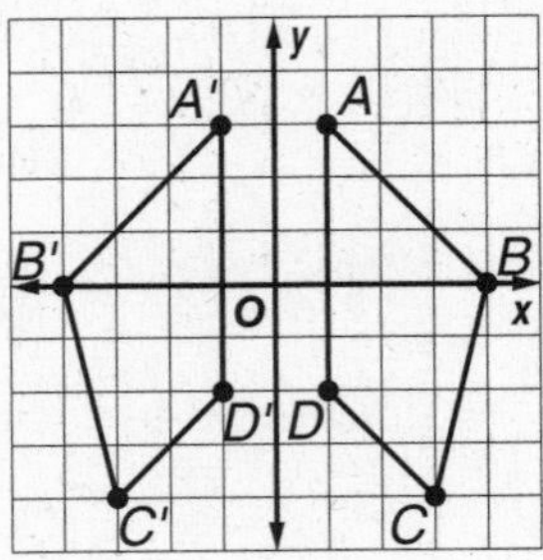

The coordinates of the vertices of the image are *A'* ______, *B'* ______, *C'* ______, and *D'* ______.

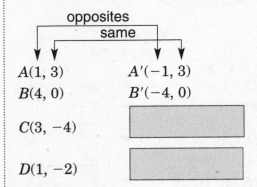

opposites
same

A(1, 3) *A'*(−1, 3)

B(4, 0) *B'*(−4, 0)

C(3, −4) ______

D(1, −2) ______

Notice that the *x*-coordinate of a point reflected over the *y*-axis is the opposite of the *x*-coordinate of the ______ point.

Your Turn Graph quadrilateral *ABCD* with vertices *A*(2, 2), *B*(5, 0), *C*(4, −2), and *D*(2, −1). Then graph the image of ABCD after a reflection over the *y*-axis, and write the coordinates of its vertices.

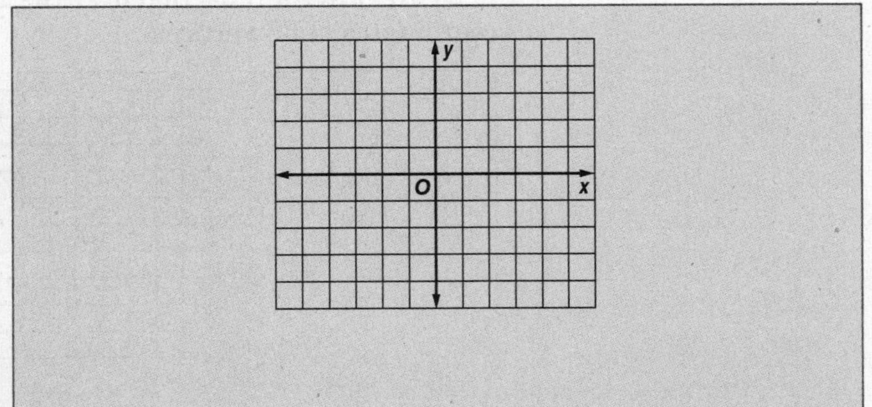

HOMEWORK ASSIGNMENT

Page(s):

Exercises:

© Glencoe/McGraw-Hill

6–8 Translations

BUILD YOUR VOCABULARY (page 135)

A **translation** (sometimes called a *slide*) is the ________ of a figure from one position to another ________ turning it.

WHAT YOU'LL LEARN

- Graph translations on a coordinate plane.

KEY CONCEPT

Properties of Translations

1. Every point on the original figure is moved the same distance and in the same direction.
2. The image is congruent to the original figure, and the orientation of the image is *the same* as that of the original figure.

EXAMPLE Draw a Translation

1 **Draw the image of $\triangle EFG$ after a translation of 3 units right and 2 units up.**

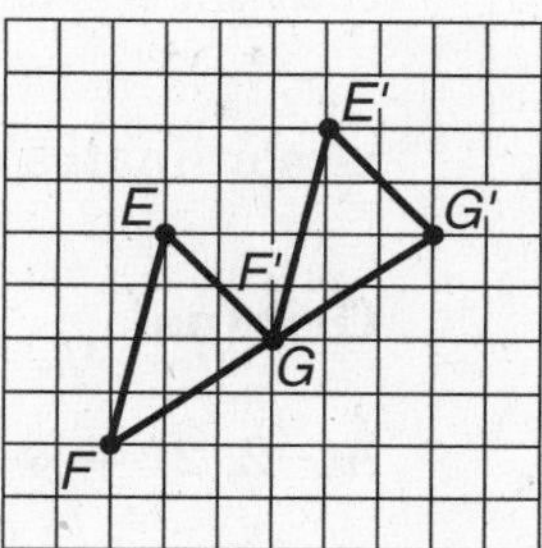

Step 1 Move each vertex of the triangle ________ units right and ________ units up.

Step 2 Connect the new vertices to form the ________.

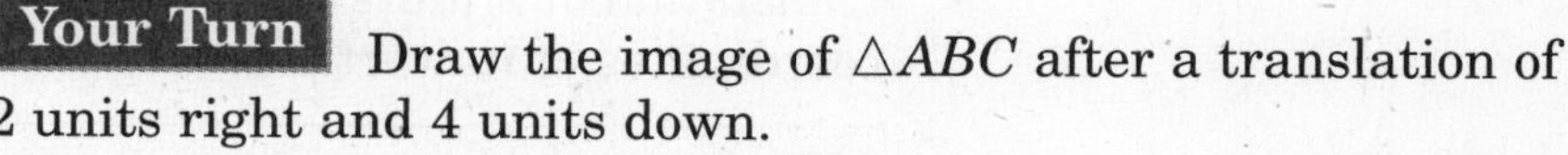

Your Turn Draw the image of $\triangle ABC$ after a translation of 2 units right and 4 units down.

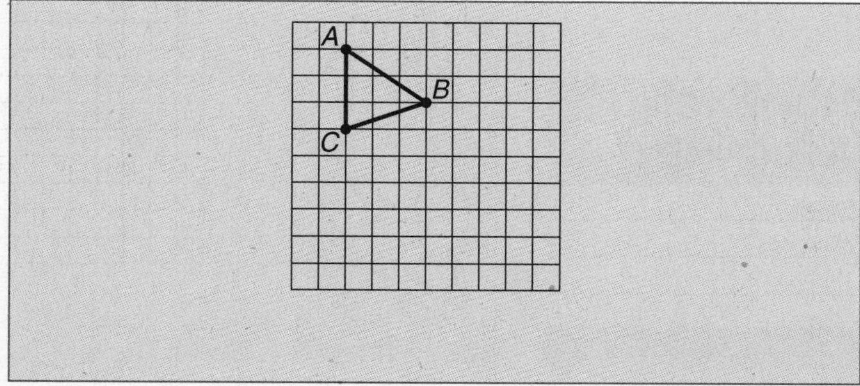

© Glencoe/McGraw-Hill

EXAMPLE Translation in the Coordinate Plane

2 **Graph $\triangle ABC$ with vertices $A(-2, 2)$, $B(3, 4)$, and $C(4, 1)$. Then graph the image of $\triangle ABC$ after a translation of 2 units left and 5 units down. Write the coordinates of its vertices.**

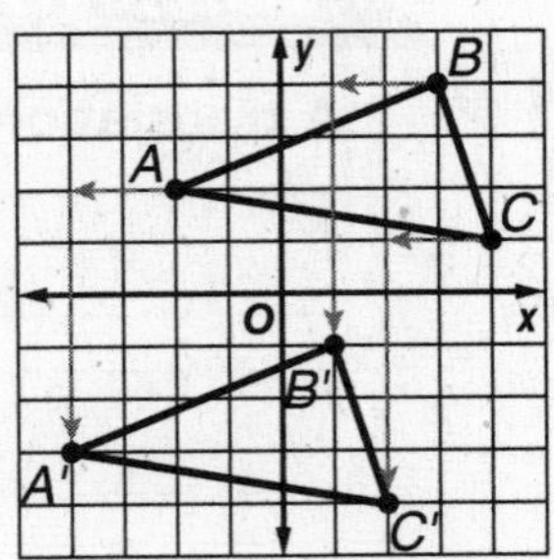

The coordinates of the vertices of the image are

A' ______, B' ______, and C' ______. Notice that these vertices can also be found by adding ______ to the x-coordinates and ______ to the y-coordinates, or $(-2, -5)$.

Original	**Add (−2, −5)**	**Image**
$A(-2, 2)$ →	$(-2 + (-2), 2 + (-5))$	→ ______
$B(3, 4)$ →	$(3 + (-2), 4 + (-5))$	→ ______
$C(4, 1)$ →	$(4 + (-2), 1 + (-5))$	→ ______

Your Turn Graph $\triangle PQR$ with vertices $P(-1, 3)$, $Q(2, 4)$, and $R(3, 2)$. Then graph the image of $\triangle PQR$ after a translation of 2 units right and 3 units down. Write the coordinates of its vertices.

y
O
x

FOLDABLES™

ORGANIZE IT

Draw a triangle or simple quadrilateral on graph paper. Then draw a translation. Show how you determined the points needed to graph the translated figure. Put your work in the "Transformations" section of your Foldable table.

	words	model
lines		
polygons		
symmetry		
trans-formations		

HOMEWORK ASSIGNMENT

Page(s):

Exercises:

© Glencoe/McGraw-Hill

6–9 Rotations

BUILD YOUR VOCABULARY (page 135)

A **rotation** (sometimes called a *turn*) is a transformation involving the ________ or spinning of a figure around a fixed point.

WHAT YOU'LL LEARN

- Graph rotations on a coordinate plane.

KEY CONCEPT

Properties of Rotations

1. Corresponding points are the same distance from *R*. The angles formed by connecting *R* to corresponding points are congruent.
2. The image is congruent to the original figure, and their orientations are *the same*.

EXAMPLE Rotations in the Coordinate Plane

1 **Graph $\triangle QRS$ with vertices $Q(1, 1)$, $R(3, 4)$, and $S(4, 1)$. Then graph the image of $\triangle QRS$ after a rotation of 180° counterclockwise about the origin, and write the coordinates of its vertices.**

Step 1 Lightly draw a line connecting point *Q* to the ________.

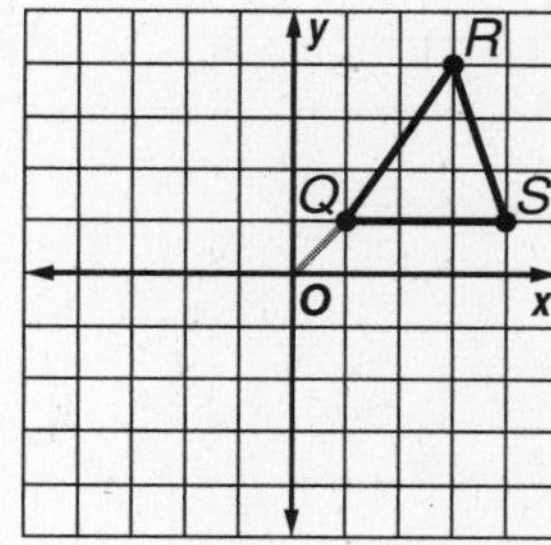

Step 2 Lightly draw $\overline{OQ'}$ so that $m\angle Q'OQ = 180°$ and $OQ' = OQ$.

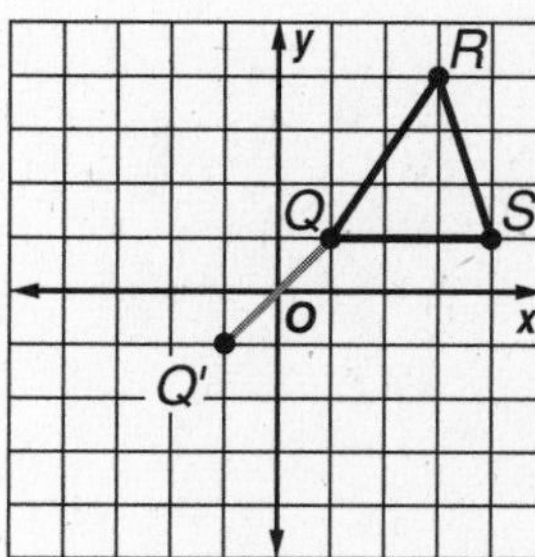

Step 3 Repeat steps 1–3 for points *R* and *Q*. Then erase all lightly drawn lines and connect the ________ to form $\triangle Q'R'S'$.

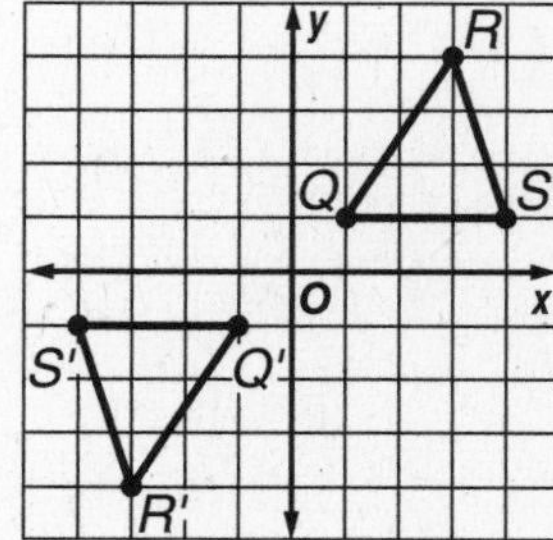

Triangle $Q'R'S'$ has vertices

Q' ________, R' ________, and S' ________.

© Glencoe/McGraw-Hill

FOLDABLES™

ORGANIZE IT

Draw a triangle on graph paper. Then draw a rotation. Label the rotation with degrees and direction of rotation. Put your work in the "Transformations" section of your Foldable table.

	words	model
lines		
polygons		
symmetry		
trans-formations		

Your Turn Graph $\triangle ABC$ with vertices $A(4, 1)$, $B(2, 1)$, and $C(2, 4)$. Then graph the image of $\triangle ABC$ after a rotation of 180° counterclockwise about the origin, and write the coordinates of its vertices.

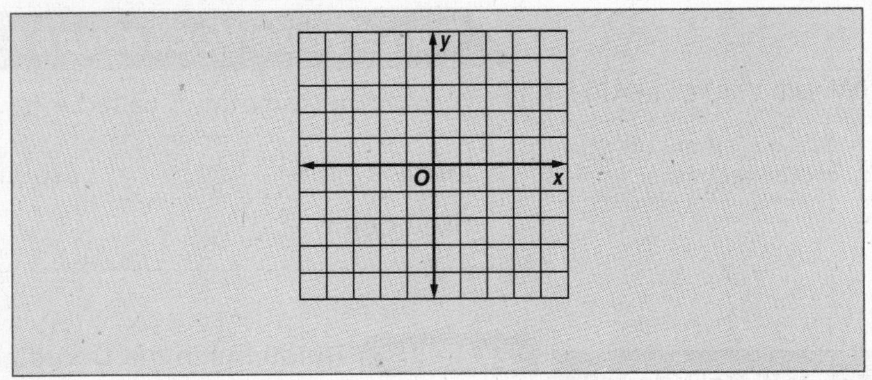

EXAMPLE Use a Rotation

2 QUILTS **Complete the quilt piece shown below so that the completed figure has rotational symmetry with 90°, 180°, and 270°, as its angles of rotation.**

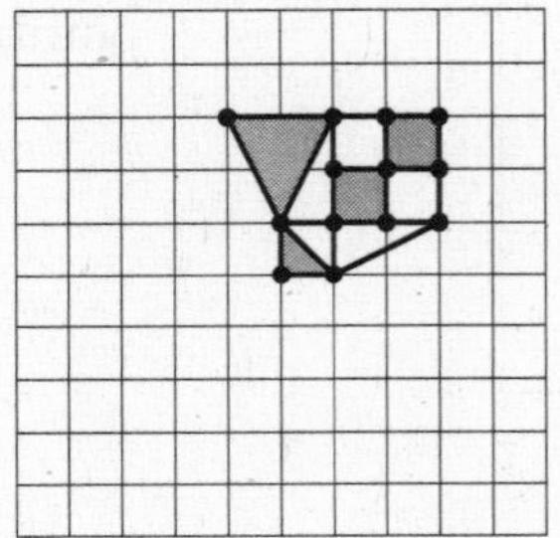

Rotate the figure 90°, 180°, and 270° counterclockwise. Use a ______ rotation *clockwise* to produce the same rotation as a 270° rotation *counterclockwise.*

90° counterclockwise

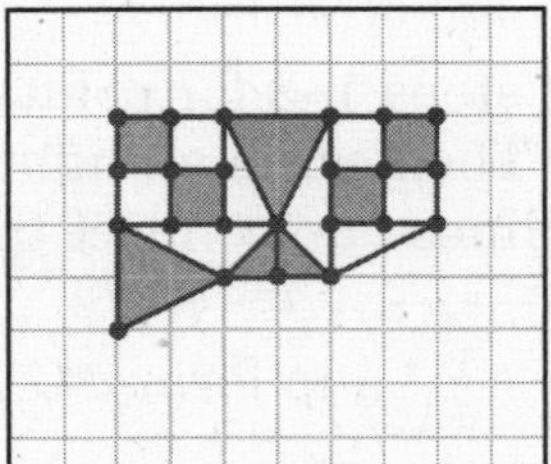

180°counterclockwise 90° clockwise

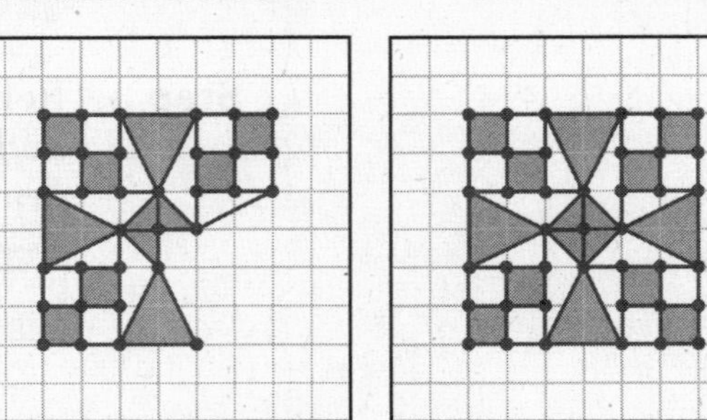

© Glencoe/McGraw-Hill

Your Turn Complete the quilt piece shown below so that the completed figure has rotational symmetry with 90°, 180°, and 270°, as its angles of rotation.

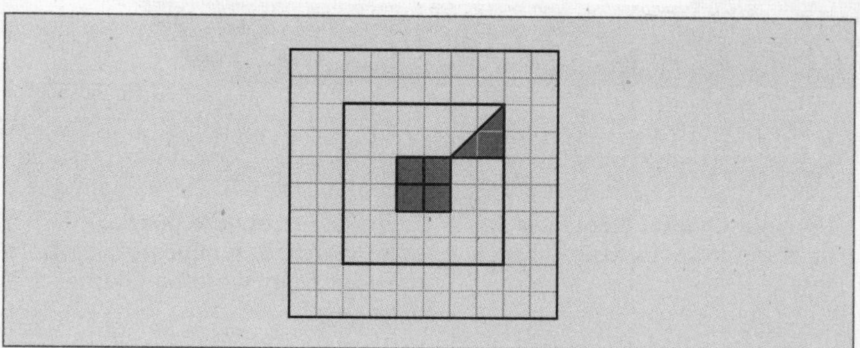

© Glencoe/McGraw-Hill

HOMEWORK ASSIGNMENT

Page(s):

Exercises:

CHAPTER 6

BRINGING IT ALL TOGETHER

STUDY GUIDE

FOLDABLES™	VOCABULARY PUZZLEMAKER	BUILD YOUR VOCABULARY
Use your **Chapter 6 Foldable** to help you study for your chapter test.	To make a crossword puzzle, word search, or jumble puzzle of the vocabulary words in Chapter 6, go to www.glencoe.com/sec/math/t_resources/free/index.php.	You can use your completed **Vocabulary Builder** (*pages 134–135*) to help you solve the puzzle.

6-1

Line and Angle Relationships

Match each figure to the words that describe it.

1. obtuse

2. right

3. acute

4. straight

5. 180°

6. 90° 

a. 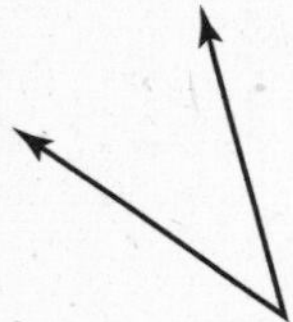

b. 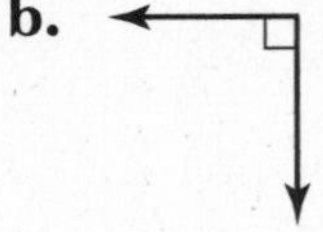

c. 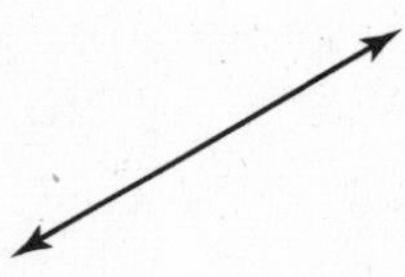

d. 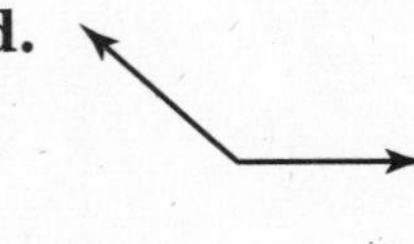

7. How do you know if two lines are perpendicular?

For Questions 8–10, use the figure at the right.

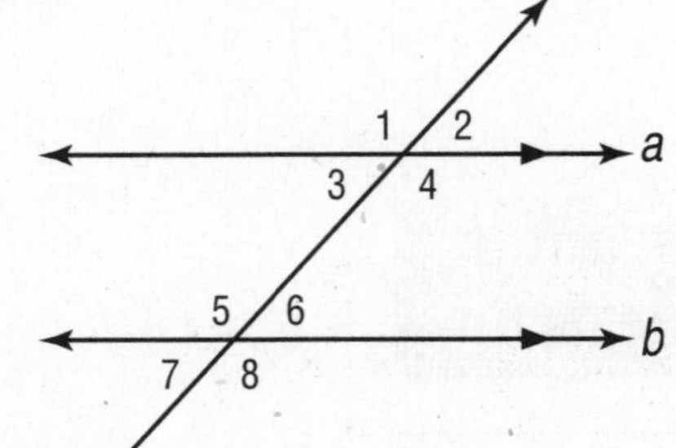

8. Look at $\angle 5$ and $\angle 6$. Classify the angle pair using all names that apply.

9. Find $m\angle 3$ if $m\angle 2 = 60°$.

10. Find $m\angle 4$ if $m\angle 2 = 60°$. 

© Glencoe/McGraw-Hill

6-2

Triangles and Angles

11. How can you determine whether a triangle is a right triangle?

12. If one angle of a triangle is either right or obtuse, what must be true of the other two angles?

13. Complete the table.

Type of Triangle	Number of Congruent Sides
Scalene	
Isosceles	
Equilateral	

14. Triangle *RST* is an equilateral triangle. What is the measure of angle *T*? How do you know?

6-3

Special Right Angles

15. The two legs in a 45°-45° right triangle are always congruent. Why?

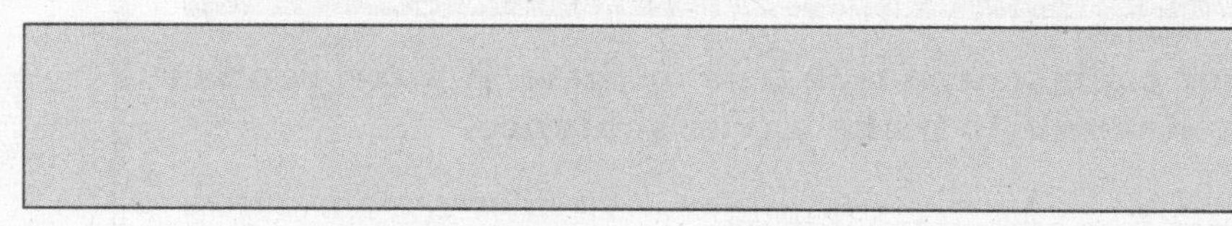

16. Triangle *FGH* is a 45°-45° right triangle. The shorter side is 18 inches. Find the length of the longest side, rounded to the nearest tenth.

© Glencoe/McGraw-Hill

6-4 Classifying Quadrilaterals

17. Refer to the quadrilateral shown. Explain how to find the value of y in quadrilateral $ABCD$.

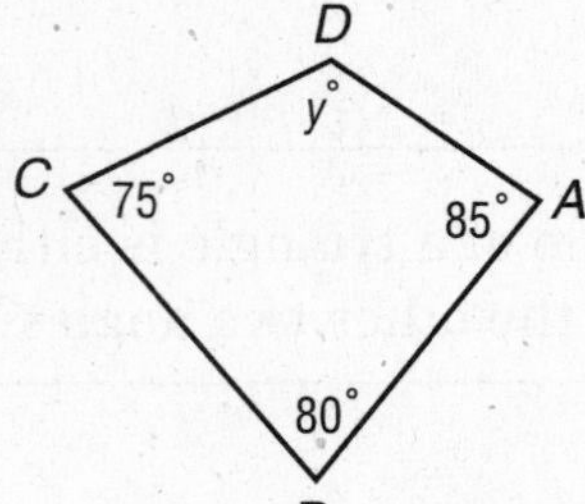

18. What is the value of y?

6-5 Congruent Polygons

19. Complete the sentence. Two polygons are congruent if their ______ sides are congruent and the corresponding angles are ______.

$\triangle ABC \cong \triangle EDF$. $m\angle A = 40°$ and $m\angle B = 50°$
$\angle E \cong \angle A$ and $\angle F \cong \angle C$

20. What is $m\angle C$?

21. What is $m\angle D$?

6-6 Symmetry

Write whether each sentence is *true* or *false*. If *false*, replace the underlined words to make a true sentence.

22. A figure has line symmetry if it can be <u>folded over a line</u> so that one half of the figure matches the other half.

23. To rotate a figure means to turn the figure from its <u>center</u>.

24. A figure has rotational symmetry if it first matches itself after being rotated <u>exactly</u> 360°.

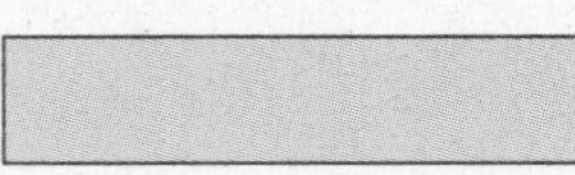

© Glencoe/McGraw-Hill

6-7 Reflections

25. Complete. A reflection is a [ ] image of a figure produced by flipping the figure over a line.

26. If you graphed quadrilateral *HIJK* reflected over the *y*-axis, what would be the coordinates of these vertices:

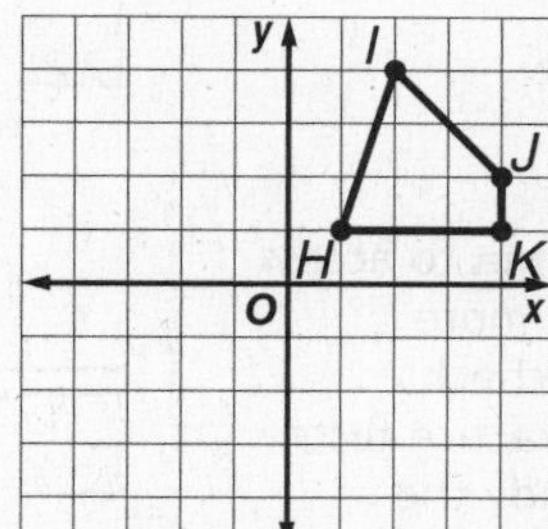

H'([ ]) J'([ ])

6-8 Translations

27. Complete. A translation is the movement of a figure from one position to another [ ] turning it.

28. If you graphed the image of quadrilateral *DEFG* after a translation 3 units right and 4 units down, what would be the coordinates of these vertices:

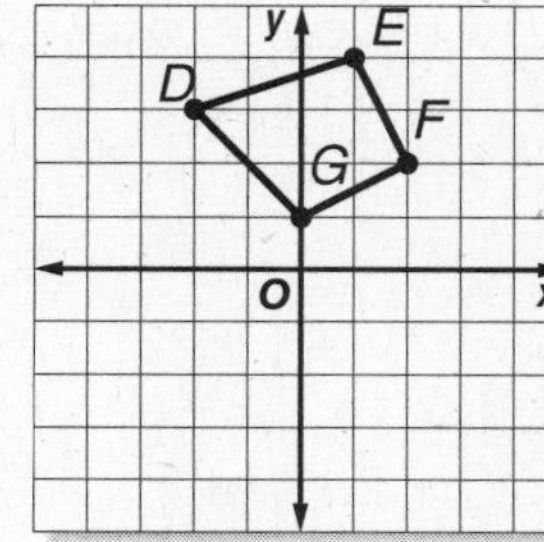

D'([ ]) F'([ ])

6-9 Rotations

29. Find the distance of each of the vertices *A, B, A'*, and *B'* from the given point. Are corresponding vertices, such as *A* and *A'*, the same distance from the point?

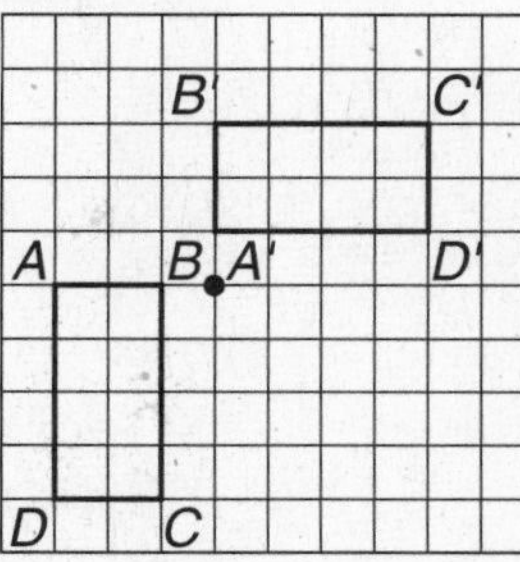

30. Is quadrilateral $A'B'C'D'$ a rotation of quadrilateral *ABCD* about the given point? Why or why not?

© Glencoe/McGraw-Hill

ARE YOU READY FOR THE CHAPTER TEST?

Checklist

Visit **msmath3.net** to access your textbook, more examples, self-check quizzes, and practice tests to help you study the concepts in Chapter 6.

Check the one that applies. Suggestions to help you study are given with each item.

- [ ] **I completed the review of all or most lessons without using my notes or asking for help.**
  - You are probably ready for the Chapter Test.
  - You may want to take the Chapter 6 Practice Test on page 309 of your textbook as a final check.

- [ ] **I used my Foldable or Study Notebook to complete the review of all or most lessons.**
  - You should complete the Chapter 6 Study Guide and Review on pages 306–308 of your textbook.
  - If you are unsure of any concepts or skills, refer to the specific lesson(s).
  - You may also want to take the Chapter 6 Practice Test on page 309.

- [ ] **I asked for help from someone else to complete the review of all or most lessons.**
  - You should review the examples and concepts in your Study Notebook and Chapter 6 Foldable.
  - Then complete the Chapter 6 Study Guide and Review on pages 306–308 of your textbook.
  - If you are unsure of any concepts or skills, refer to the specific lesson(s).
  - You may also want to take the Chapter 6 Practice Test on page 309.

Student Signature

Parent/Guardian Signature

Teacher Signature

© Glencoe/McGraw-Hill

Geometry: Measuring Area and Volume

Use the instructions below to make a Foldable to help you organize your notes as you study the chapter. You will see Foldable reminders in the margin of this Interactive Study Notebook to help you in taking notes.

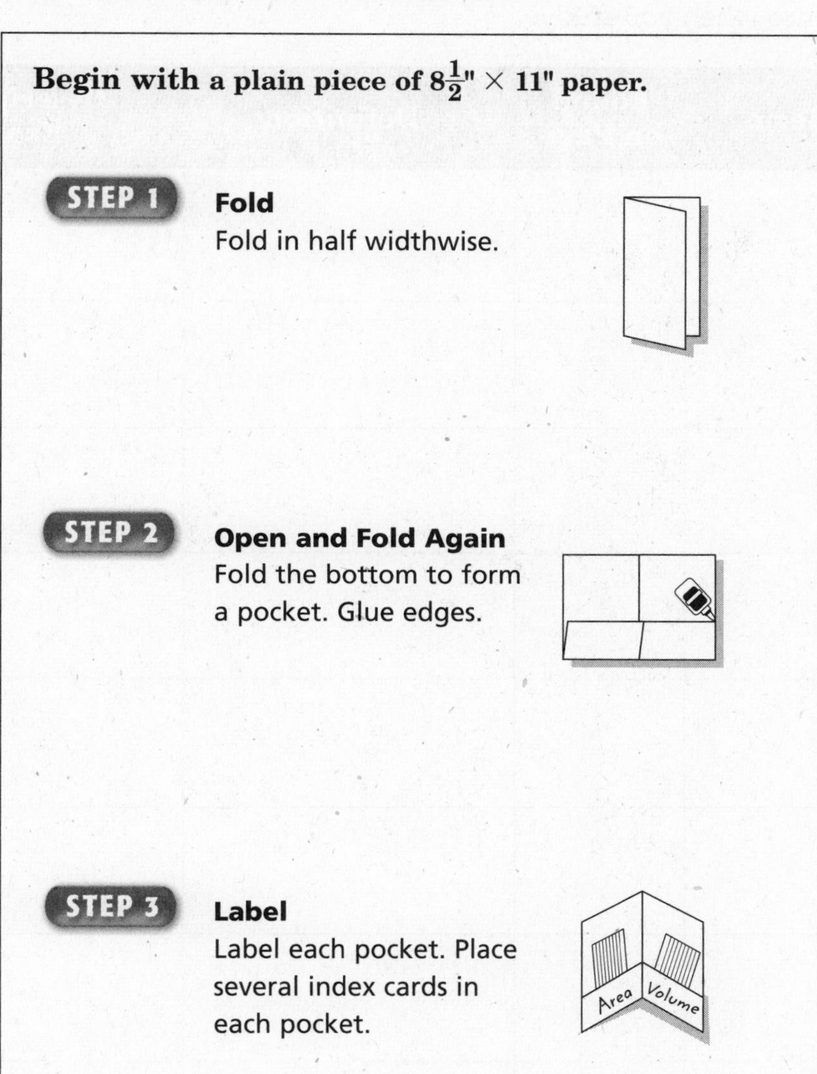

Begin with a plain piece of $8\frac{1}{2}$" × 11" paper.

STEP 1 **Fold**
Fold in half widthwise.

STEP 2 **Open and Fold Again**
Fold the bottom to form a pocket. Glue edges.

STEP 3 **Label**
Label each pocket. Place several index cards in each pocket.

NOTE-TAKING TIP: As you read and learn a new concept, such as how to measure area or volume, write examples and explanations showing the main ideas of the concept.

© Glencoe/McGraw-Hill

BUILD YOUR VOCABULARY

This is an alphabetical list of new vocabulary terms you will learn in Chapter 7. As you complete the study notes for the chapter, you will see Build Your Vocabulary reminders to complete each term's definition or description on these pages. Remember to add the textbook page number in the second column for reference when you study.

Vocabulary Term	Found on Page	Definition	Description or Example
circumference			
complex figure			
cone			
cylinder			
diameter			
edge			
face			
lateral area			
lateral face			

© Glencoe/McGraw-Hill

Vocabulary Term	Found on Page	Definition	Description or Example
polyhedron			
precision			
prism			
pyramid			
radius			
significant digits			
slant height			
surface area			
vertex			
volume			

© Glencoe/McGraw-Hill

7-1 Area of Parallelograms, Triangles, and Trapezoids

WHAT YOU'LL LEARN

- Find the areas of parallelograms, triangles, and trapezoids.

KEY CONCEPT

Area of a Parallelogram, Triangle, and Trapezoid

Area of a Parallelogram The area A of a parallelogram is the product of any base b and its height h.

Area of a Triangle The area A of a triangle is half the product of any base b and its height h.

Area of a Trapezoid The area A of a trapezoid is half the product of the height h, and the sum of the bases b_1 and b_2.

EXAMPLE Find the Area of a Parallelogram

1 **Find the area of the parallelogram.**

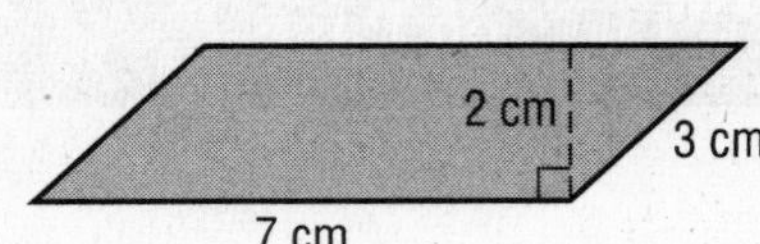

The base is ____ centimeters. The height is ____ centimeters.

$A = bh$ Area of a ____

$A =$ ____ · ____ Replace b with ____ and h with ____.

$A =$ ____ Multiply.

The area is ____ square centimenters.

EXAMPLE Find the Area of a Triangle

2 **Find the area of the triangle.**

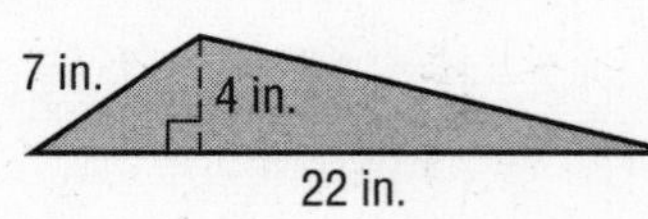

The base is ____ inches. The height is ____ inches.

$A = \frac{1}{2}bh$ Area of a ____

$A = \frac{1}{2}($ ____ $)($ ____ $)$ Replace b with ____ and h with ____.

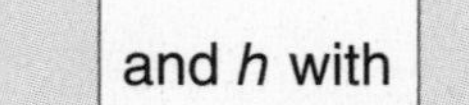

$A = \frac{1}{2}($ ____ $)$ Multiply. $22 \times 4 =$ ____

$A =$ ____ Multiply. $\frac{1}{2} \times 88 =$ ____

The area is ____ square inches.

© Glencoe/McGraw-Hill

FOLDABLES™

ORGANIZE IT

On index cards, write the formula for the area of a parallelogram, a triangle, and a trapezoid. Sketch each figure and label the parts. Place your cards in the "Area" pocket of your Foldable.

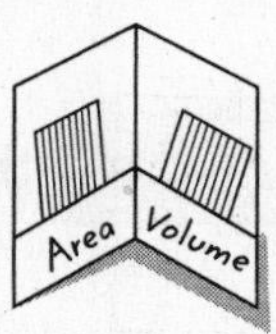

Your Turn **Find the area of each figure.**

a.

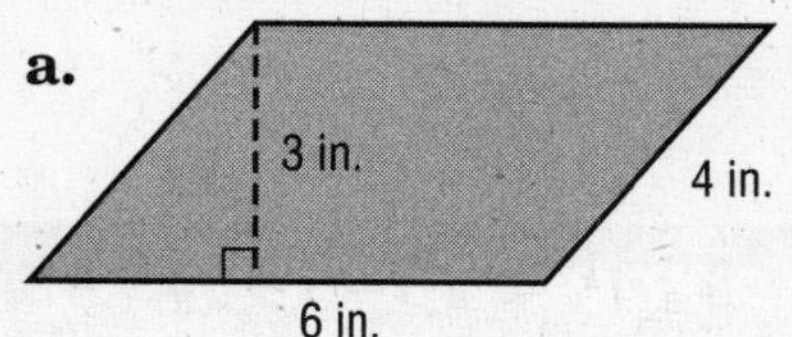

b.

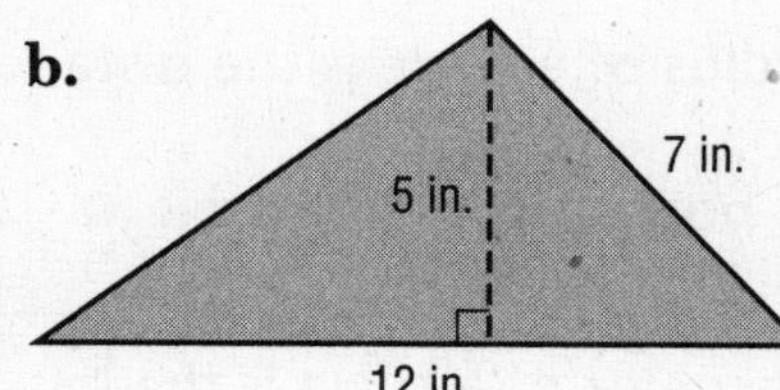

EXAMPLE **Find the Area of a Trapezoid**

3 Find the area of the trapezoid.

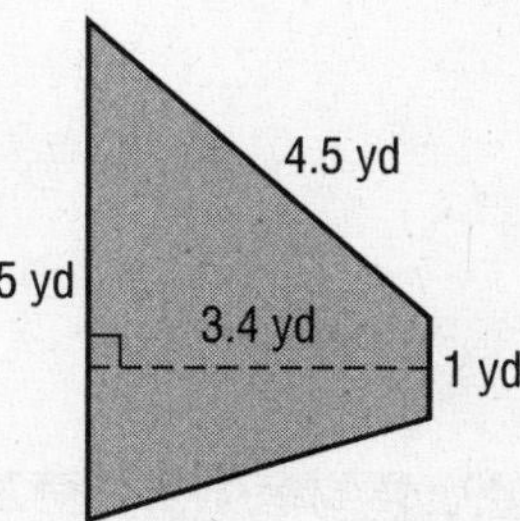

The height is ______ yards. The lengths 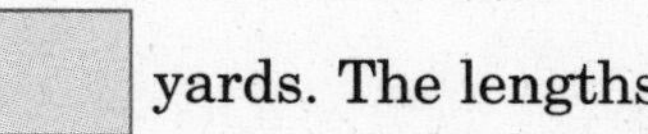

of the bases are ______ yards and ______ yard.

$A = \frac{1}{2}h(b_1 + b_2)$ Area of a ______

$A = \frac{1}{2}(\quad)(5 + 1)$ Replace h with ______, b_1 with 5, and b_2 with 1.

$A = \frac{1}{2}(\quad)(\quad)$ or ______ Simplify.

The area of the trapezoid is ______ square yards.

Your Turn Find the area of the trapezoid.

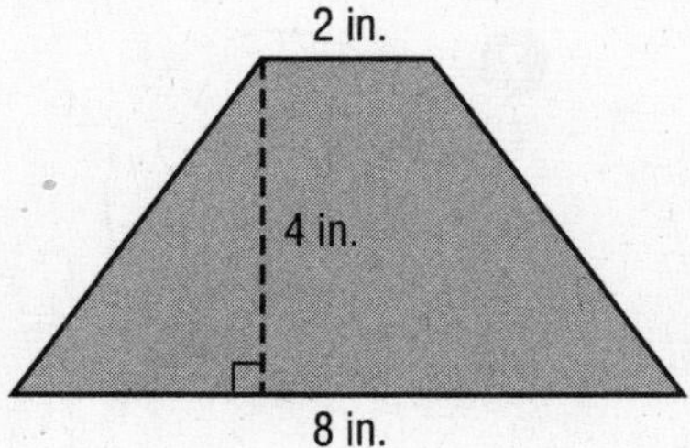

HOMEWORK ASSIGNMENT

Page(s):

Exercises:

© Glencoe/McGraw-Hill

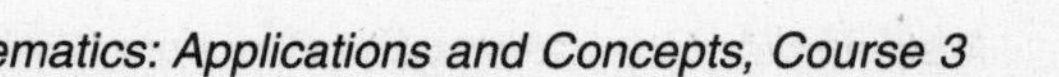

7–2 Circumference and Area of Circles

WHAT YOU'LL LEARN

- Find the circumference and the area of circles.

BUILD YOUR VOCABULARY (pages 164–165)

The **radius** of a circle is the distance from the ______ to any point ______ the circle.

The **diameter** of a circle is the ______ the circle through the center.

The **circumference** of a circle is the ______ the circle.

KEY CONCEPTS

Circumference of a Circle The circumference C of a circle is equal to its diameter d times π, or 2 times its radius r times π.

Area of a Circle The area A of a circle is equal to π times the square of the radius r.

EXAMPLES Find the Circumferences of Circles

Find the circumference of each circle. Round to the nearest tenth.

1

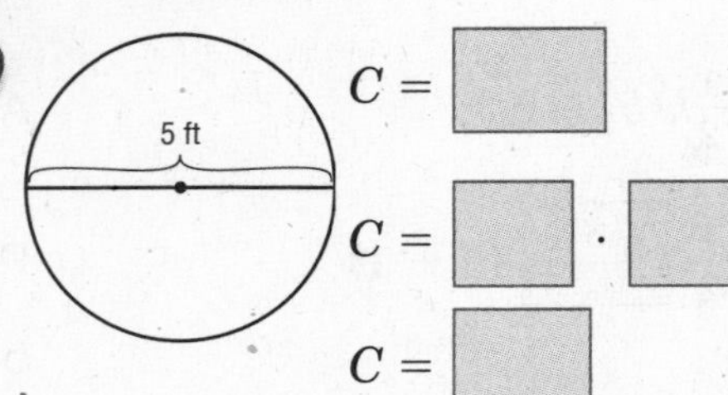

$C =$ ______ Circumference of a circle

$C =$ ______ $\cdot$ ______ Replace d with ______.

$C =$ ______ This is the exact circumference.

Use a calculator to find 5π.

5 [×] [π] [ENTER =] 15.70796327

The circumference is about ______.

2

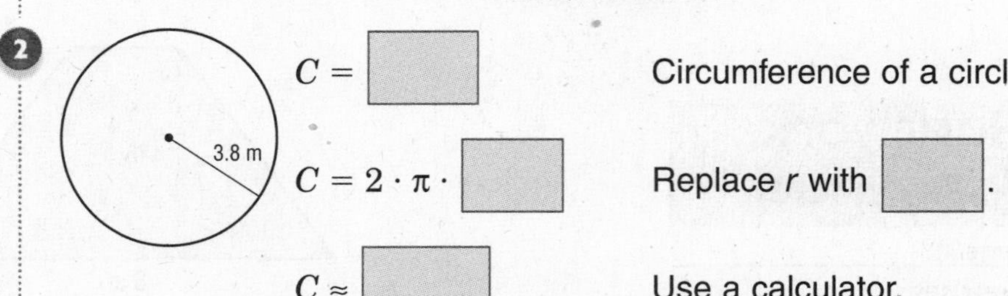

$C =$ ______ Circumference of a circle

$C = 2 \cdot \pi \cdot$ ______ Replace r with ______.

$C \approx$ ______ Use a calculator.

The circumference is about ______.

© Glencoe/McGraw-Hill

Your Turn **Find the circumference of each circle. Round to the nearest tenth.**

a.

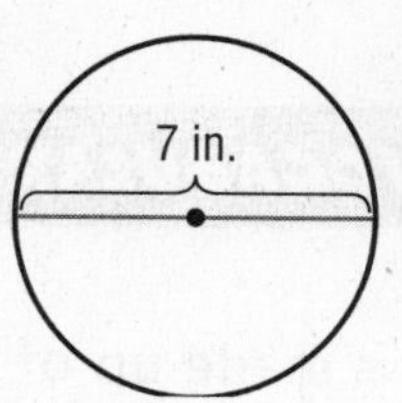

b.

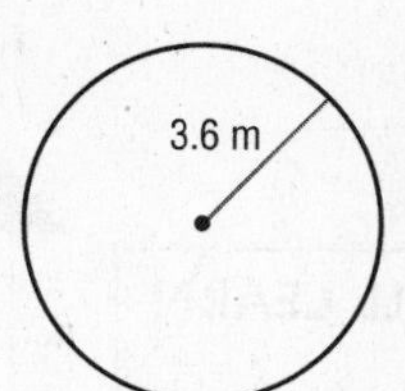

EXAMPLES Find the Areas of Circles

Find the area of each circle. Round to the nearest tenth.

FOLDABLES™

ORGANIZE IT

On index cards, write the formulas for finding the circumference and area of a circle. Sketch a circle and label its parts. Place your cards in the "Area" pocket of your Foldable.

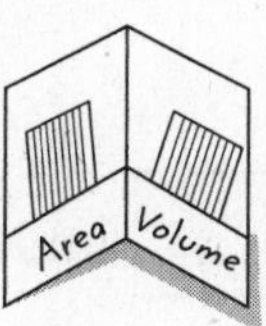

3

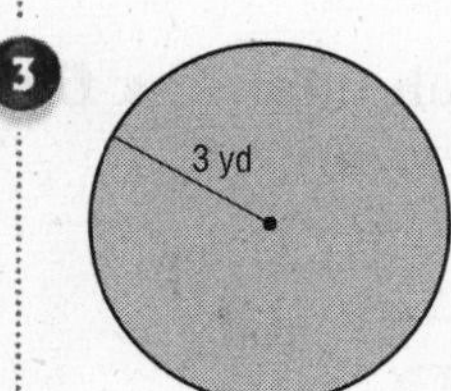

$A =$ ______ Area of a circle

$A = \pi \cdot$ ______^{2} Replace r with ______.

$A = \pi \cdot$ ______ Evaluate 3^2.

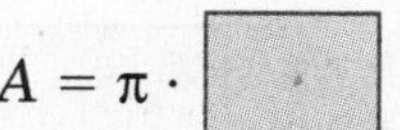

$A \approx$ ______ Use a calculator.

The area is about ______.

4

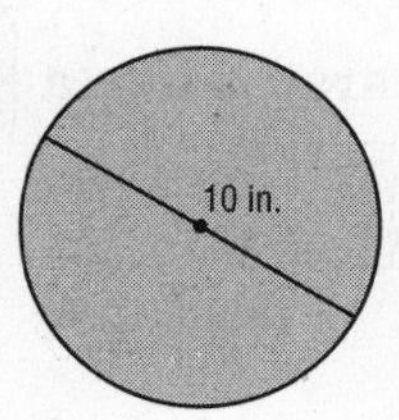

$A = \pi r^2$ Area of a circle

$A = \pi \cdot$ ______^{2} $r = \frac{1}{2}$ of 10

$A = \pi \cdot$ ______ Evaluate 5^2.

$A \approx$ ______ Use a calculator.

The area is about ______.

Your Turn **Find the area of each circle. Round to the nearest tenth.**

a.

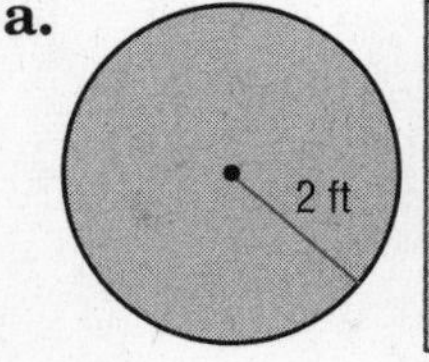

b.

HOMEWORK ASSIGNMENT

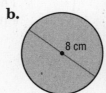

Page(s):

Exercises:

© Glencoe/McGraw-Hill

7-3 Area of Complex Figures

WHAT YOU'LL LEARN

- Find the area of complex figures.

BUILD YOUR VOCABULARY (pages 164–165)

A **complex figure** is made up of ______ shapes.

EXAMPLES Find the Areas of Complex Figures

Find the area of each complex figure. Round to the nearest tenth if necessary.

1

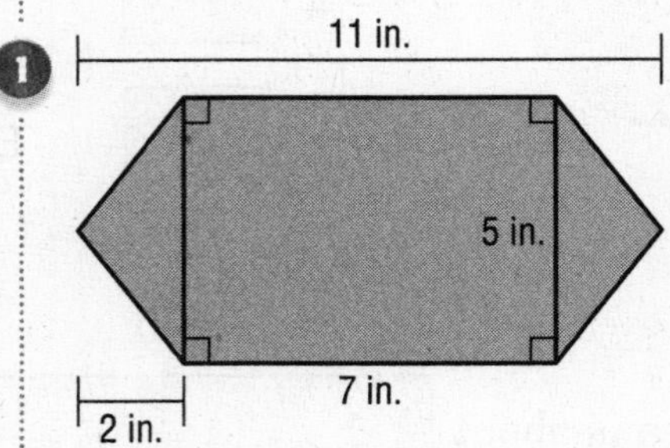

The figure can be separated into a ______ and two congruent ______.

Area of rectangle

$A = \ell w$

$A =$ ______

$A =$ ______

Area of one triangle

$A = \frac{1}{2}bh$

$A =$ ______

$A =$ ______

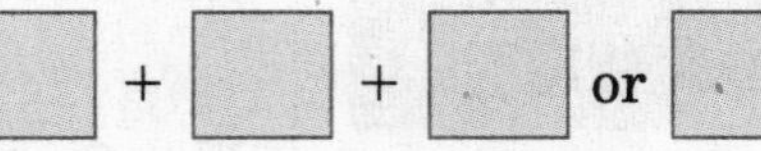

The area of the figure is ______ + ______ + ______ or ______ square inches.

© Glencoe/McGraw-Hill

2

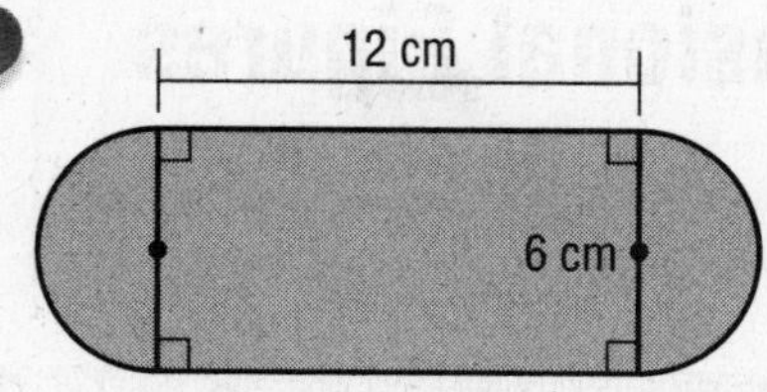

The figure can be separated into two ________ and a ________.

Area of one semicircle

$A = \frac{1}{2}\pi r^2$

$A =$ ________

$A \approx$ ________

Area of rectangle

$A = \ell w$

$A =$ ________

$A =$ ________

The area of the figure is about 14.1 + ________ + ________ or 100.3 square centimeters.

Your Turn **Find the area of each complex figure. Round to the nearest tenth if necessary.**

a.

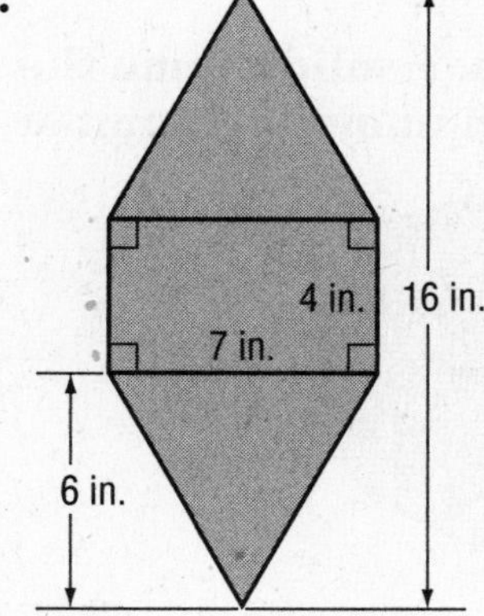

b.

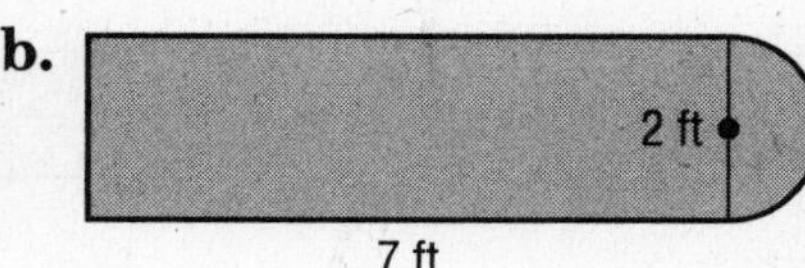

HOMEWORK ASSIGNMENT

Page(s):

Exercises:

© Glencoe/McGraw-Hill

7–4 Three-Dimensional Figures

What You'll Learn

- Identify and draw three-dimensional figures.

Key Concept

Common Polyhedrons

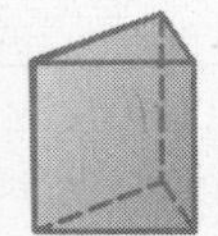
triangular prism

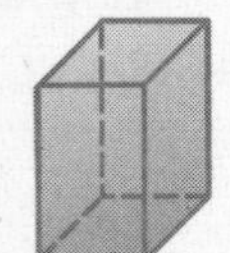
rectangular prism

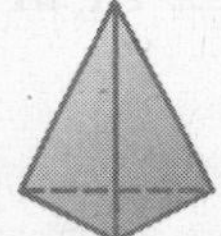
triangular pyramid

rectangular pyramid

Build Your Vocabulary (pages 164–165)

A **polyhedron** is a solid with ______ surfaces that are ______.

An **edge** is where two planes ______ in a line.

A **face** is a ______ surface.

A **vertex** is where three or more planes ______ at a point.

A **prism** is a polyhedron with two ______ faces, or bases.

A **pyramid** is a polyhedron with one base that is a ______ and faces that are ______.

EXAMPLE Identify Prisms and Pyramids

Identify each solid. Name the number and shapes of the faces. Then name the number of edges and vertices.

1

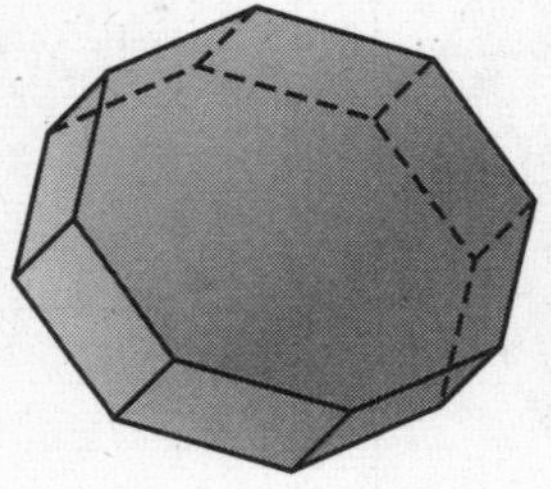

The figure has two parallel ______ bases that are ______, so it is an ______ prism. The other faces are rectangles. It has a total of ______ faces, ______ edges, and ______ vertices.

© Glencoe/McGraw-Hill

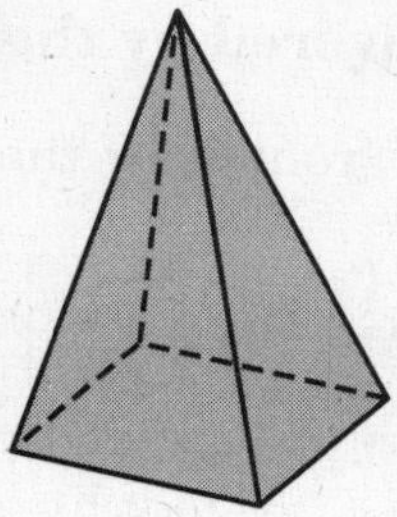

2 The figure has one base that is a ______, so it is a ______. The other faces are triangles. It has a total of ______ faces, ______ edges, and ______ vertices.

Your Turn **Identify each solid. Name the number and shapes of the faces. Then name the number of edges and vertices.**

a.

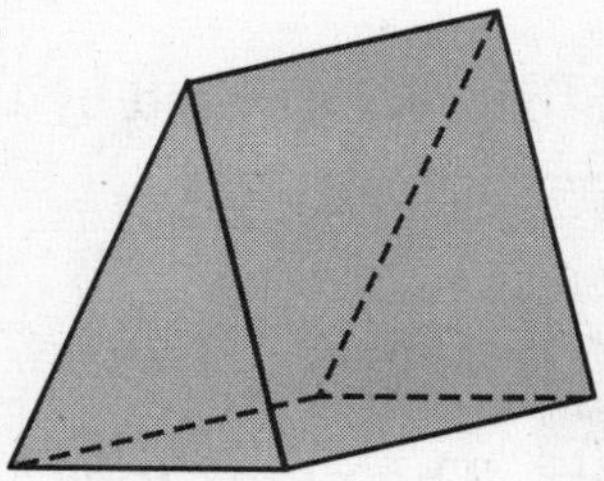

b.

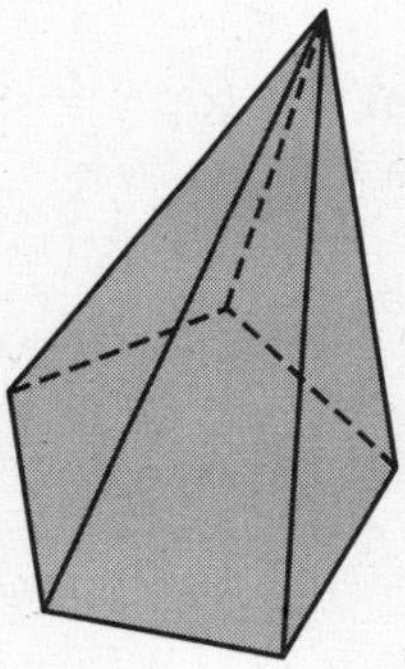

EXAMPLES Analyze Real-Life Drawings

ARCHITECTURE **The plans for a hotel fireplace are shown at the right.**

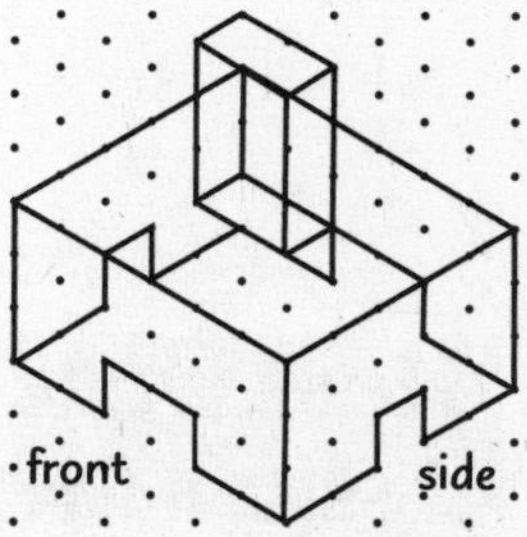

front

side

3 **Draw and label the top, front, and side views.**

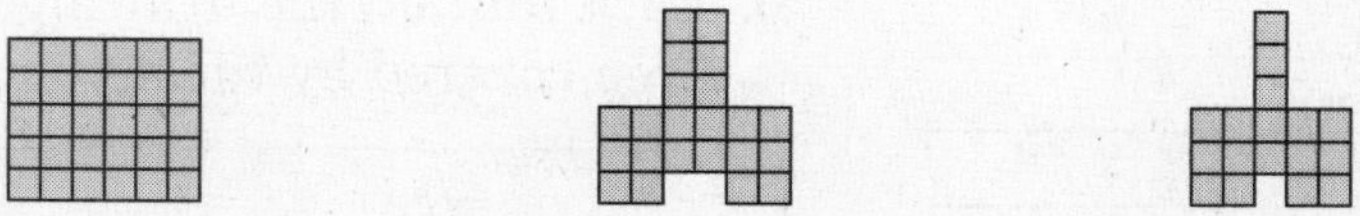

______ view ______ view ______ view

© Glencoe/McGraw-Hill

4 **Each unit on the drawing represents 1.5 feet. Find the area of the floor covered by the fireplace.**

You can see from the front view that the floor is a rectangle

that is [ ] units wide by [ ] units long. The actual

dimensions are [ ](1.5) feet by [ ](1.5) feet or 7.5 feet

by 9 feet.

$A = 7.5 \cdot 9$ $\quad$ $A = \ell \cdot w$

$A = 67.5$ $\quad$ Simplify.

The area of the floor covered by the fireplace is

[ ].

Your Turn **The plans for a building are shown to the right.**

a. Draw and label the top, front, and side views.

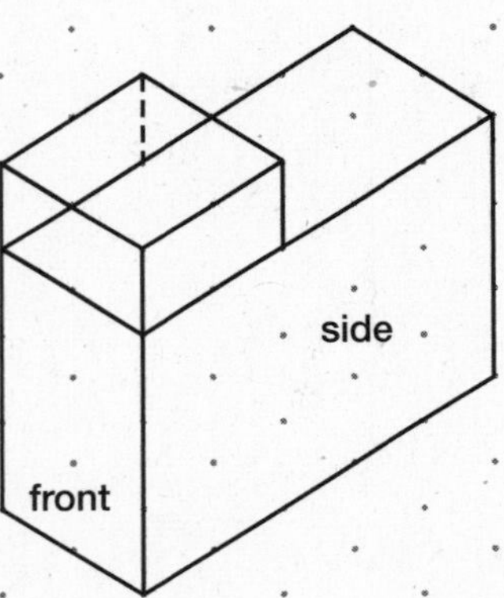

b. Each unit on the drawing represents 15 feet. Find the area covered by the second floor.

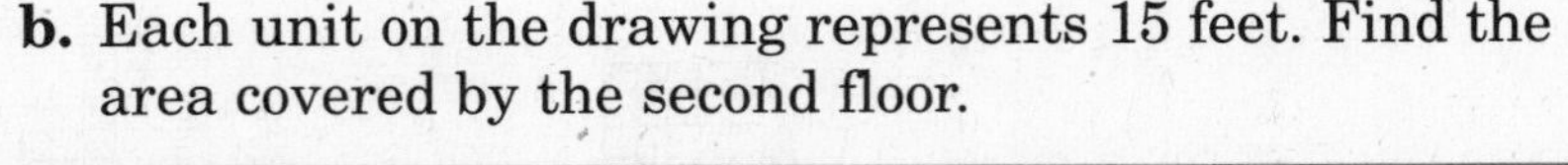

HOMEWORK ASSIGNMENT

Page(s):

Exercises:

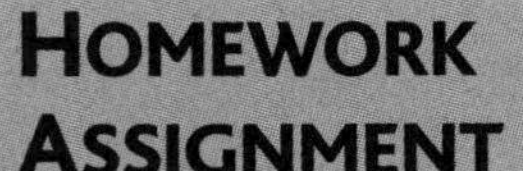

© Glencoe/McGraw-Hill

Volume of Prisms and Cylinders

What You'll Learn

- Find the volumes of prisms and cylinders.

BUILD YOUR VOCABULARY (pages 164–165)

Volume is the measure of the _____ occupied by a solid. Volume is measured in cubic units.

Key Concept

Volume of a Prism The volume V of a prism is the area of the base B times the height h.

EXAMPLE Find the Volume of a Rectangular Prism

1 Find the volume of the prism.

5 in.
7 in.
11 in.

$V = Bh$ Volume of a prism

$V = (\_\_\_\_\_)h$ The base is a rectangle, so $B =$ _____.

$V = (5 \cdot 7)11$ $\ell = 5, w = 7, h = 11$

$V =$ _____ Simplify.

The volume is 385 _____ inches.

EXAMPLE Find the Volume of a Triangular Prism

2 Find the volume of the prism.

9 ft
15 ft
4 ft

$V = Bh$ Volume of a prism

$V = \left(\frac{1}{2} \cdot 9 \cdot 15\right)h$ The base is a _____, so $B = \frac{1}{2} \cdot 9 \cdot 15$.

$V = \left(\frac{1}{2} \cdot 9 \cdot 15\right)4$ The height of the prism is _____.

$V =$ _____ Simplify.

The volume is _____ cubic feet.

© Glencoe/McGraw-Hill

Your Turn **Find the volume of each prism.**

a.

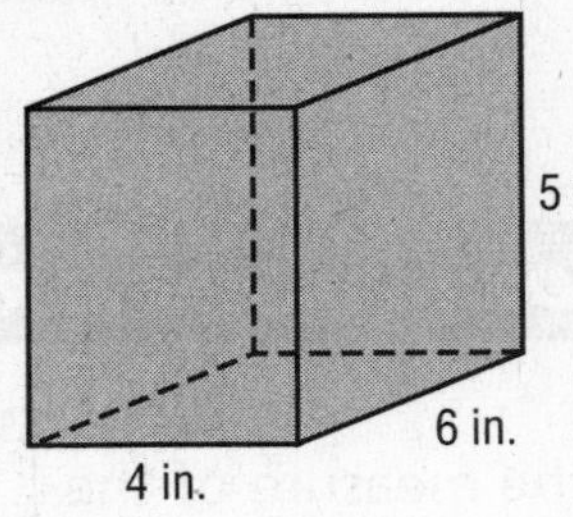

b.

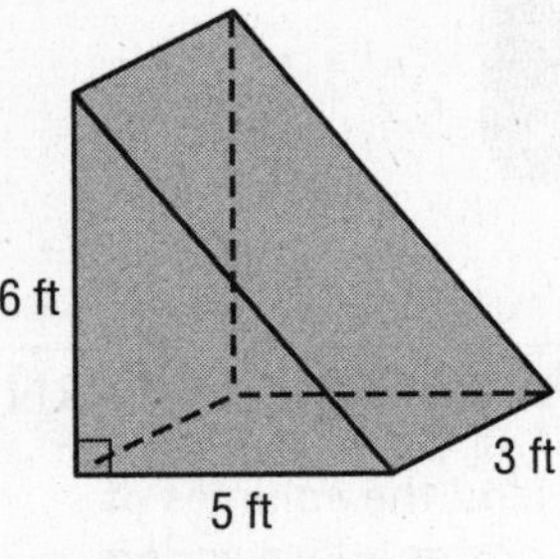

BUILD YOUR VOCABULARY (pages 164–165)

A **cylinder** is a solid whose bases are congruent, parallel,

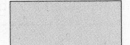

, connected with a  side.

EXAMPLES **Find the Volumes of Cylinders**

Find the volume of each cylinder. Round to the nearest tenth if necessary.

KEY CONCEPT

Volume of a Cylinder The volume V of a cylinder with radius r is the area of the base B times the height h.

3

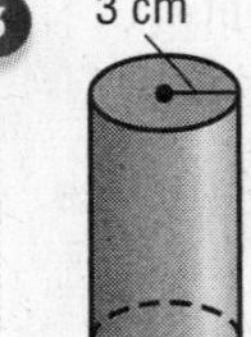

$V = \pi r^2 h$ Volume of a cylinder

$V = \pi \cdot \square^2 \cdot \square$  $r = \square, h = \square$ 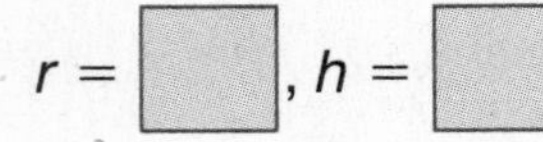

$V \approx$  Simplify.

The volume is about 339.3 ______ centimeters.

4 **diameter of base, 18 yd; height, 25.4 yd**

Since the diameter is 18 yards, the radius is ______ yards.

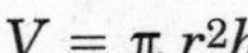

$V = \pi r^2 h$ Volume of a cylinder

$V = \pi \cdot \square^2 \cdot \square$ $r = \square, h = \square$

$V \approx$ ______ Simplify.

The volume is ______.

© Glencoe/McGraw-Hill

FOLDABLES

ORGANIZE IT

On index cards, write the formula for the volume of a rectangular prism, a triangular prism, and a cylinder. Sketch each figure and label its parts. Place your cards in the "Volume" pocket of your Foldable.

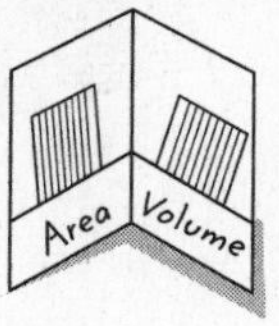

Your Turn **Find the volume of each cylinder. Round to the nearest tenth if necessary.**

a.

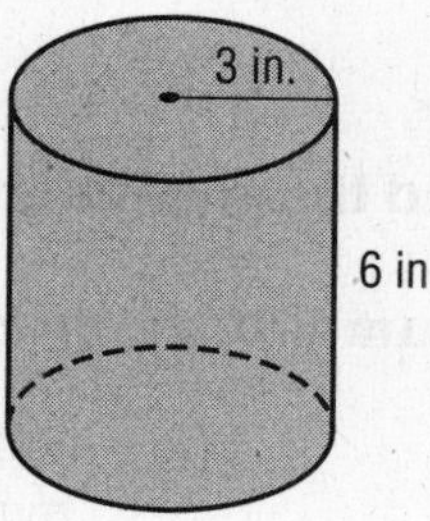

b. diameter of base, 8 yd; height, 10 yd

EXAMPLE **Find the Volume of a Complex Solid**

5 **TOYS** **A wooden block has a single hole drilled entirely though it. What is the volume of the block? Round to the nearest hundredth.**

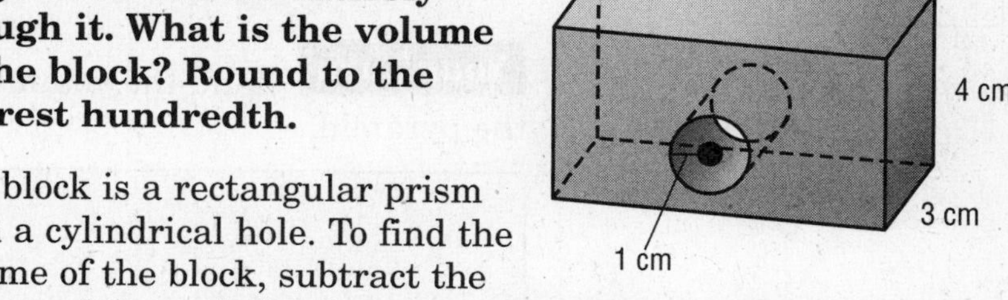

The block is a rectangular prism with a cylindrical hole. To find the volume of the block, subtract the volume of the cylinder from the volume of the prism.

Rectangular Prism	**Cylinder**
$V =$ ______	$V =$ ______
$V = (6 \cdot 3)4$ or 72	$V = \pi(1)^2(3)$ or 9.42

The volume of the box is about 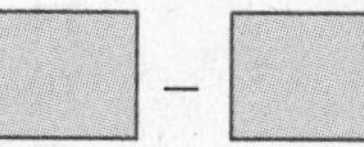 –  or ______ 62.58 cubic centimeters.

Your Turn A small wooden cube has been glued to a larger wooden block for a whittling project. What is the volume of the wood to be whittled?

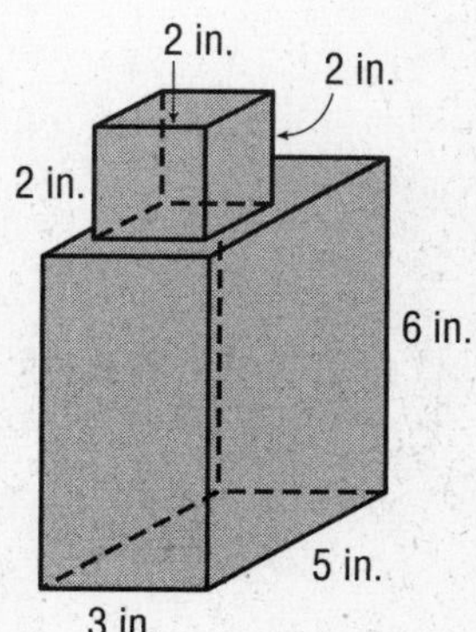

© Glencoe/McGraw-Hill

HOMEWORK ASSIGNMENT

Page(s):

Exercises:

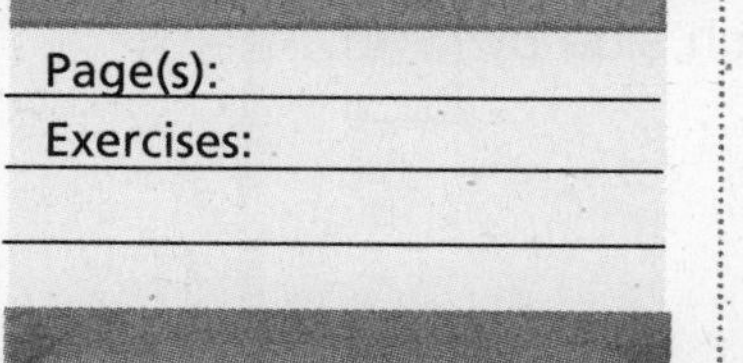

7–6 Volume of Pyramids and Cones

WHAT YOU'LL LEARN

- Find the volumes of pyramids and cones.

KEY CONCEPT

Volume of a Pyramid The volume *V* of a pyramid is one-third the area of the base *B* times the height *h*.

EXAMPLE Find the Volume of a Pyramid

1 **Find the volume of the pyramid.**

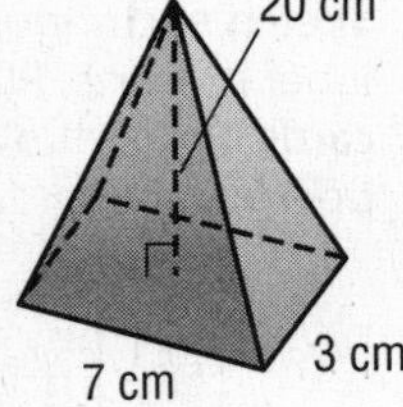

$V = \frac{1}{3}Bh$ Volume of a pyramid

$V = \frac{1}{3}(\square \cdot \square)\square$ $B = \square \cdot \square$, $h = \square$

$V = 140$ Simplify.

The volume is ______.

Your Turn Find the volume of the pyramid.

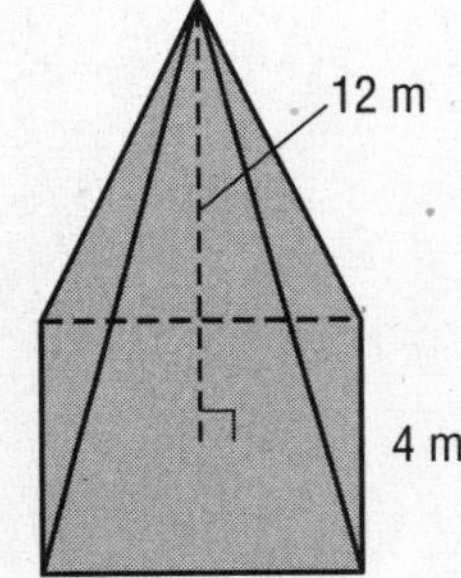

EXAMPLE Use Volume to Solve a Problem

2 **SOUVENIRS A novelty souvenir company wants to make snow globes shaped like a pyramid. It decides that the most cost-effective maximum volume of water for the pyramids is 12 cubic inches. If a pyramid globe measures 4 inches in height, find the area of its base.**

$V = \frac{1}{3}Bh$ Volume of a pyramid

$\square = \frac{1}{3} \cdot B \cdot 4$ Replace *V* with $\square$ and *h* with $\square$.

$12 = \frac{4}{3} \cdot B$ Simplify.

$\square \cdot 12 = \square \cdot \frac{4}{3} \cdot B$ Multiply each side by $\square$.

$\square = B$

The area of the base of the snow globe is 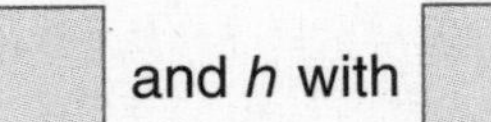.

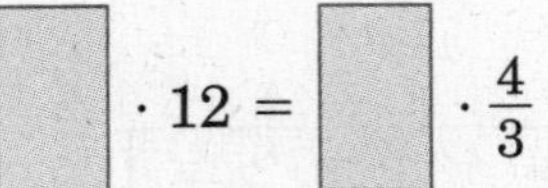
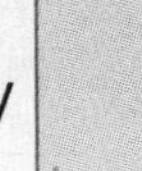

© Glencoe/McGraw-Hill

Your Turn A company is designing pyramid shaped building blocks with a square base. They want the volume of the blocks to be 18 cubic inches. If the length of the side of the base is 3 inches, what should be the height of the blocks?

BUILD YOUR VOCABULARY (pages 164–165)

A **cone** is a three-dimensional figure with one ______ base. A curved surface connects the base and the ______.

KEY CONCEPT

Volume of a Cone
The volume V of a cone with radius r is one-third the area of the base B times the height h.

EXAMPLE Find the Volume of a Cone

3 Find the volume of the cone. Round to the nearest tenth.

$V = \frac{1}{3}\pi r^2 h$ Volume of a cone

$V = \frac{1}{3} \cdot \pi \cdot \square^2 \cdot \square$ Replace r with ______ and h with ______.

$V \approx$ ______ Simplify.

The volume is ______.

FOLDABLES ORGANIZE IT

On index cards, write the formula for the volume of a pyramid and a cone. Sketch each figure and label its parts. Place your cards in the "Volume" pocket of your Foldable.

Your Turn Find the volume of the cone. Round to the nearest tenth.

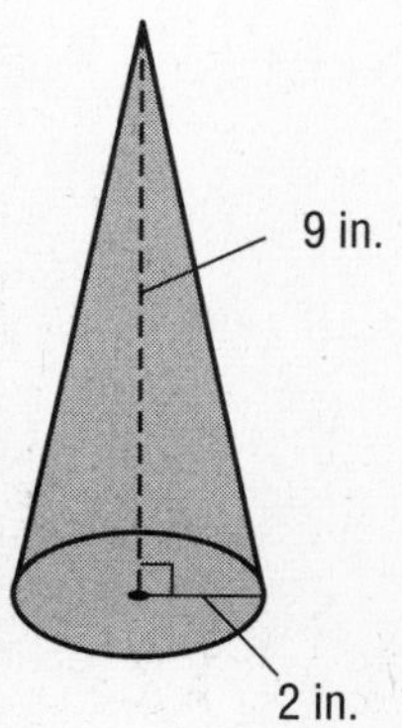

HOMEWORK ASSIGNMENT

Page(s):
Exercises:

© Glencoe/McGraw-Hill

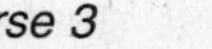

7–7 Surface Area of Prisms and Cylinders

BUILD YOUR VOCABULARY (pages 164–165)

The **surface area** of a solid is the ______ of the ______ of all its ______, or faces.

WHAT YOU'LL LEARN

- Find the surface areas of prisms and cylinders.

KEY CONCEPT

Surface Area of a Rectangular Prism The surface area S of a rectangular prism with length ℓ, width w, and height h is the sum of the areas of the faces.

EXAMPLE Surface Area of a Rectangular Prism

1 **Find the surface area of the rectangular prism.**

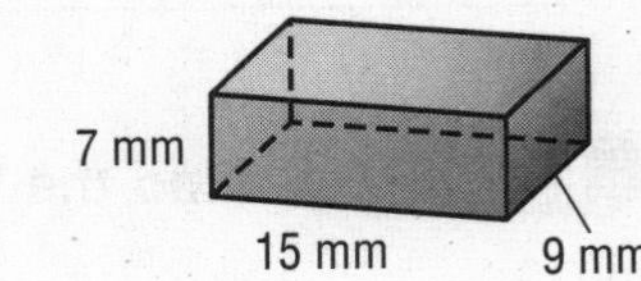

$S = 2$ ______ $+ 2$ ______ $+ 2$ ______ Write the formula.

$S = 2(15)(9) + 2(15)(7) + 2(9)(7)$ Substitution

$S =$ ______ Simplify.

The surface area is ______.

Your Turn Find the surface area of the rectangular prism.

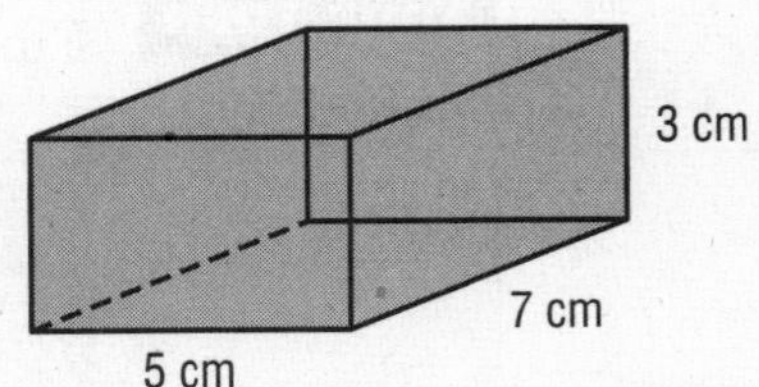

© Glencoe/McGraw-Hill

EXAMPLE Surface Area of a Triangular Prism

REVIEW IT

What is the formula for finding the area of a triangle? How does this relate to finding the surface area of a triangular prism? *(Lesson 7-1)*

____

____

____

____

2 CAMPING **A family wants to reinforce the fabric of its tent with a waterproofing treatment. Find the surface area, including the floor, of the tent below.**

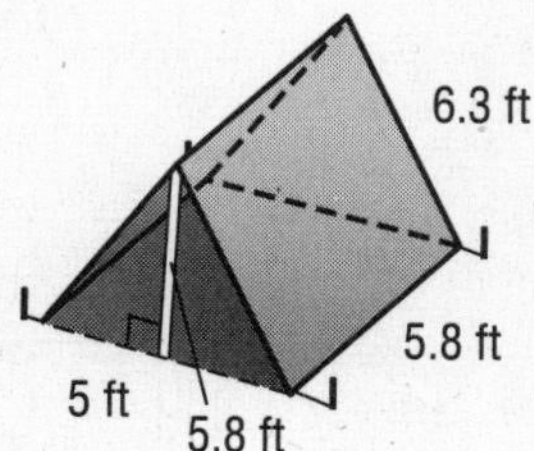

A triangular prism consists of two congruent ________ faces and three ________ faces.

Draw and label a net of this prism. Find the area of each face.

bottom ____ · ____ = 29

left side ____ · ____ = 36.54

right side ____ · ____ = 36.54

two bases $2\left(\frac{1}{2} \cdot 5 \cdot \text{____}\right) = 29$

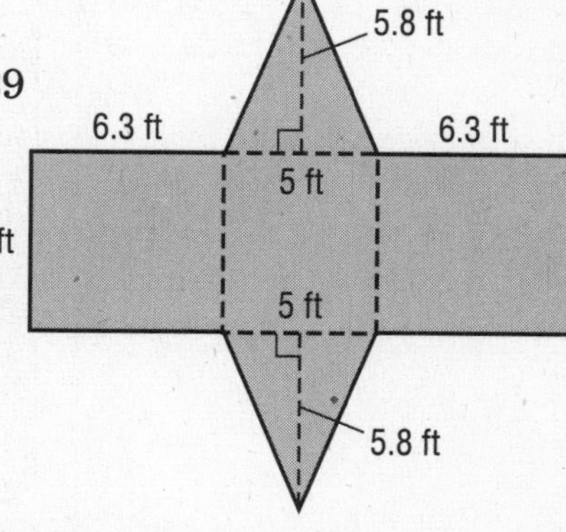

The surface area of the tent is 29 + 36.54 + 36.54 + 29 or about ________.

Your Turn Julia is painting triangular prisms to use as decoration in her garden. Find the surface area of the prism.

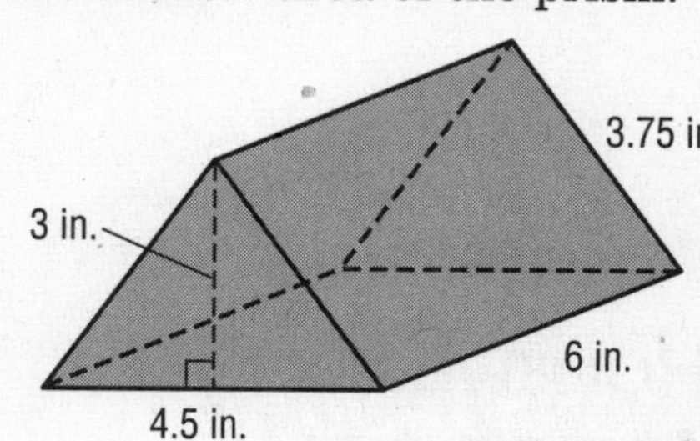

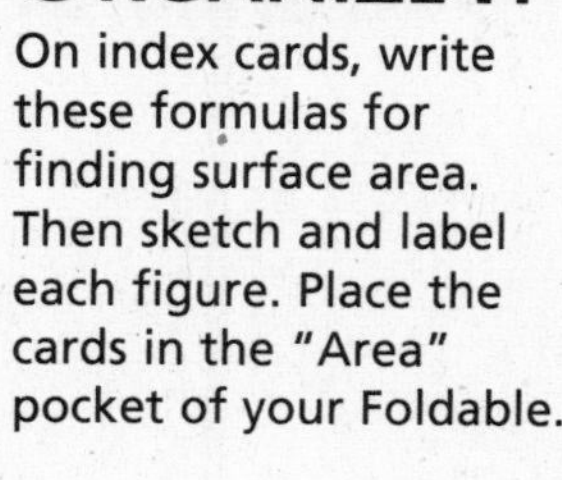

ORGANIZE IT

On index cards, write these formulas for finding surface area. Then sketch and label each figure. Place the cards in the "Area" pocket of your Foldable.

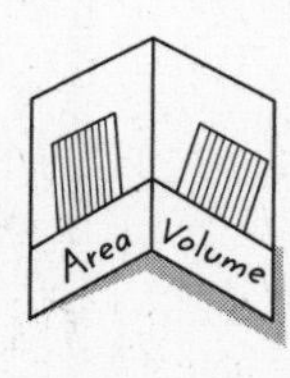

© Glencoe/McGraw-Hill

EXAMPLE Surface Area of a Cylinder

KEY CONCEPT

Surface Area of a Cylinder The surface area S of a cylinder with height h and radius r is the area of the two bases plus the area of the curved surface.

3 **Find the surface area of the cylinder. Round to the nearest tenth.**

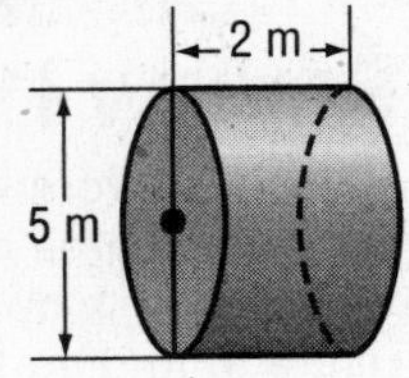

$S = 2\pi r^2 + 2\pi rh$ Surface area of a cylinder

$S = 2\pi(\quad)^2 + 2\pi(\quad)(\quad)$ Replace r with ____ and h with ____.

$S \approx$ ____ Simplify.

The surface area is about ____ square meters.

Your Turn Find the surface area of the cylinder. Round to the nearest tenth.

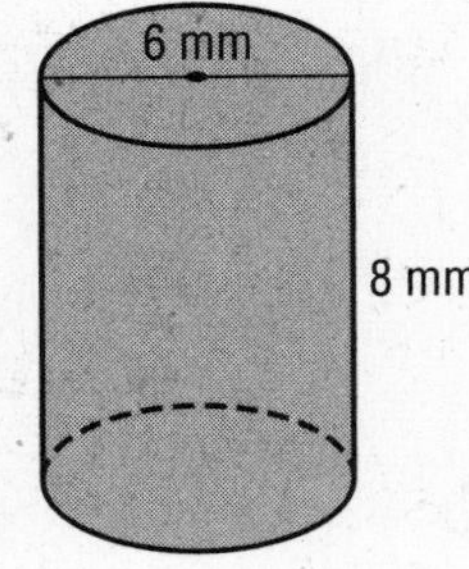

HOMEWORK ASSIGNMENT

Page(s):

Exercises:

© Glencoe/McGraw-Hill

7–8 Surface Area of Pyramids and Cones

What You'll Learn

- Find the surface areas of pyramids and cones.

BUILD YOUR VOCABULARY (pages 164–165)

The ______ of a pyramid are called **lateral faces.**

The altitude or ______ of each ______ is called the **slant height.**

The sum of the ______ of the ______ is the **lateral area.**

FOLDABLES

ORGANIZE IT

On a card, write the formula for finding the surface area of a pyramid. Then sketch a pyramid and label the parts. Place the card in the "Area" pocket of your Foldable.

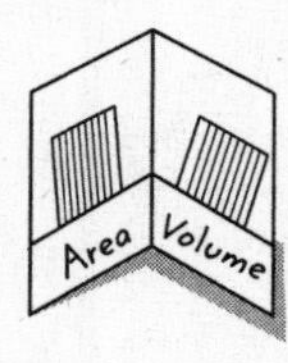

EXAMPLE Surface Area of a Pyramid

1 **Find the surface area of the triangular pyramid.**

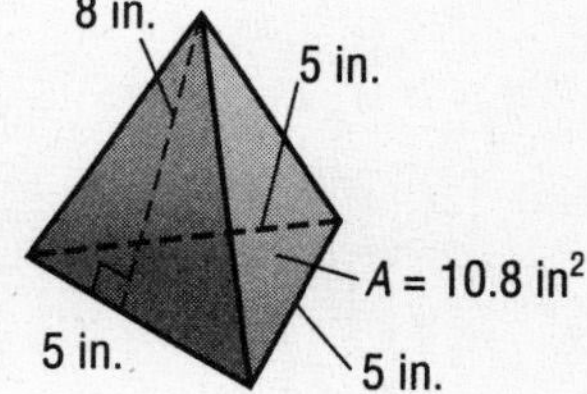

Find the lateral area and the area of the base.

Area of each lateral face

$A =$ ______ Area of a triangle

$A = \frac{1}{2}(\ \ )(\ \ )$ or ______ Replace b with ______ and h with ______.

There are 3 faces, so the lateral area is 3( ) or ______ square inches.

Area of base

$A =$ ______

The surface area of the pyramid is ______ + ______ or ______ square inches.

© Glencoe/McGraw-Hill

Your Turn Find the surface area of the square pyramid.

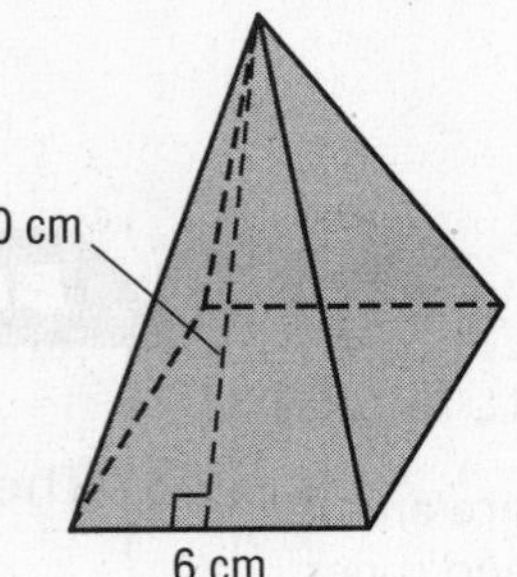

EXAMPLE **Surface Area of a Cone**

KEY CONCEPT

Surface Area of a Cone The surface area S of a cone with slant height ℓ and radius r is the lateral area plus the area of the base.

2 Find the surface of the cone. Round to the nearest tenth.

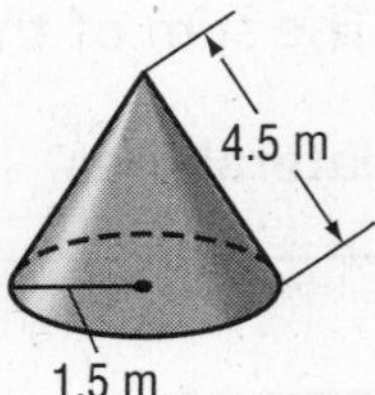

$S = \pi r \ell +$ ______ Surface area of a cone

$S = \pi($ ______ $)($ ______ $) = \pi($ ______ $)^2$ $r =$ ______, $\ell =$ ______

$S \approx$ ______ Simplify.

The surface area of the cone is about ______ square meters.

Your Turn Find the surface area of the cone. Round to the nearest tenth.

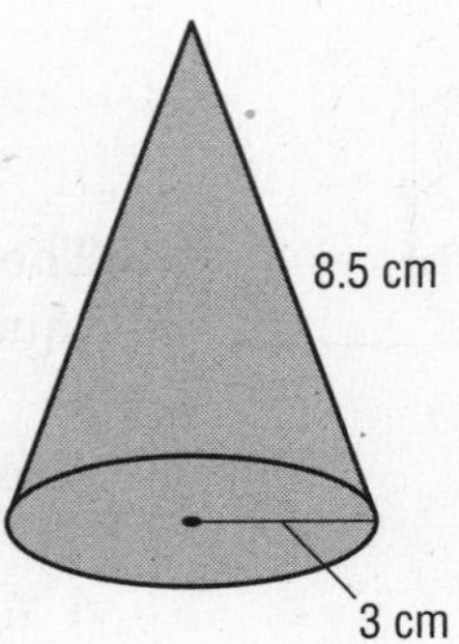

HOMEWORK ASSIGNMENT

Page(s):

Exercises:

© Glencoe/McGraw-Hill

7–9 Precision and Significant Digits

WHAT YOU'LL LEARN

- Analyze measurements.

BUILD YOUR VOCABULARY (pages 164–165)

The **precision** of a measurement is the ______ to which a measurement is made.

Significant digits are all of the digits of a measurement that are known to be ______ plus one ______ digit.

EXAMPLE Identify Precision Units

1 Identify the precision unit of the thermometer.

There are ______ spaces between each 10° mark, so the precision unit is ______ of 10 degrees or ______ degrees.

REMEMBER IT 

The precision unit of a measuring instrument determines the number of significant digits.

Your Turn Identify the precision unit of the ruler.

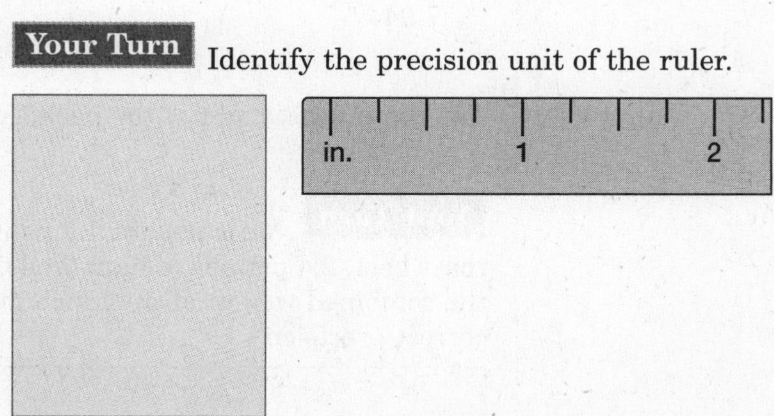

© Glencoe/McGraw-Hill

EXAMPLES Identify Significant Digits

Determine the number of significant digits in each measure.

2 **5,000 ft**

[ ] significant digit

3 **315.05 ounces**

[ ] significant digits

Your Turn **Determine the number of significant digits in each measure.**

a. 21.0 miles

b. 35.006 centimeters

EXAMPLE Add Measurements

4 MAIL **Venita took four packages to the post office. They weighed 1.27 pounds, 3.45 pounds, 0.524 pound, and 2.7 pounds. Write the combined weight of her mail using correct precision.**

1.27 ← [ ] decimal places

3.45 ← [ ] decimal places

0.524 ← [ ] decimal places

+2.7 ← 1 decimal place ← The least precise measurement has 1 decimal place, so round the sum to 1 decimal place.

7.944

The combined weight of the packages is about [ ] pounds.

Your Turn Veric bought 1.5 pounds of turkey, 0.75 pound of roast beef, 2.4 pounds of ham, and 0.5 pound of salami. Write the combined weight of the lunch meat that Veric bought using correct precision.

© Glencoe/McGraw-Hill

EXAMPLE **Multiply Measurements**

5 **Use the correct number of significant digits to find the volume of the cylinder. Use 3.14 for pi.**

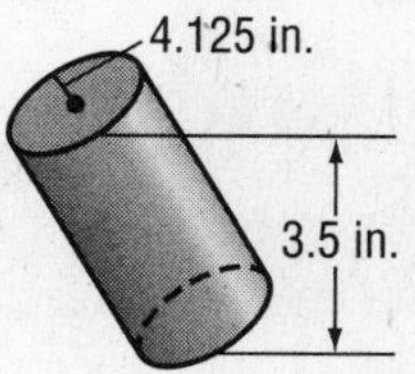

$V = \pi r^2 h$ Volume of a cylinder

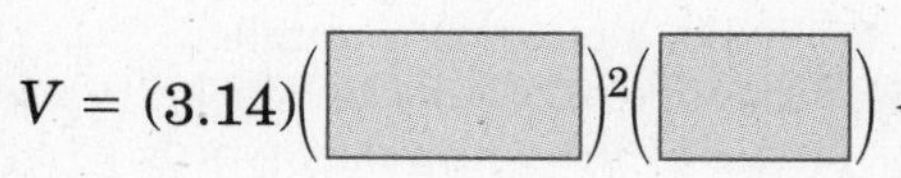

The height of the cylinder has the least number of significant digits, 2.

$V = 187.0017188$

Round the answer, 187.0017188, so that it has 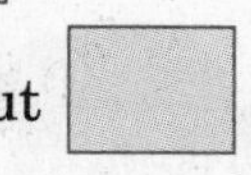 significant digits. The volume of the cylinder is about ______ cubic inches.

REVIEW IT

What is the formula for finding the volume of a rectangular prism? *(Lesson 7-5)*

______

Your Turn Use the correct number of significant digits to find the volume of the rectangular prism.

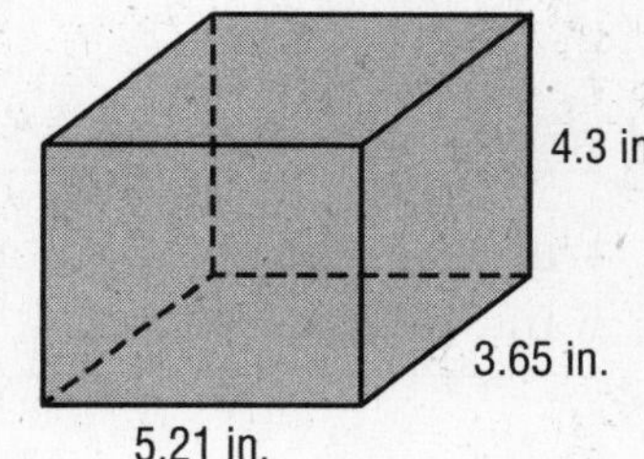

HOMEWORK ASSIGNMENT

Page(s):
Exercises:

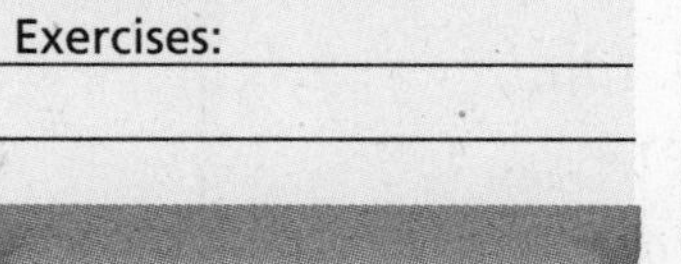

© Glencoe/McGraw-Hill

CHAPTER 7

BRINGING IT ALL TOGETHER

STUDY GUIDE

FOLDABLES™	VOCABULARY PUZZLEMAKER	BUILD YOUR VOCABULARY
Use your **Chapter 7 Foldable** to help you study for your chapter test.	To make a crossword puzzle, word search, or jumble puzzle of the vocabulary words in Chapter 7, go to: www.glencoe.com/sec/math/t_resources/free/index.php	You can use your completed **Vocabulary Builder** *(pages 164–165)* to help you solve the puzzle.

7-1

Area of Parallelograms, Triangles, and Trapezoids

Find the area of each figure.

1. triangle: base, 14 m; height, 4 m

2. parallelogram: base, 5 ft; height, 9.2 ft

3. trapezoid: bases, 2 in. and 6 in.; height, 4.3 in.

4. trapezoid: bases, 3 yd and 12 yd; height 10 yd

7-2

Circumference and Area of Circles

Complete.

5. The distance from the center of a circle to any point on the circle is called the ______, while the distance around the circle is called the ______.

Find the circumference and area of each circle. Round to the nearest tenth.

6. The radius is 14 miles.

7. The diameter is 17.4 in.

© Glencoe/McGraw-Hill

7-3

Area of Complex Figures

8. What is a complex figure?

9. What is the first step in finding the area of a complex figure?

10. Explain how to divide up the figure shown.

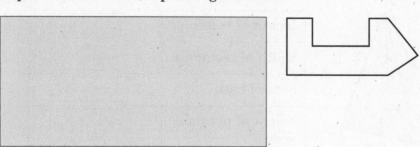

7-4

Three-Dimensional Figures

Match each description with the word it describes.

a. vertex
b. edge
c. face
d. base
e. prism
f. pyramid

11. a flat surface

12. a polyhedron with one base that is a polygon and faces that are triangles

13. where three or more planes intersect at a point

14. where two planes intersect in a line

15. a polyhedron with two parallel, congruent faces

© Glencoe/McGraw-Hill

7-5 Volume of Prisms and Cylinders

Find the volume of each solid. Round to the nearest tenth if necessary.

16.

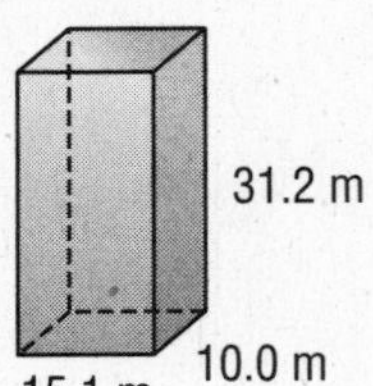

17.

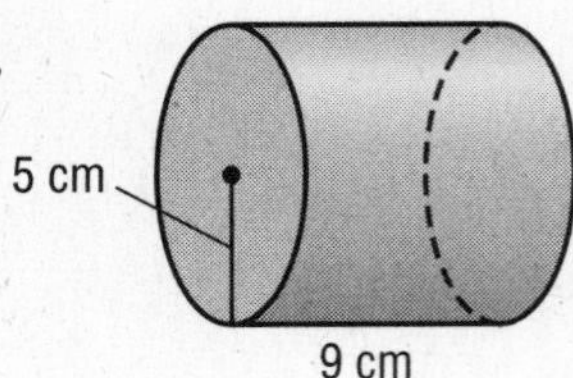

18.

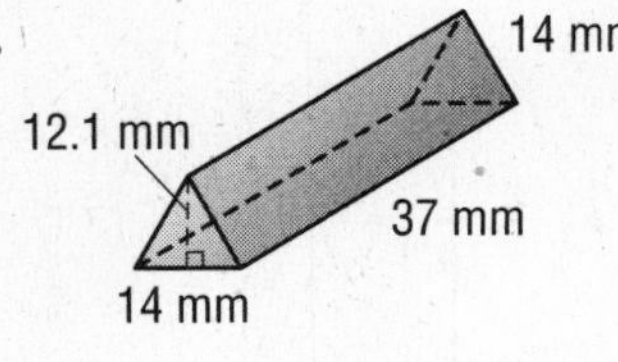

7-6 Volume of Pyramids and Cones

19. Fill in the table about what you know from the diagram. Then complete the volume of the pyramid.

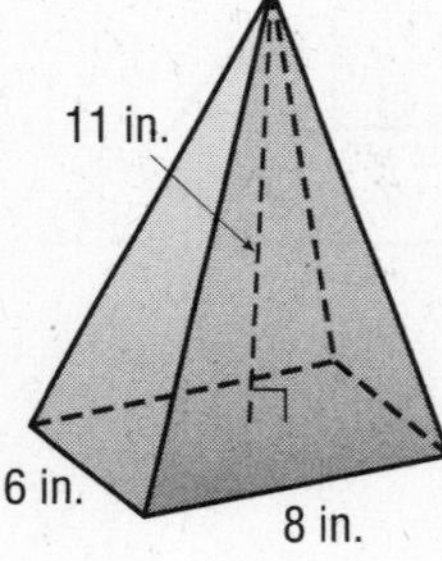

length of rectangle	
width of rectangle	
area of base	
height of pyramid	
volume of pyramid	

7-7 Surface Area of Prisms and Cylinders

20. Complete the sentence with the correct numbers. When you draw a net of a triangular prism, there are ☐ congruent triangular faces and ☐ rectangular faces.

21. If you unroll a cylinder, what does the net look like?

22. Find the surface area of the cylinder. Round the nearest tenth.

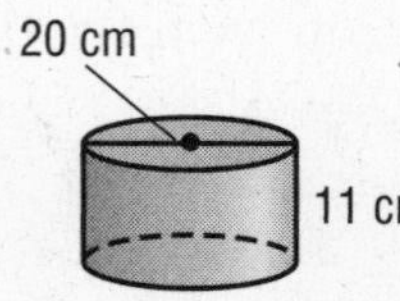

© Glencoe/McGraw-Hill

7-8

Surface Area of Pyramids and Cones

23. Complete the steps in finding the surface area of a square pyramid.

Area of each lateral face

$A = \frac{1}{2}bh$

$A = \frac{1}{2}(9)(16)$

$A = 72$

There are ____ faces, so the lateral area is 4(72) = ____ square inches.

Area of base

$A = s^2$

$A = 9^2$ or 81

The surface area of the square pyramid is ____ + ____ or ____ square inches.

24. What two areas are needed to calculate the surface area of a cone?

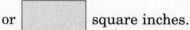

7-9

Precision and Significant Digits

25. How do you determine that a measurement is accurate?

26. What determines the number of significant digits?

© Glencoe/McGraw-Hill

ARE YOU READY FOR THE CHAPTER TEST?

Visit **msmath3.net** to access your textbook, more examples, self-check quizzes, and practice tests to help you study the concepts in Chapter 7.

Check the one that applies. Suggestions to help you study are given with each item.

☐ **I completed the review of all or most lessons without using my notes or asking for help.**

- You are probably ready for the Chapter Test.
- You may want to take the Chapter 7 Practice Test on page 367 of your textbook as a final check.

☐ **I used my Foldable or Study Notebook to complete the review of all or most lessons.**

- You should complete the Chapter 7 Study Guide and Review on pages 363–366 of your textbook.
- If you are unsure of any concepts or skills, refer to the specific lesson(s).
- You may also want to take the Chapter 7 Practice Test on page 367.

☐ **I asked for help from someone else to complete the review of all or most lessons.**

- You should review the examples and concepts in your Study Notebook and Chapter 7 Foldable.
- Then complete the Chapter 7 Study Guide and Review on pages 363–366 of your textbook.
- If you are unsure of any concepts or skills, refer to the specific lesson(s).
- You may also want to take the Chapter 7 Practice Test on page 367.

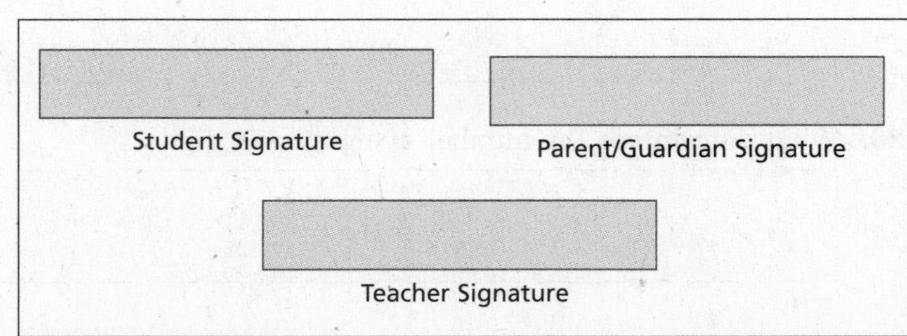

© Glencoe/McGraw-Hill

Probability

Use the instructions below to make a Foldable to help you organize your notes as you study the chapter. You will see Foldable reminders in the margin of this Interactive Study Notebook to help you in taking notes.

Begin with two sheets of 8.5" × 11" unlined paper.

STEP 1 **Fold in Quarters**
Fold each sheet in quarters along the width.

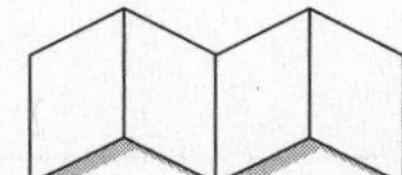

STEP 2 **Tape**
Unfold each sheet and tape to form one long piece.

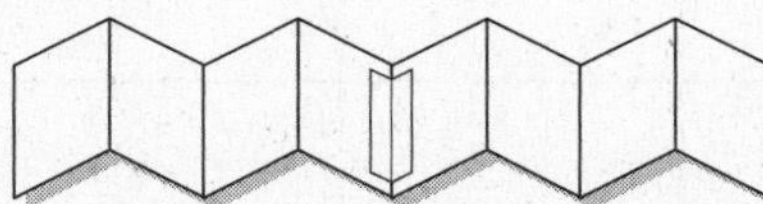

STEP 3 **Label**
Label each page with the lesson number to form one long piece.

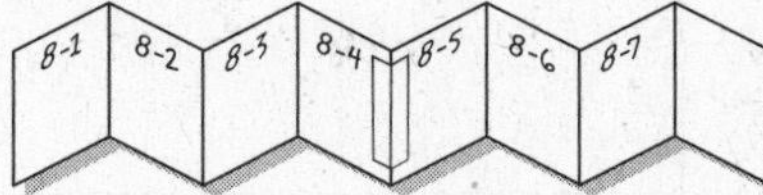

NOTE-TAKING TIP: It helps to take notes as you progress through studying a subject. New concepts often build upon concepts you have just learned in a previous lesson. If you take notes as you go, you will know what you need to know for the concept you are now learning.

© Glencoe/McGraw-Hill

BUILD YOUR VOCABULARY

This is an alphabetical list of new vocabulary terms you will learn in Chapter 8. As you complete the study notes for the chapter, you will see Build Your Vocabulary reminders to complete each term's definition or description on these pages. Remember to add the textbook page number in the second column for reference when you study.

Vocabulary Term	Found on Page	Definition	Description or Example
biased sample			
combination			
complementary events			
compound events			
dependent events			
experimental probability			
factorial			
Fundamental Counting Principle			
independent events			
outcome			

© Glencoe/McGraw-Hill

Vocabulary Term	Found on Page	Definition	Description or Example
permutation			
population			
probability			
random			
sample			
stratified random sample			
systematic random sample			
theoretical probability			
tree diagram			
unbiased sample			

© Glencoe/McGraw-Hill

8–1 Probability of Simple Events

WHAT YOU'LL LEARN

- Find the probability of a simple event.

KEY CONCEPT

Probability The probability of an event is a ratio that compares the number of favorable outcomes to the number of possible outcomes.

BUILD YOUR VOCABULARY (pages 194–195)

An **outcome** is one possible ______ of a probability event.

If all outcomes occur by chance the outcomes happen at **random**.

Probability is the ______ that an event will happen.

EXAMPLES Find Probabilities

A bag contains 5 blue marbles, 10 red marbles, and 10 yellow marbles. A marble is picked at random.

1 What is the probability the marble is yellow?

There are 5 + 10 + 10 or 25 marbles in the bag.

$$P(\text{yellow}) = \frac{\underline{\quad} \text{ marbles}}{\underline{\quad} \text{ number of marbles}}$$

$= \underline{\quad}$ or $\underline{\quad}$ There are ______ yellow marbles out of ______ marbles.

The probability the marble is yellow is ______.

2 What is the probability the marble is blue or red?

$$P(\text{blue or red}) = \frac{\underline{\quad} \text{ marbles} + \underline{\quad} \text{ marbles}}{\text{total number of marbles}}$$

$= \frac{5 + 10}{25}$ There are 5 blue marbles and 10 red marbles.

$= \underline{\quad}$ or $\underline{\quad}$ Simplify.

The probability the marble is blue or red is ______.

© Glencoe/McGraw-Hill

Your Turn **A bag contains 3 green marbles, 7 purple marbles, and 15 black marbles. A marble is picked at random.**

a. What is the probability the marble is black?

b. What is the probability the marble is green or purple?

c. What is the probability the marble is red?

FOLDABLES™

ORGANIZE IT

Under Lesson 8-1, explain how to find probabilities. Include examples and diagrams. On the last page in your Foldable, write the key terms in the lesson and their definitions.

BUILD YOUR VOCABULARY (page 194)

The events of one outcome happening and that outcome ______ happenning are **complementary events.**

EXAMPLE Probability of a Complementary Event

3 **PHONE LISTINGS** **One town has 10,000 phone numbers in use. Of these, 500 are not listed in the local phone book. What is the probability that the phone number you are looking for is listed in the phone book?**

______ − ______ or ______ phone numbers are listed.

$P(\text{listed}) = \dfrac{\text{listed numbers}}{\text{total number of phone numbers}}$ Definition of probability

$= \dfrac{\_\_\_}{\_\_\_}$ or ______ There are 9,500 listed phone numbers.

The probability is ______.

HOMEWORK ASSIGNMENT

Page(s):

Exercises:

Your Turn Oakdale Junior High School has a total enrollment of 670 students. Of these, 45 are absent today. Suppose a student's name is picked at random. What is the probability that the student picked is absent today?

© Glencoe/McGraw-Hill

8–2 Counting Outcomes

What You'll Learn

- Count outcomes by using a tree diagram or the Fundamental Counting Principle.

Build Your Vocabulary (pages 194–195)

A **tree diagram** is a diagram used to show the ______ number of ______ in a probability experiment.

The **Fundamental Counting Principle** uses ______ of the number of ways each event in an experiment can occur to find the number of ______ in a sample space.

Write It

How is using a tree diagram to find total number of outcomes like using a factor tree to find prime factors? (see factor trees in Prerequisite Skills page 609)

______
______
______
______

EXAMPLE Use a Tree Diagram

1 BOOKS **A flea market vendor sells new and used books for adults and teens. Today she has fantasy novels and poetry collections to choose from. Draw a tree diagram to determine the number of categories of books.**

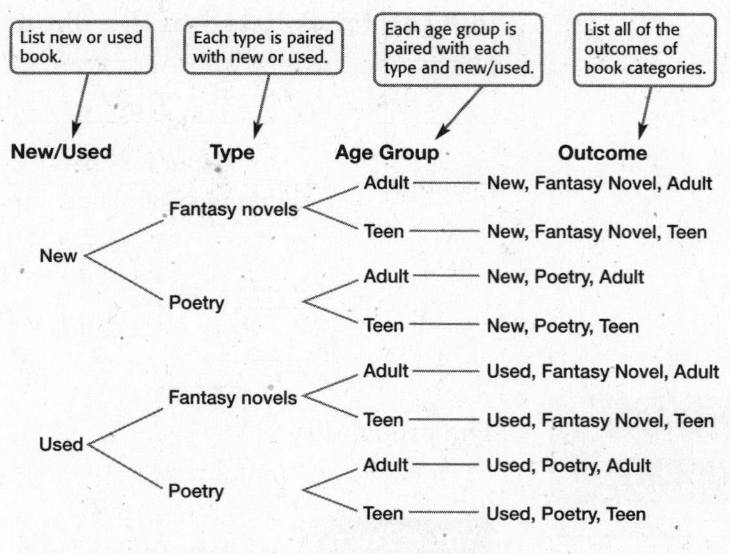

There are ______ different categories.

© Glencoe/McGraw-Hill

KEY CONCEPT

Fundamental Counting Principle If event *M* an occur in *m* ways and is followed by event *N* that can occur in n ways, then the event *M* followed by the event *N* can occur in $m \cdot n$ ways.

Your Turn A store has spring outfits on sale. You can choose either striped or solid pants. You can also choose green, pink, or orange shirts. Finally, you can choose either long-sleeved shirts or short-sleeved shirts. Draw a tree diagram to determine the number of possible outfits.

EXAMPLE Use the Fundamental Counting Principle

2 RESTAURANTS **A manager assigns different codes to all the tables in a restaurant to make it easier for the wait staff to identify them. Each code consists of the vowel A, E, I, O, or U, followed by two digits from 0 through 9. How many codes could the manager assign using this method?**

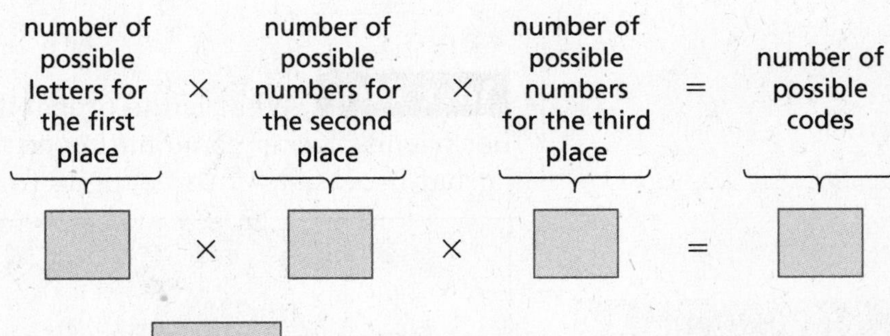

There are ______ possible codes.

© Glencoe/McGraw-Hill

Your Turn A middle school assigns each student a code to use for scheduling. Each code consists of a letter, followed by two digits from 0 though 9. How many codes are possible?

FOLDABLES™

ORGANIZE IT

Under Lesson 8-2, write notes on what you learned about counting outcomes by using a tree diagram and by using the Counting Principle. Include examples of each. On the last page of your Foldable, write the key terms in the lesson and their definitions.

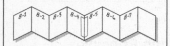

EXAMPLE Find Probability

3 COMPUTERS **What is the probability that Liana will guess her friend's computer password on the first try if all she knows is that it consists of three letters?**

Find the number of possible outcomes. Use the Fundamental Counting Principle.

choices for the first letter	×	choices the second letter	×	choices for the third letter	=	total number of outcomes
☐	×	☐	×	☐	=	☐

There are ☐ possible outcomes. There is ☐ correct password. So, the probability of guessing on the first try is ☐.

Your Turn What is the probability that Shauna will guess her friend's locker combination on the first try if all she knows is that it consists of three digits from 0 through 9?

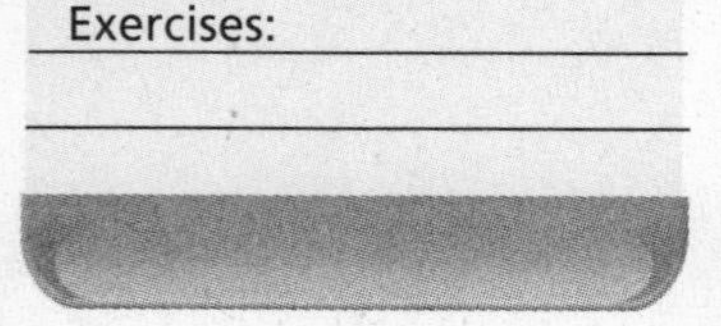

HOMEWORK ASSIGNMENT

Page(s):

Exercises:

© Glencoe/McGraw-Hill

8-3 Permutations

WHAT YOU'LL LEARN

- Find the number of permutations of objects.

BUILD YOUR VOCABULARY (pages 194–195)

A **permutation** is an arrangement or listing in which ______.

A **factorial** is a mathematical expression in which $n!$ is the ______ of all counting numbers beginning with n and counting backward to ______.

EXAMPLE Find a Permutation

1 SOFTBALL **There are 10 players on a softball team. In how many ways can the manager choose three players for first, second, and third base?**

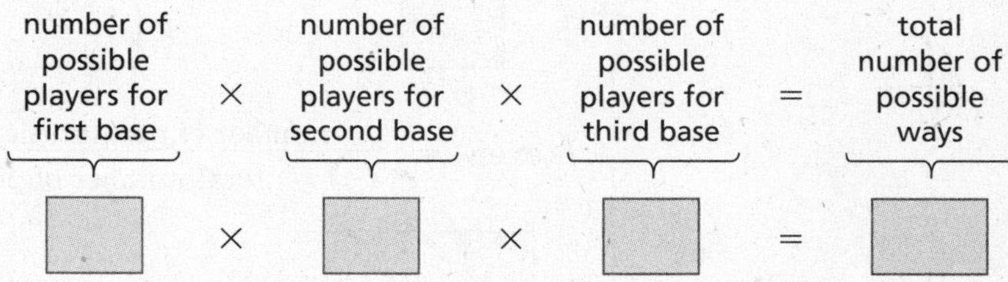

number of possible players for first base
×
number of possible players for second base
×
number of possible players for third base
=
total number of possible ways
×
×
=

There are ______ different ways the manager can pick players for first, second, and third base.

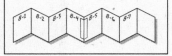

Your Turn There are 15 students on student council. In how many ways can Mrs. Sommers choose three students for president, vice president, and secretary?

______

FOLDABLES™

ORGANIZE IT

Under Lesson 8-3, explain how to find the number of permutations of objects. Use words and symbols and include examples.

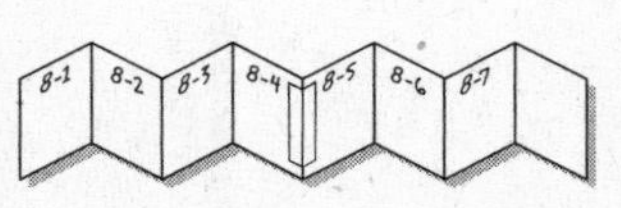

EXAMPLE Use Permutation Notation

2 **Find each value.**

***P*(7, 2)**

$P(7, 2) =$ ______ · ______ or ______ 7 things taken ______ at a time.

© Glencoe/McGraw-Hill

Your Turn **Find each value.**

a. $P(8, 4)$

b. $P(12, 5)$

EXAMPLE **Find Probability**

3 **Consider all of the five-digit numbers that can be formed using the digits 1, 2, 3, 4, and 5 where no digit is used twice. Find the probability that one of these numbers picked at random is an even number.**

Find the number of possible five-digit numbers.

$P(5, 5) = \square!$

In order for a number to be even, the ones digit must be 2 or 4.

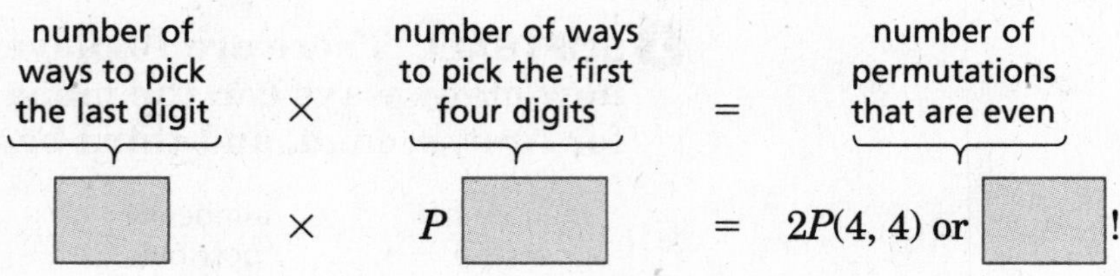

number of ways to pick the last digit
×
number of ways to pick the first four digits
=
number of permutations that are even
×
P
= 2P(4, 4) or
!

$P(\text{even}) = \dfrac{\text{number of permutations that are even}}{\text{total number of permutations}}$

$= \square$ Substitute.

$= \dfrac{2 \cdot \overset{1}{\cancel{4}} \cdot \overset{1}{\cancel{3}} \cdot \overset{1}{\cancel{2}} \cdot \overset{1}{\cancel{1}}}{5 \cdot \underset{1}{\cancel{4}} \cdot \underset{1}{\cancel{3}} \cdot \underset{1}{\cancel{2}} \cdot \underset{1}{\cancel{1}}}$ Definition of factorial

$= \square$ or $\square$ Simplify.

The probability is

Your Turn Consider all of the five-digit numbers that can be formed using the digits 1, 2, 3, 4, and 5 where no digit is used twice. Find the probability that one of these numbers picked at random is an odd number.

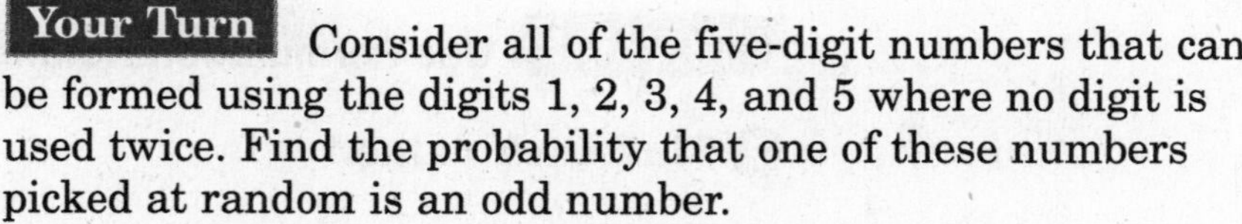

REMEMBER IT

Remember that the symbol ! does not always represent an exclamation. Sometimes it is used to represent factorials, such as 3! for 3 · 2 · 1.

HOMEWORK ASSIGNMENT

Page(s):

Exercises:

© Glencoe/McGraw-Hill

8–4 Combinations

WHAT YOU'LL LEARN

- Find the number of combinations of objects.

BUILD YOUR VOCABULARY (page 194)

A **combination** is an arrangement or listing in which __________.

EXAMPLE Find a Combination

1 TOURNAMENTS **Five teams are playing each other in a tournament. If each team plays every other team once, how many games are played?**

Method 1

Let A, B, C, D, and E represent the five teams. First, list all of the possible permutations of A, B, C, D, and E taken  at a time. Then cross out the letter pairs that are the same as one another.

AB	AC	AD	AE	~~BA~~
BC	BD	BE	~~CA~~	~~CB~~
CD	CE	~~DA~~	~~DB~~	~~DC~~
DE	~~EA~~	~~EB~~	~~EC~~	~~ED~~

Team A playing Team B is the same as Team B playing Team A, so cross off one of them.

There are only ____ different games

Method 2

Find the number of permutations of 5 teams taken  at a time.

$P(5, 2) = 5 \cdot 4$ or ____

Since order is not important, divide the number of permutations by the number of ways 2 things can be arranged.

$\frac{20}{2!} = \frac{20}{\_\_\_\_ \cdot \_\_\_\_}$ or ____

There are ____ games that can be played.

© Glencoe/McGraw-Hill

Your Turn Six teams are playing each other in a tournament. If each team plays every other team once, how many games are played?

EXAMPLE **Use Combination Notation**

2 Find $C(8, 5)$.

$C(8, 5) = \frac{P(8, 5)}{\square}$ Definition of $C(8, 5)$

$= \frac{8 \cdot 7 \cdot \overset{2}{\cancel{6}} \cdot \overset{1}{\cancel{5}} \cdot \overset{1}{\cancel{4}}}{\underset{1}{\cancel{5}} \cdot \underset{1}{\cancel{4}} \cdot \underset{1}{\cancel{3}} \cdot \underset{1}{\cancel{2}} \cdot 1}$ or $\square$ $P(8, 5) = \square$

and $5! = \square$

Your Turn Find $C(6, 3)$.

EXAMPLES **Combinations and Permutations**

SCHOOL An eighth grade teacher needs to select 4 students from a class of 22 to help with sixth grade orientation.

3 Does this represent a combination or a permutation? How many possible groups could be selected to help out the new students?

This is a ______ problem since the order is not important.

$C(22, 4) = \frac{P(22, 4)}{4!}$ 22 students taken 4 at a time.

$= \frac{\overset{11}{\cancel{22}} \cdot \overset{7}{\cancel{21}} \cdot \overset{5}{\cancel{20}} \cdot 19}{\underset{1}{\cancel{4}} \cdot \underset{1}{\cancel{3}} \cdot \underset{1}{\cancel{2}} \cdot 1}$ or $\square$

There are ______ different groups of eighth grade students that could help the new students.

Explain the difference between combinations and *permutations*. *(Lesson 8-3)*

______

______

______

______

______

© Glencoe/McGraw-Hill

4 SCHOOL One eighth grade student will be assigned to sixth grade classes on the first floor, another student will be assigned to classes on the second floor, another student will be assigned to classes on the third floor, and still another student will be assigned to classes on the fourth floor. Does this represent a combination or a permutation? In how many possible ways can the eighth graders be assigned to help with the sixth grade orientation?

Since it makes a difference which student goes to which floor, order is important. This is a ______.

$P(22, 4) = 22 \cdot 21 \cdot 20 \cdot$ ______ Definition of $P(22, 4)$

$= 175{,}560$

There are ______ for the eighth grade students to be selected to help with sixth grade orientation.

FOLDABLES™

ORGANIZE IT

Under Lesson 8-4, write notes on what you learned about finding the number of combinations of objects. Include examples. On the last page in your Foldable, write the key terms in the lesson and their definitions.

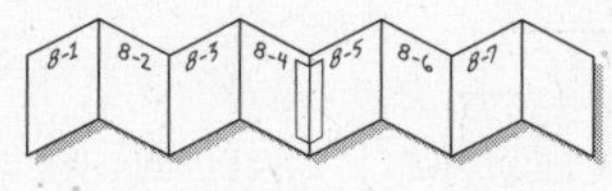

Your Turn **A teacher needs to select 5 students from a class of 26 to help with parent teacher conferences.**

a. Does this represent a combination or a permutation? How many possible groups could be selected to help?

b. One student will be assigned to fifth grade parents, another student will be assigned to sixth grade parents, another student will be assigned to seventh grade parents, another student will be assigned to eighth grade parents and still another student will be assigned to ninth grade parents. Does this represent a combination or a permutation? In how many possible ways can the students be assigned to help with the parent teacher conferences?

HOMEWORK ASSIGNMENT

Page(s):

Exercises:

© Glencoe/McGraw-Hill

8–5 Probability of Compound Events

BUILD YOUR VOCABULARY (page 194)

WHAT YOU'LL LEARN

- Find the probability of independent and dependent events.

A **compound event** consists of ______ simple events.

Independent events are ______ events in which the outcome of one event ______ affect the outcome of the other events.

KEY CONCEPT

Probability of Two Independent Events The probability of two independent events can be found by multiplying the probability of the first event by the probability of the second event.

EXAMPLE Probability of Independent Events

1 **The two spinners below are spun. What is the probability that both spinners will show a number greater than 6?**

P(first spinner is greater than 6) = 

P(second spinner is greater than 6) = 

P(both spinners are greater than 6) = $\frac{3}{10} \cdot \frac{3}{10}$ or 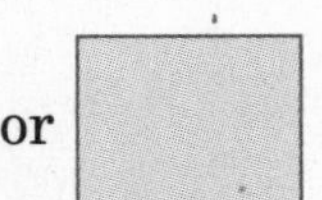

Your Turn The two spinners below are spun. What is the probability that both spinners will show a number less than 4?

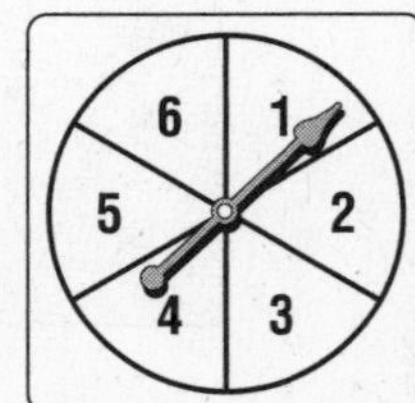

© Glencoe/McGraw-Hill

EXAMPLE Use a Probability to Solve a Problem

2 POPULATION Use the information below. What is the probability that a student picked at random will be an eighth-grade girl?

Cross River Middle School	
Demographic Group	**Fraction of the Population**
Grade 6	$\frac{4}{10}$
Grade 7	$\frac{3}{10}$
Grade 8	$\frac{3}{10}$
Boys	$\frac{4}{9}$
Girls	$\frac{5}{9}$

P(8th grade) = ____

P(girl) = ____

P(8th grade and girl) =

$\frac{\overset{1}{\cancel{3}}}{\underset{2}{\cancel{10}}} \cdot \frac{\overset{1}{\cancel{5}}}{\underset{3}{\cancel{9}}}$ or ____

The probability that the ____ events will occur is $\frac{1}{6}$.

Your Turn Use the information in the table. What is the probability that a student picked at random will be a sixth grade boy?

Monterey Middle School	
Demographic Group	**Fraction of the Population**
Grade 6	$\frac{2}{9}$
Grade 7	$\frac{4}{9}$
Grade 8	$\frac{1}{3}$
Boys	$\frac{7}{10}$
Girls	$\frac{3}{10}$

KEY CONCEPT

Probability of Two Dependent Events If two events, *A* and *B*, are dependent, then the probability of both events occurring is the product of the probability of *A* and the probability of *B* after *A* occurs.

BUILD YOUR VOCABULARY (page 194)

If the outcome of one event does ____ the outcome of another event, the compound events are called **dependent events**.

© Glencoe/McGraw-Hill

FOLDABLES™

ORGANIZE IT

Under Lesson 8-5, write what you learned about how to find the probability of independent and dependent events. On the last page in your Foldable, write the key terms in the lesson and their definitions.

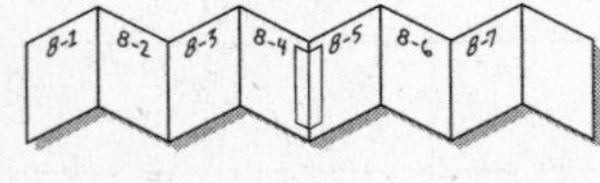

EXAMPLE Probability of Dependent Events

3 There are 4 red, 8 yellow, and 6 blue socks in a drawer. Once a sock is selected, it is not replaced. Find the probability that two blue socks are chosen.

Since the first sock ______ replaced, the first event affects the second event. These are dependent events.

$P(\text{first sock is blue}) =$ ______ ← number of blue socks / ← total number of socks

$P(\text{second sock is blue}) =$ ______ ← number of blue socks after one blue sock is removed / ← total number of socks after one blue sock is removed

$P(\text{two blue socks}) =$ ______ or ______

Your Turn There are 6 green, 9 purple, and 3 orange marbles in a bag. Once a marble is selected, it is not replaced. Find the probability that two purple marbles are chosen.

HOMEWORK ASSIGNMENT

Page(s):

Exercises:

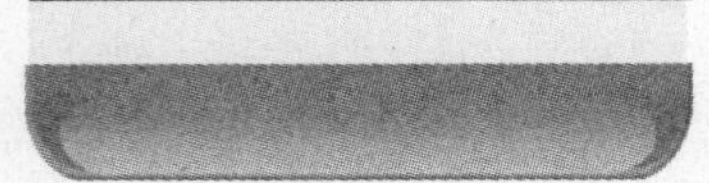

© Glencoe/McGraw-Hill

8–6 Experimental Probability

BUILD YOUR VOCABULARY (pages 194–195)

A probability that is based on ______ obtained by conducting an ______ is called an **experimental probability**.

A probabililty that is based on ______ ______ is called a **theoretical probability**.

WHAT YOU'LL LEARN

- Find experimental probability.

EXAMPLES Experimental Probability

Nikki is conducting an experiment to find the probability of getting various results when three coins are tossed. The results of her experiment are given in the table.

Result	Number of Tosses
all heads	6
two heads	32
one head	30
no heads	12

1 According to the experimental probability, is Nikki more likely to get all heads or no heads on the next toss?

Based on the results so far, ______ heads is more likely.

2 How many possible outcomes are there for tossing three coins if order is important?

There are $2 \cdot 2 \cdot 2$ or ______ possible outcomes.

© Glencoe/McGraw-Hill

Your Turn **Marcus is conducting an experiment to find the probability of getting various results when four coins are tossed. The results of his experiment are given in the table.**

Result	Number of Tosses
all heads	6
three heads	12
two heads	20
one head	7
no heads	5

a. According to the experiment probability, is Marcus more likely to get all heads or no heads on the next toss?

b. How many possible outcomes are there for tossing four coins if order is important?

EXAMPLE Theoretical Probability

3 **Nikki is conducting an experiment to find the probability of getting various results when three coins are tossed. The results of her experiment are given in the table. Is the theoretical probability greater for tossing all heads or no heads? What is the theoretical probability of each?**

Result	Number of Tosses
all heads	6
two heads	32
one head	30
no heads	12

The theoretical probability of ______ heads is

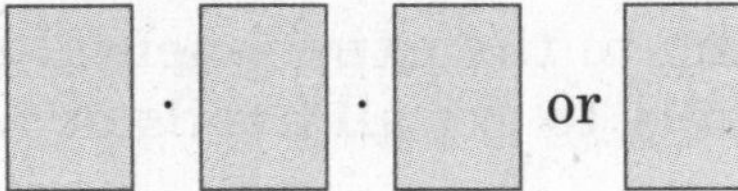
· · or .

The theoretical probability of ______ heads is

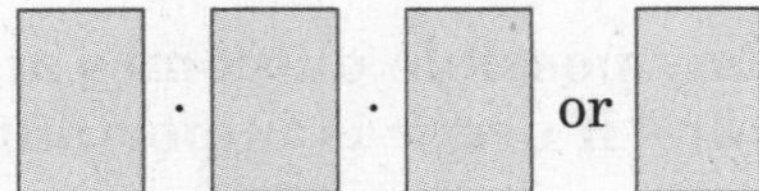
· · or .

The theoretical probabilities are .

© Glencoe/McGraw-Hill

Your Turn Marcus is conducting an experiment to find the probability of getting various results when four coins are tossed. The results of his experiment are given below. Is the theoretical probability greater for tossing all heads or no heads? What is the theoretical probability of each?

Result	Number of Tosses
all heads	6
three heads	12
two heads	20
one head	7
no heads	5

EXAMPLE Experimental Probability

FOLDABLES

ORGANIZE IT

Under Lesson 8-6, write a few words to compare and contrast experimental and theoretical probabilities. On the last page in your Foldable, write the key terms in the lesson and their definitions.

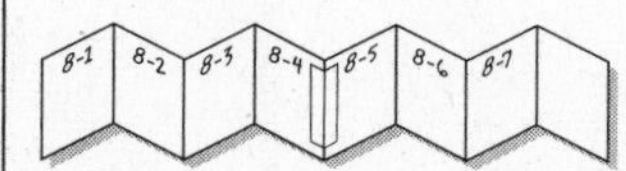

4 MARKETING **Eight hundred adults were asked whether they were planning to stay home for winter vacation. Of those surveyed, 560 said that they were. What is the experimental probability that an adult planned to stay home for winter vacation?**

There were ______ people surveyed and ______ said that they were staying home.

The experimental probability is ______ or ______.

Your Turn Five hundred adults were asked whether they were planning to stay home for New Year's Eve. Of those surveyed, 300 said that they were. What is the experimental probability that an adult planned to stay home for New Year's Eve?

© Glencoe/McGraw-Hill

EXAMPLES Use Probability to Predict

MATH TEAM Over the past three years, the probability that the school math team would win a meet is $\frac{3}{5}$.

5 Is this probability experimental or theoretical? Explain.

This is an experimental probability since it is based on what

happened in the ______.

6 If the team wants to win 12 more meets in the next 3 years, how many meets should the team enter?

This problem can be solved using a proportion.

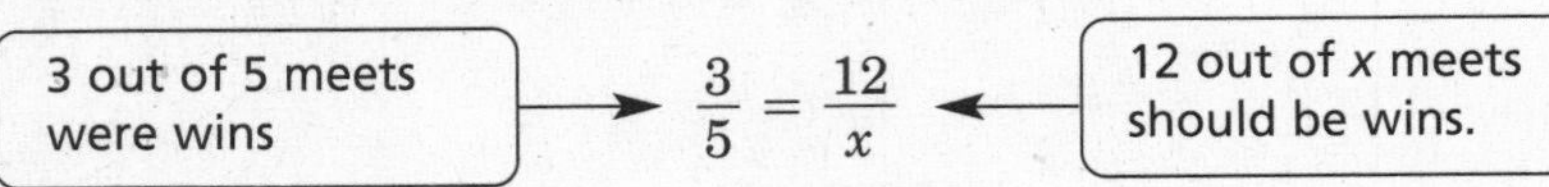

Solve the proportion.

$\frac{3}{5} = \frac{12}{x}$ Write the proportion.

______ = ______ Find the cross products.

______ = ______ Multiply.

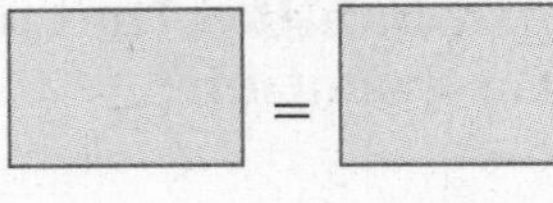

______ = ______ Divide each side by ______.

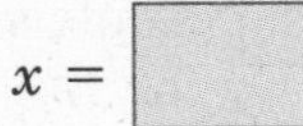

$x =$ ______

They should enter ______ meets.

Your Turn **Over the past three years, the probability that the school speech and debate team would win a meet is $\frac{4}{5}$.**

a. Is this probability experimental or theoretical? Explain.

b. If the team wants to win 20 more meets in the next 3 years, how many meets should the team enter?

REVIEW IT

Explain what a proportion is and how you can solve a proportion. *(Lesson 4-4)*

HOMEWORK ASSIGNMENT

Page(s):

Exercises:

© Glencoe/McGraw-Hill

8–7 Using Sampling to Predict

WHAT YOU'LL LEARN

- Predict the actions of a larger group by using a sample.

BUILD YOUR VOCABULARY (pages 194–195)

A **sample** is a ________ selected group chosen for the purpose of collecting data.

The **population** is the ________ from which the samples under consideration are taken.

An **unbiased sample** is selected so that it is ________ of the entire population.

In a **stratified random sample**, the population is divided into ________, nonoverlapping groups.

In a **systematic random sample**, the items or people are selected according to a specific ________ or item interval.

In a **biased sample**, one or more parts of the population are ________ over others.

EXAMPLES Describe Samples

Describe each sample.

1 **To determine which school lunches students like most, every twentieth student to walk into the cafeteria is surveyed.**

Since the population is the students entering the cafeteria, the sample is a ________. It is an ________ sample.

© Glencoe/McGraw-Hill

2 **To determine what sports teenagers like, the student athletes on the girls' field hockey team are surveyed.**

Teenagers on the field hockey team are more likely to choose field hockey. This is a ______ sample. The sample is a ______ sample because the people are easily accessed.

Your Turn **Describe each sample.**

a. To determine which CDs customers like most, every tenth customer to walk into the music store is surveyed.

______

b. To determine what restaurant teenagers like, the teenagers eating at Pete's Diner are surveyed.

______

EXAMPLES Using Sampling to Predict

BOOKS The student council is trying to decide what types of books to sell at its annual book fair to help raise money for the eighth-grade trip. It surveys 40 students at random. The books they prefer are in the table.

Book Type	Number of Students
mystery	12
adventure novel	9
sports	11
short stories	8

3 **What percent of the students prefer mysteries?**

______ out of ______ students prefer mysteries.

______ ÷ ______ = ______

So, ______ of the students prefer mysteries.

© Glencoe/McGraw-Hill

4 **If 220 books are to be sold at the book fair, how many should be mysteries?**

Find 30% of ____.

0.30 × ____ = ____

About ____ books should be mysteries.

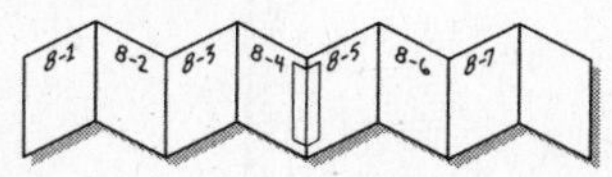

FOLDABLES™

ORGANIZE IT

Under Lesson 8-7, list the different types of samples and how to use them to make predictions. Give examples. On the last page of your Foldable, write the key terms in the lesson and their definitions.

Your Turn **The student shop sells pens. It surveys 50 students at random. The pens they prefer are in the table.**

Type	Number
gel pens	22
ball point	8
glitter pens	10
roller balls	10

a. What percent of the students prefer gel pens?

b. If 300 pens are to be sold at the student shop, how many should be gel pens?

© Glencoe/McGraw-Hill

HOMEWORK ASSIGNMENT

Page(s):

Exercises:

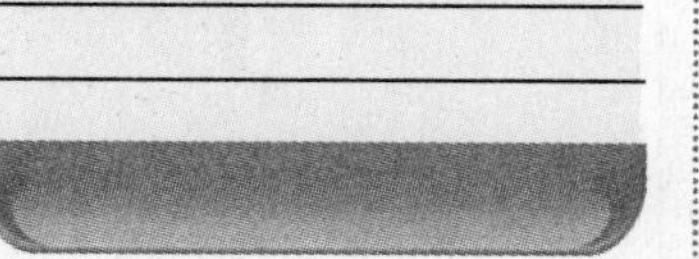

CHAPTER 8

BRINGING IT ALL TOGETHER

STUDY GUIDE

FOLDABLES™	VOCABULARY PUZZLEMAKER	BUILD YOUR VOCABULARY
Use your **Chapter 8 Foldable** to help you study for your chapter test.	To make a crossword puzzle, word search, or jumble puzzle of the vocabulary words in Chapter 8, go to: www.glencoe.com/sec/math/t_resources/free/index.php	You can use your completed **Vocabulary Builder** *(pages 194–195)* to help you solve the puzzle.

8-1 Probability of Simple Events

1. Complete the sentence. Probability is a ratio that compares the number of ________ outcomes to the number of ________ outcomes.

A spinner is divided into 8 equal parts, and one of the first eight letters of the alphabet is placed in each section such that all the letters are used.

2. What is the sample space?

3. What is the probability of spinning a vowel?

4. If an event has a probability of 0.1, what does this mean?

5. If you are given the probability of an event happening, how can you find the probability of the event *not* happening?

© Glencoe/McGraw-Hill

8-2

Counting Outcomes

6. Complete the tree diagram shown below for how many boys and and how many girls are likely to be in a family of three children.

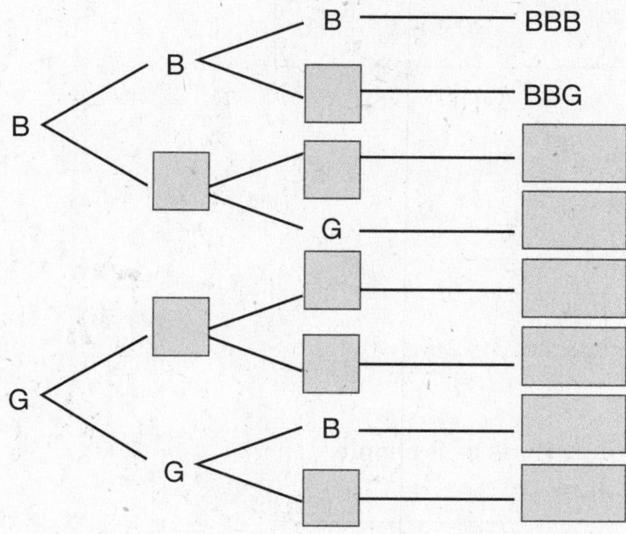

7. Use the Fundamental Counting Principle to find the number of possible outcomes if there are 4 true-false questions on a test.

☐ × ☐ × ☐ × ☐ = ☐

8-3

Permutations

8. What does the notation $P(14, 4)$ represent?

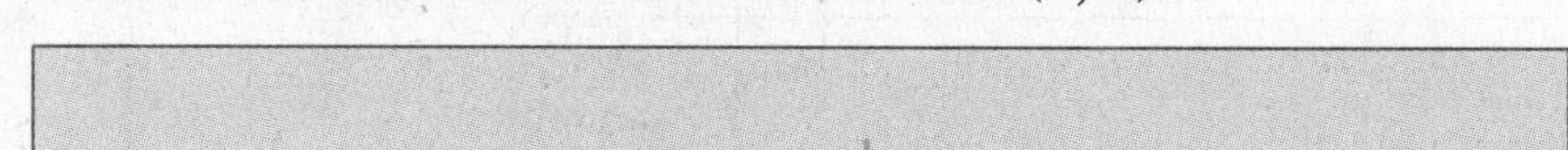

9. What is the difference between 6! and $P(6, 4)$?

A security system has a number pad with 9 digits.

10. How many three-number passwords are available if a digit cannot be repeated?

11. If a digit can be repeated, how many passwords are available?

© Glencoe/McGraw-Hill

8-4

Combinations

12. What is the difference between a permutation and a combination?

13. Fill in the blanks to find $C(9, 4)$.

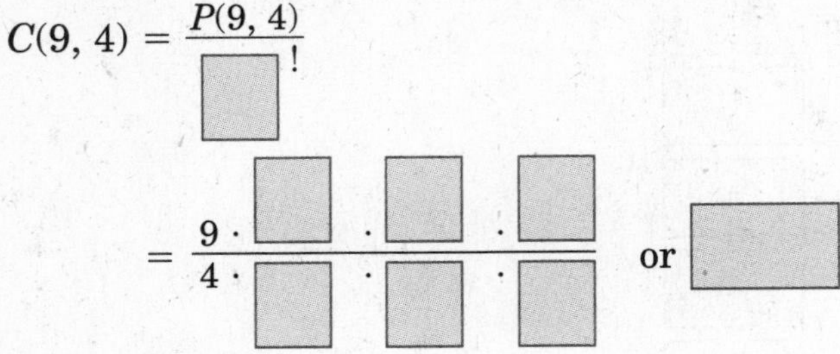

$C(9, 4) = \frac{P(9, 4)}{\square!}$

$= \frac{9 \cdot \square \cdot \square \cdot \square}{4 \cdot \square \cdot \square \cdot \square}$ or $\square$

14. Are there more combinations or permutations of 3 people chosen from a group of 6 people? Explain.

8-5

Probability of Compound Events

15. What is a compound event?

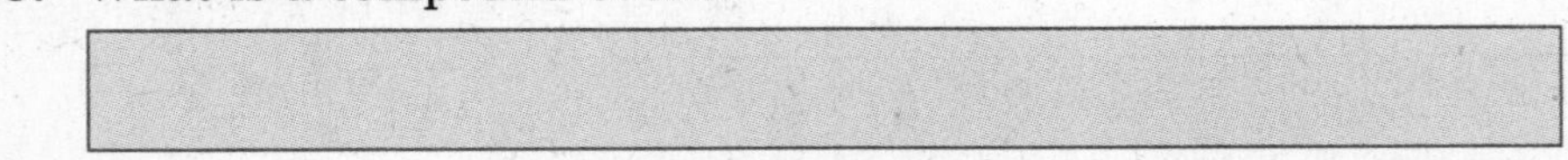

16. Are the events of spinning a spinner and rolling a number cube independent events? Why or why not?

A number cube is rolled and a penny is tossed. Find each probability.

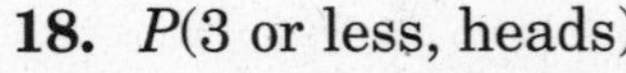

17. P(4 and tails)

18. P(3 or less, heads)

© Glencoe/McGraw-Hill

8-6

Experimental Probability

The table at the right shows the results of a survey.

Item	Number of People
balloons	75
cards	15
decorations	25
cake	50

19. How many people bought balloons?

20. How many people were surveyed?

21. What is the experimental probability that a person surveyed preferred balloons?

22. A bag contains 15 red marbles, 25 purple marbles, and 10 yellow marbles. Describe an experiment that you could conduct with the marbles to find an experimental probability.

8-7

Using Sampling to Predict

23. What is the first step in conducting a survey?

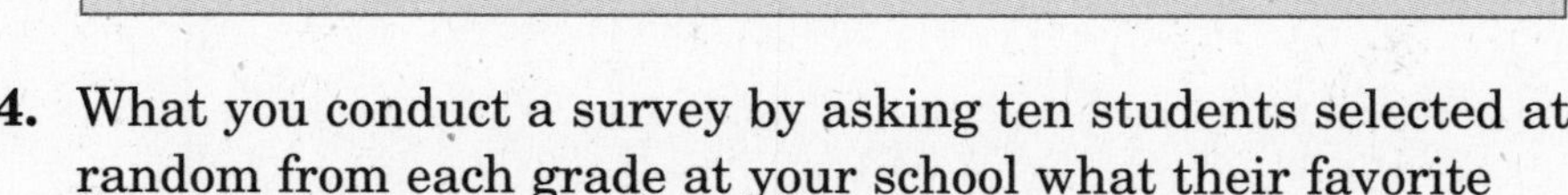

24. What you conduct a survey by asking ten students selected at random from each grade at your school what their favorite class is, what type of random sample have you taken?

25. A grocery store owner asks the shoppers in his store where they prefer to shop for groceries. What type of sample has he conducted?

© Glencoe/McGraw-Hill

ARE YOU READY FOR THE CHAPTER TEST?

Visit **msmath3.net** to access your textbook, more examples, self-check quizzes, and practice tests to help you study the concepts in Chapter 8.

Check the one that applies. Suggestions to help you study are given with each item.

☐ **I completed the review of all or most lessons without using my notes or asking for help.**

- You are probably ready for the Chapter Test.
- You may want to take the Chapter 8 Practice Test on page 413 of your textbook as a final check.

☐ **I used my Foldable or Study Notebook to complete the review of all or most lessons.**

- You should complete the Chapter 8 Study Guide and Review on pages 410–412 of your textbook.
- If you are unsure of any concepts or skills, refer back to the specific lesson(s).
- You may also want to take the Chapter 8 Practice Test on page 413.

☐ **I asked for help from someone else to complete the review of all or most lessons.**

- You should review the examples and concepts in your Study Notebook and Chapter 8 Foldable.
- Then complete the Chapter 8 Study Guide and Review on pages 410–412 of your textbook.
- If you are unsure of any concepts or skills, refer back to the specific lesson(s).
- You may also want to take the Chapter 8 Practice Test on page 413.

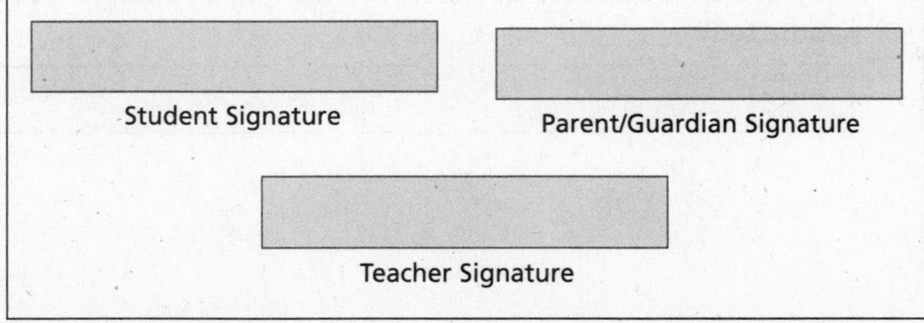

© Glencoe/McGraw-Hill

Statistics and Matrices

Use the instructions below to make a Foldable to help you organize your notes as you study the chapter. You will see Foldable reminders in the margin of this Interactive Study Notebook to help you in taking notes.

Begin with four pieces of $8\frac{1}{2}$" by 11" paper.

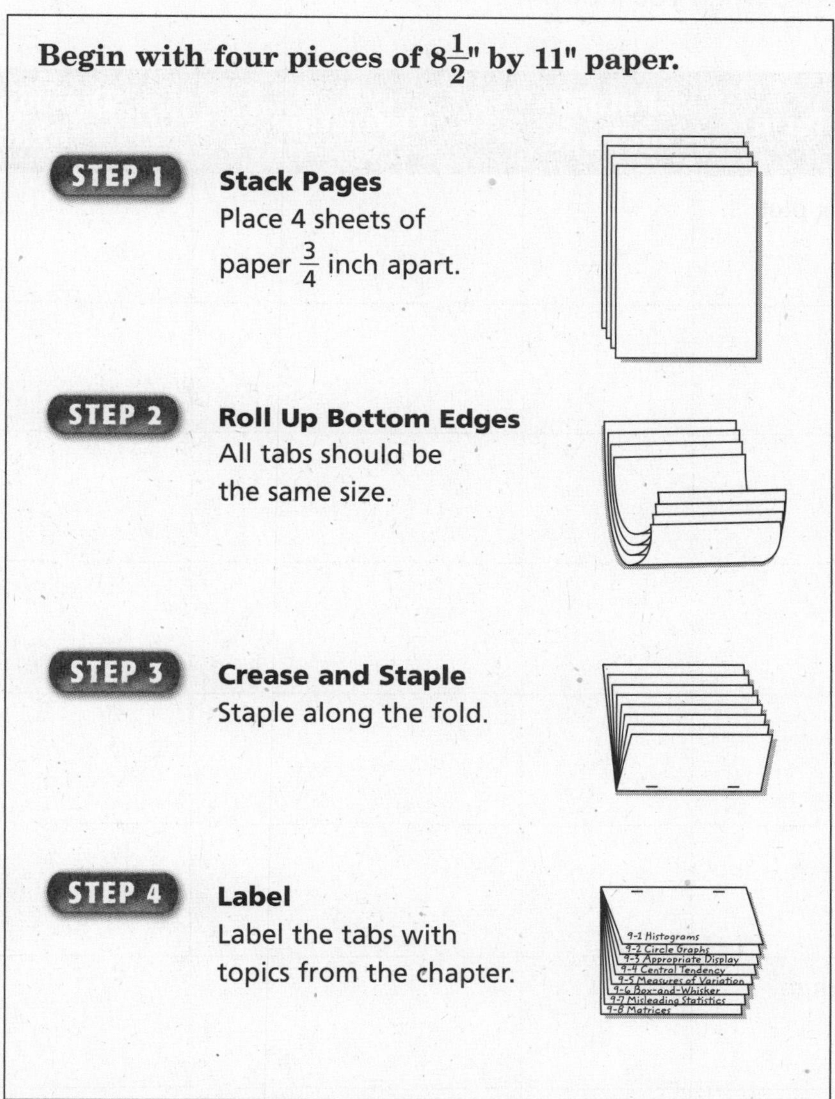

Stack Pages
Place 4 sheets of paper $\frac{3}{4}$ inch apart.

Roll Up Bottom Edges
All tabs should be the same size.

Crease and Staple
Staple along the fold.

Label
Label the tabs with topics from the chapter.

NOTE-TAKING TIP: As you take notes on a topic, it helps to write how the subject relates to your life. For example, as you learn about different kinds of statistical measures and graphs, you will understand how to evaluate statistical information presented in such places as advertisements and persuasive articles in magazines.

© Glencoe/McGraw-Hill

BUILD YOUR VOCABULARY

This is an alphabetical list of new vocabulary terms you will learn in Chapter 9. As you complete the study notes for the chapter, you will see Build Your Vocabulary reminders to complete each term's definition or description on these pages. Remember to add the textbook page number in the second column for reference when you study.

Vocabulary Term	Found on Page	Definition	Description or Example
box-and-whisker plot			
circle graph			
column			
dimensions			
element			
histogram			
interquartile range			
lower quartile			
matrix [MAE-triks]			

© Glencoe/McGraw-Hill

Vocabulary Term	Found on Page	Definition	Description or Example
mean			
measures of central tendency			
measures of variation			
median			
mode			
outlier			
quartiles			
range			
row			
upper quartile			

© Glencoe/McGraw-Hill

9–1 Histograms

BUILD YOUR VOCABULARY (page 222)

A **histogram** is a type of ______ graph used to display numerical data that have been organized into ______ intervals.

WHAT YOU'LL LEARN

- Construct and interpret histograms.

FOLDABLES™

ORGANIZE IT

Under the tab for Lesson 9–1, explain the difference between a bar graph and a histogram. Describe a type of statistics that could be displayed with a histogram.

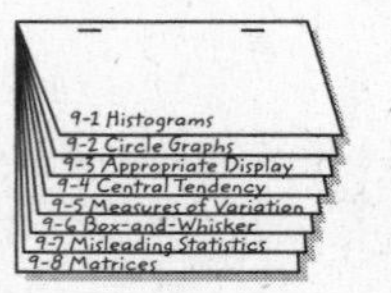

EXAMPLE Draw a Histogram

1 FOOD **The frequency table below shows the amount of caffeine in certain types of tea. Draw a histogram to represent the data.**

Caffeine Content of Certain Types of Tea		
Caffeine (mg)	Tally	Frequency
1–20	𝍸 II	7
21–40	𝍸 III	8
41–60	II	2
61–80		0
81–100	II	2

Step 1 Draw and label a horizontal and vertical axis. Include a title.

Step 2 Show the ______ from the frequency table on the ______ axis.

Step 3 For each caffeine interval, draw a bar whose height is given by the frequency.

© Glencoe/McGraw-Hill

Your Turn The frequency table below shows the amount of caffeine in certain drinks. Draw a histogram to represent the data.

Caffeine Content of Certain Types of Drink		
Caffeine (mg)	**Tally**	**Frequency**
0–50	III	3
51–100	IIII	4
101–150	𝍸 I	6
151–200	𝍸 II	7

EXAMPLE Interpret Data

2 WEATHER **How many months had 6 or more days of rain?**

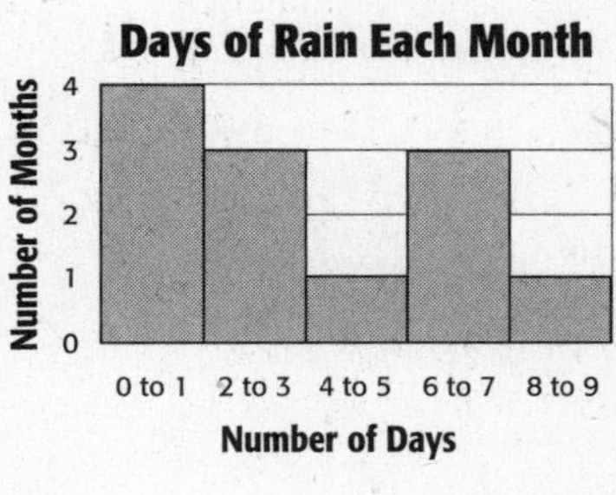

Three months had 6 to 7 days of rain, and one month had 8 to 9 days of rain.

Therefore, ☐ + ☐ or ☐ months had 6 or more days of rain.

Your Turn How many months had 6 or more days of snow?

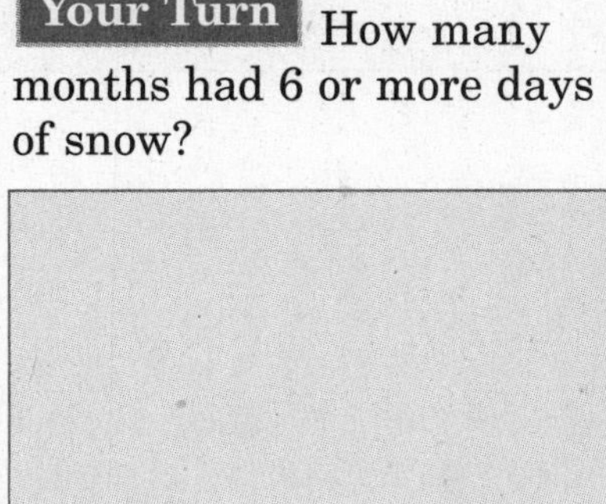

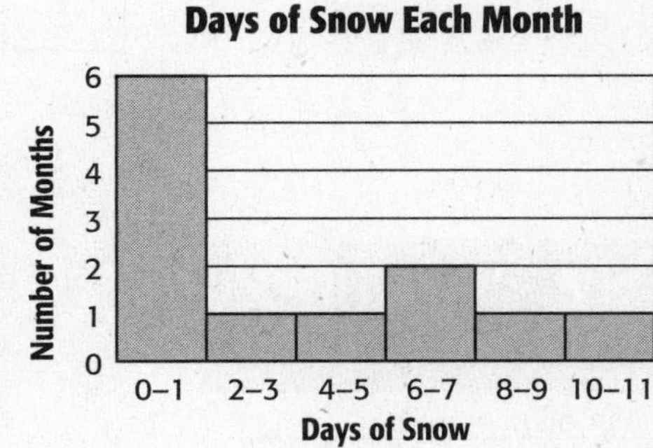

© Glencoe/McGraw-Hill

EXAMPLE Compare Two Sets of Data

> **REMEMBER IT**
>
> When you compare two histograms, be sure to first check to make sure the scales and intervals are the same.

3 GRADES **Determine which test had the greater number of students scoring 86 or higher.**

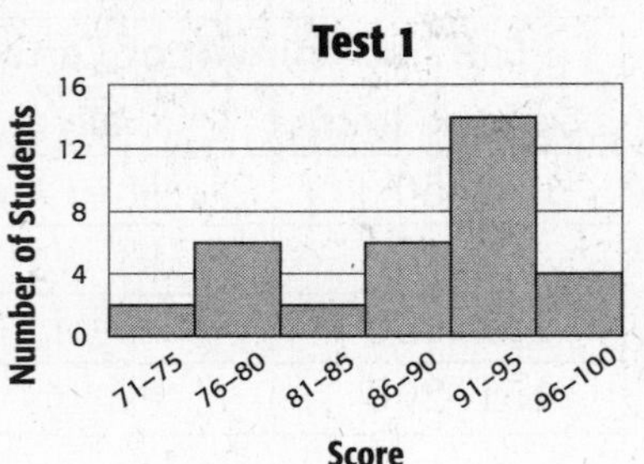

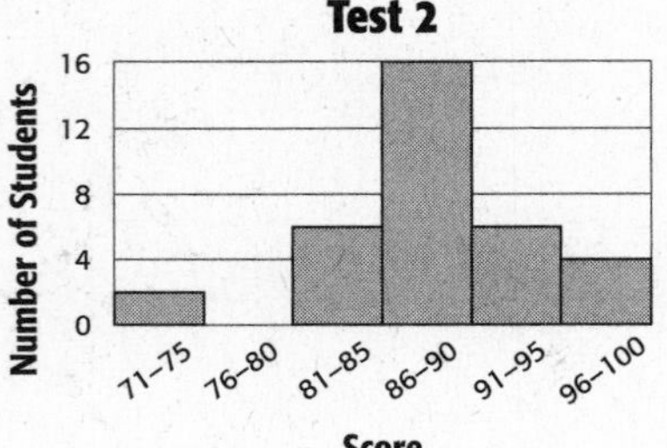

On test 1, 6 + 14 + ______ or 24 students scored 86 or higher.

On test 2, ______ + 6 + 4 or 26 students scored 86 or higher.

A greater number of students scored 86 or higher on test 2.

Your Turn Determine which test had the greater number of students scoring 71 or higher.

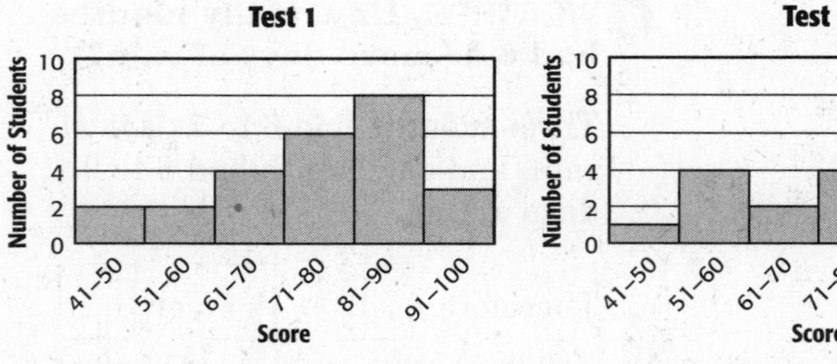

______

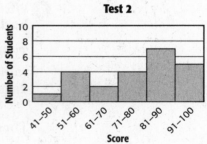

HOMEWORK ASSIGNMENT

Page(s):

Exercises:

© Glencoe/McGraw-Hill

9–2 Circle Graphs

What You'll Learn

- Construct and interpret circle graphs.

Build Your Vocabulary (page 222)

A **circle graph** is used to compare parts of a ________.

The entire ________ represents that whole.

FOLDABLES™

Organize It

Under the tab for Lesson 9–2, find an example of a circle graph from a newspaper or magazine. Explain what the graph shows.

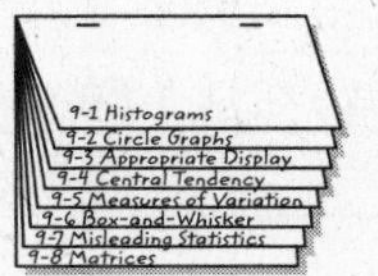

EXAMPLE Draw a Circle Graph

1 TORNADOES **The table shows when tornadoes occurred in the United States from 1999 to 2001. Make a circle graph using this information.**

Tornadoes in the United States, 1999–2001	
January–March	15%
April–June	53%
July–September	21%
October–December	11%

Source: spc.noaa.gov/

Step 1 There are ________ in a circle. So, multiply each percent by 360 to find the number of degrees for each ________ of the graph.

Jan–Mar:

15% of 360 = ________ · 360 or ________

Apr–Jun:

53% of 360 = ________ · 360 or about ________

Jul–Sept:

21% of 360 = ________ · 360 or about ________

Oct–Dec:

11% of 360 = ________ · 360 or about ________

© Glencoe/McGraw-Hill

Step 2 Use a compass to draw a circle and a radius. Then use a protractor to draw a ______ angle. This section represents January–March. From the new radius, draw the next angle. Repeat for each of the remaining angles. Label each ______. Then give the graph a ______.

Your Turn The table shows when hurricanes or tropical storms occurred in the Atlantic Ocean during the hurricane season of 2002. Make a circle graph using this information.

Hurricanes in the United States, 2002	
Month	**Percent**
July	7%
August	21%
September	64%
October	8%

Source: nhc.noaa.gov/

© Glencoe/McGraw-Hill

EXAMPLES Use Circle Graphs to Interpret Data

2 BASKETBALL Make a circle graph using the information in the histogram below.

Average Points Per Basketball
Game for Top 25 Scorers

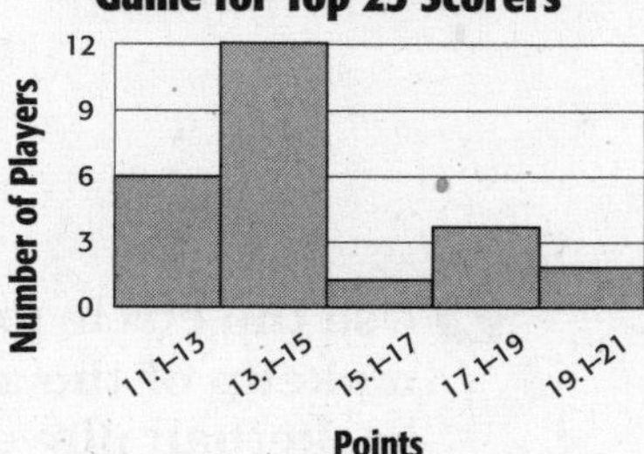

Step 1 Find the total number of players.

$6 + \square + 1 + \square + 2 = \square$

Step 2 Find the ratio that compares the number in each point range to the total number of players. Round to the nearest hundredth.

11.1 to 13 : $6 \div 25 = \square$

13.1 to 15 : $12 \div 25 = \square$

15.1 to 17 : $1 \div 25 = \square$

17.1 to 19 : $4 \div 25 = \square$

19.1 to 21 : $2 \div 25 = \square$

Step 3 Use these ratios to find the number of degrees of each section. Round to the nearest degree if necessary.

11.1 to 13 : $\square \cdot 360 = \square$ or about $\square$

13.1 to 15 : $\square \cdot 360 = \square$ or about 173

15.1 to 17 : $\square \cdot 360 = \square$ or about $\square$

17.1 to 19 : $\square \cdot 360 = \square$ or about $\square$

19.1 to 21 : $\square \cdot 360 = \square$ or about 29

© Glencoe/McGraw-Hill

Step 4 Use a compass and protractor to draw a circle and the appropriate sections. Label each section and give the graph a title. Write the ratios as percents.

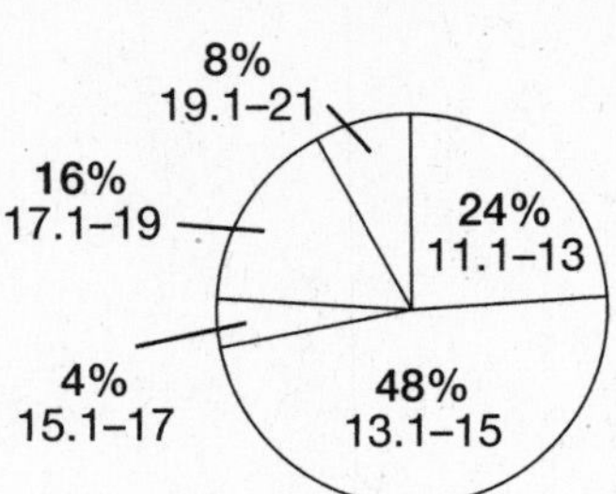

3 Use the circle graph from Example 2 to describe the makeup of the average game scores of the 25 top-scoring basketball players.

Almost $\frac{3}{4}$ of the players had average game scores between 11.1 and 15 points. Fewer than $\frac{1}{4}$ had average game scores greater than ☐ points.

Your Turn

a. Make a circle graph using the information in the histogram at right.

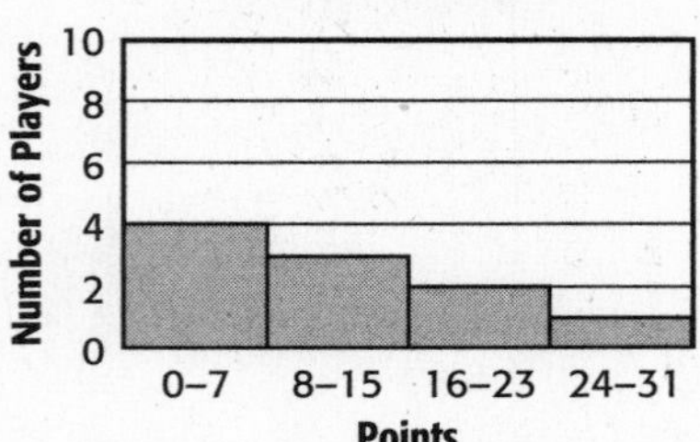

b. Use the graph to describe the makeup of the average game scores of the 10 top-scoring football players.

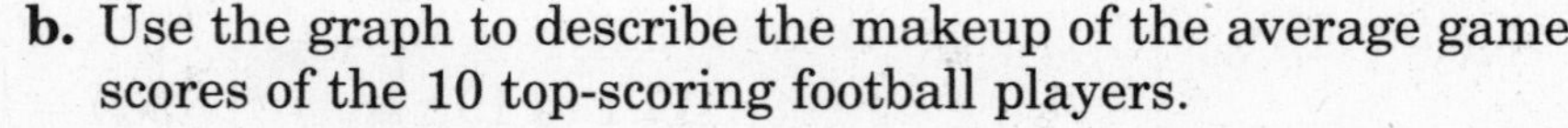

HOMEWORK ASSIGNMENT

Page(s): ________

Exercises: ________

© Glencoe/McGraw-Hill

9–3 Choosing an Appropriate Display

WHAT YOU'LL LEARN

- Choose an appropriate display for a set of data.

EXAMPLE Choose an Appropriate Display

Choose an appropriate type of display for each situation. Then make a display.

1 FARMS **The table shows farm acres in Maine. Choose an appropriate display for this situation. Then make a display.**

Farms in Maine by Size	
1–99 acres	46.8%
100–499 acres	43.8%
500–999 acres	6.9%
1,000 or more acres	2.5%

Source: ers.usda.gov

This data deals with percents that have a sum of ________.

A ________ would be a good way to show percents.

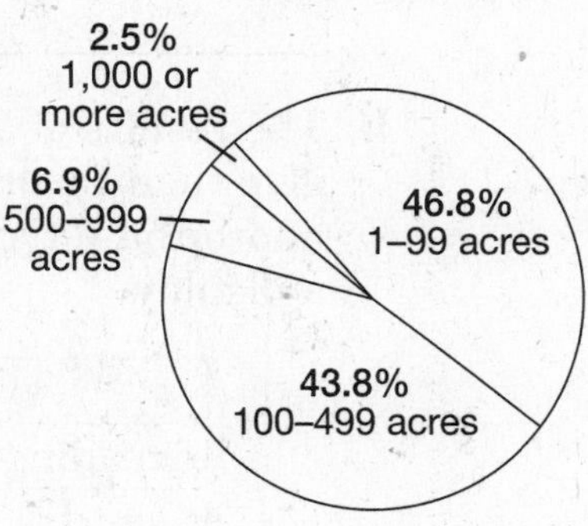

FOLDABLES™

ORGANIZE IT

Under the tab for Lesson 9-3, make a table of data from your science or social studies textbook. Draw a circle graph and bar graph displaying the data. Discuss which graph is most appropriate.

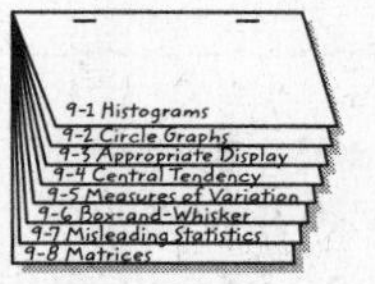

2 SCHOOL **The results of a survey of a group of students asked to give their favorite school subject are shown in the table. Choose an appropriate type of display for this situation. Then make a display.**

Favorite School Subject	
math	~~IIII~~ ~~IIII~~ ~~IIII~~ II
history	~~IIII~~ III
science	~~IIII~~ ~~IIII~~
English	~~IIII~~ II
other	~~IIII~~ I

In this case, there are specific categories. If you want to show the specific number, use a ________ or a ________.

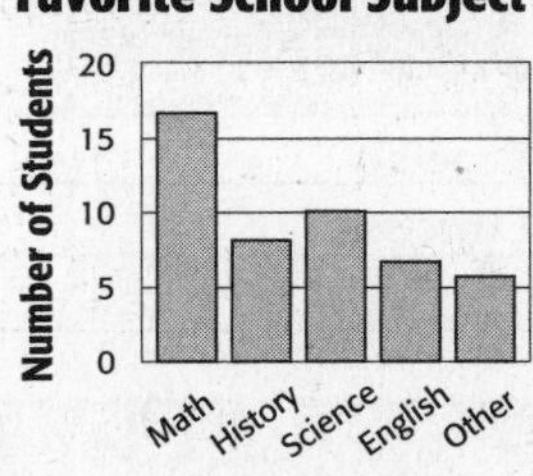

© Glencoe/McGraw-Hill

Your Turn

REMEMBER IT

There are many ways to display the same data. However, often one of those ways makes the data easier to understand than do the other ways.

a. The table shows the favorite type of television program of students at Walnut Junior High. Choose an appropriate type of display for this situation. Then make a display.

Favorite Type of Television Program	
sitcom	54%
reality	22%
news	10%
game show	8%
cartoon	6%

b. The results of a survey of a group of students asked to give their favorite hobby are shown below. Choose an appropriate type of display for this situation. Then make a display.

Hobby	Number of Students
reading	10
sports	5
listening to music	10
photography	7
other	18

HOMEWORK ASSIGNMENT

Page(s):

Exercises:

© Glencoe/McGraw-Hill

9–4 Measures of Central Tendency

BUILD YOUR VOCABULARY (page 223)

Measures of central tendency are numbers that ______ a set of data.

The **mean** of a set of data is the ______ of the data ______ the number of items in the data set.

The **median** of a set of data is the ______ number of the data ordered from least to greatest, or the mean of the ______ numbers.

The **mode** of a set of data is the number or numbers that occur ______ often.

WHAT YOU'LL LEARN

- Find the mean, median, and mode of a set of data.

WRITE IT

The words *central* and *middle* have similar definitions. If mean, median, and mode are measures of central tendency, what do they measure?

______

______

______

______

EXAMPLE Find Measures of Central Tendency

1 Find the mean, median, and mode of the set of data.
4, 16, 32, 19, 27, 32

Mean $\frac{4 + 16 + 32 + 19 + 27 + 32}{\square} = \square$

$\approx \square$

Median Arrange the numbers in order from ______ to ______.

4 16 19 27 32 32

$\frac{\square + \square}{\square} = \square$

Mode The data has a mode of ______.

© Glencoe/McGraw-Hill

Your Turn Find the mean, median, and mode of the set of data. 3, 5, 3, 7, 6, 4

EXAMPLES Using Appropriate Measures

OLYMPICS Use the table to answer each question.

Gold Medals Won by the United States at the Winter Olympics, 1924–2002			
Event	**Gold Medals**	**Event**	**Gold Medals**
Alpine skiing	10	Luge	2
Bobsleigh	6	Short track speed skating	3
Cross country	0	Skeleton	3
Figure skating	13	Ski jumping	0
Freestyle skiing	4	Snowboarding	2
Ice hockey	3	Speed skating	26

FOLDABLES™

ORGANIZE IT

Under the tab for Lesson 9–4, record how to find the mean, median, and mode of a set of data. Explain *measures of central tendency*, *mean*, *median*, and *mode* in your own words and with examples.

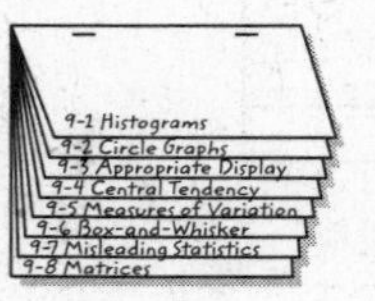

2 What is the mean, median, and mode of the data?

Mean $\dfrac{10 + 6 + 0 + 13 + 4 + 3 + 2 + 3 + 3 + 0 + 2 + 26}{\square} = \dfrac{\square}{\square}$

$= \square$

The mean is ☐ medals.

Median Arrange the numbers from least to greatest.

0, 0, 2, 2, 3, 3, 3, 4, 6, 10, 13, 26

The median is the mean of the middle two numbers, or ☐ medals.

Mode There is one mode, ☐.

3 Which measure of central tendency is most representative of the data?

The median and the mode; the mean is affected by the extreme value of ☐. The mode is the same as the median. So, both the median and the mode are good choices.

© Glencoe/McGraw-Hill

Your Turn **Use the table to answer each question.**

Country	Gold Medals (1896–2000 Summer)
United States	872
Great Britain	180
France	188
Italy	179
Sweden	136
Hungary	150
Australia	102
Finland	101
Japan	97
Romania	74
Brazil	12
Ethiopia	12

Source: infoplease.com

a. What is the mean, median, and mode of the data?

b. Which measure of central tendency is the most representative of the data?

© Glencoe/McGraw-Hill

HOMEWORK ASSIGNMENT

Page(s):

Exercises:

9–5 Measures of Variation

BUILD YOUR VOCABULARY (pages 222–223)

Measures of variation are used to describe the ______ of a set of data.

The **range** indicates how ______ the data are.

Quartiles are the values that divide the data into ______ equal parts.

The ______ of the lower half of a set of data is the **lower quartile**.

The median of the ______ of the set of data is the **upper quartile**.

Data that are more than ______ times the value of the interquartile range beyond the quartiles are called **outliers**.

WHAT YOU'LL LEARN

- Find the range and quartiles of a set of data.

KEY CONCEPTS

Range The range of a set of data is the difference between the greatest and the least numbers in the set.

Interquartile Range The interquartile range is the range of the middle half of the data. It is the difference between the upper quartile and the lower quartile.

EXAMPLES Find Measures of Variation

BASKETBALL Use the table at the right.

1 Find the range of the scores.

The greatest number of scores is ______.

The least number of scores is ______.

The range is ______ – ______ or ______ points.

Points Scored by Top Ten Teams During the NBA Playoffs, 2002	
Team	Points Scored
Dallas	109
Minnesota	102
Sacramento	101.1
L.A. Lakers	97.8
Charlotte	96.1
New Jersey	95.4
Orlando	93.8
Indiana	91.6
Boston	91.3
Portland	91.3

Source: nba.com

© Glencoe/McGraw-Hill

2 **Find the median and the upper and lower quartiles of the scores.**

Arrange the numbers in order from least to greatest.

lower quartile ↓ median ↓ upper quartile ↓

91.3 91.3 [ ] 93.8 95.4 96.1 97.8 [ ] 102 109

$\frac{95.4 + 96.1}{2} =$ [ ]

The median is [ ], the lower quartile is [ ], and the upper quartile is [ ].

REMEMBER IT

A small interquartile range means that the data in the middle of the set are close in value. A large interquartile range means that the data in the middle are spread out.

3 **Find the interquartile range of the scores.**

Interquartile Range: [ ] – [ ] or [ ]

Your Turn **Use the table at the right.**

a. Find the range of the batting averages.

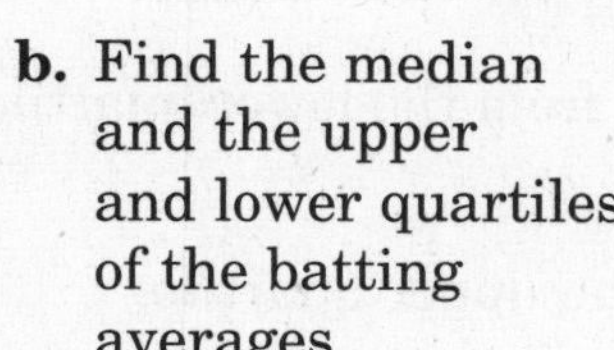

b. Find the median and the upper and lower quartiles of the batting averages.

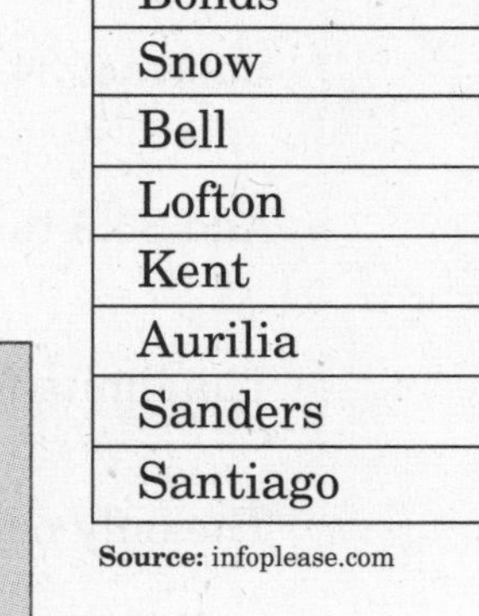

Giants Batting Average Against Anaheim in the World Series 2002	
Player	**Batting Average**
Rueter	0.500
Bonds	0.471
Snow	0.407
Bell	0.304
Lofton	0.290
Kent	0.276
Aurilia	0.250
Sanders	0.238
Santiago	0.231

Source: infoplease.com

c. Find the interquartile range of the batting averages.

© Glencoe/McGraw-Hill

EXAMPLE Find Outliers

FOLDABLES

ORGANIZE IT

Under the tab for Lesson 9–5, write what you learn about finding the range and quartiles of a set of data.

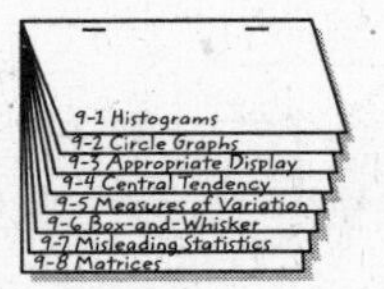

4 CONCESSION SALES **Find any outliers for the data in the table at the right.**

Items Sold at Football Game Concession Stand	
Item	**Number Sold**
Colas	196
Diet colas	32
Water	46
Coffee	18
Candy bars	39
Hotdogs	23
Hamburgers	16
Chips	41
Popcorn	24

First arrange the numbers in order from least to greatest. Then find the median, upper quartile, and lower quartile.

16 18 23 24 32 39 41 46 196

$\frac{18 + 23}{2} =$ ______ 32 $\frac{41 + 46}{2} =$ ______

Interquartile Range = ______ − ______ or 23

Multiply the interquartile range, 23, by 1.5. ______ × ______ = 34.5

Find the limits for the outliers.

Subtract 34.5 from the lower quartile. ______ − 34.5 = ______

Add 34.5 to the upper quartile. ______ + 34.5 = ______

The limits for the outliers are ______ and ______.

The only outlier is ______.

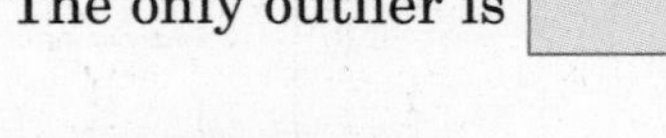

Your Turn Find any outliers for the data in the table at right.

Items Sold at School Bookstore	
Item	**Number Sold**
Pens	35
Pencils	15
Erasers	20
Candy bars	93
Folders	17
School Pennants	18
Calculators	2

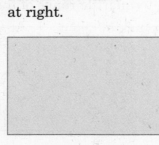

HOMEWORK ASSIGNMENT

Page(s):

Exercises:

© Glencoe/McGraw-Hill

9–6 Box-and-Whisker Plots

What You'll Learn

- Display and interpret data in a box-and-whisker plot.

Build Your Vocabulary (page 222)

A **box-and-whisker plot** uses a ______ to show the ______ of a set of data.

Foldables™ Organize It

Under the tab for Lesson 9–6, collect data from the Internet, such as number of homeruns hit by the players of a baseball team. Draw a box-and-whisker plot to display the data.

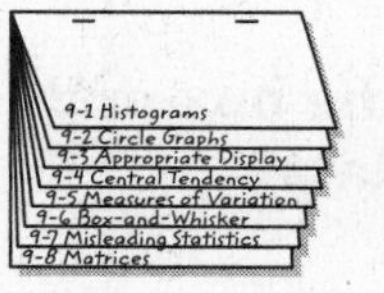

EXAMPLE Draw a Box-and-Whisker Plot

1 POPULATION **Use the data in the table at the right to draw a box-and-whisker plot.**

World's Most Populous Cities	
City	**Population (millions)**
Tokyo	34.8
New York	20.2
Seoul	19.9
Mexico City	19.8
Sao Paulo	17.9
Bombay	17.9
Osaka	17.9
Los Angeles	16.2
Cairo	14.4
Manila	13.5

Source: *Time Almanac*

Step 1 Draw a ______ that includes the least and greatest number in the data.

Step 2 Mark the extremes, the ______, and the upper and lower ______ above the number line. Since the data have an outlier, mark the greatest value that is not an ______.

Step 3 Draw the box and whiskers.

© Glencoe/McGraw-Hill

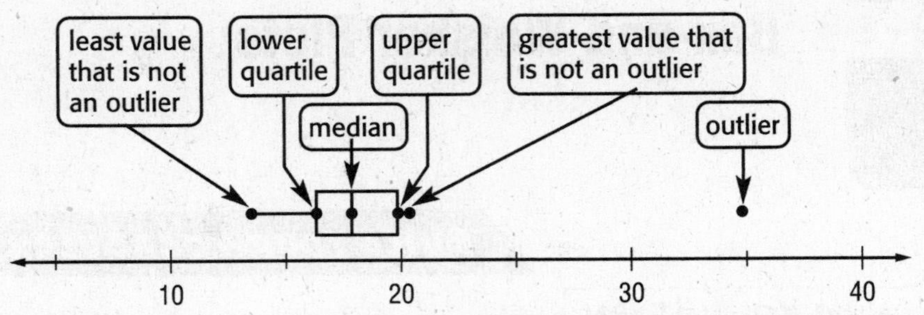

Your Turn

Use the data in the table at the right to draw a box-and-whisker plot.

Most Populous U.S. Cities	
City	**Population (in millions)**
New York	8.0
Los Angeles	3.7
Chicago	2.9
Houston	2.0
Philadelphia	1.5
Phoenix	1.3
San Diego	1.2
Dallas	1.2

Source: infoplease.com

EXAMPLE Interpret Data

2 WATERFALLS **What does the length of the box-and-whisker plot below tell you about the data?**

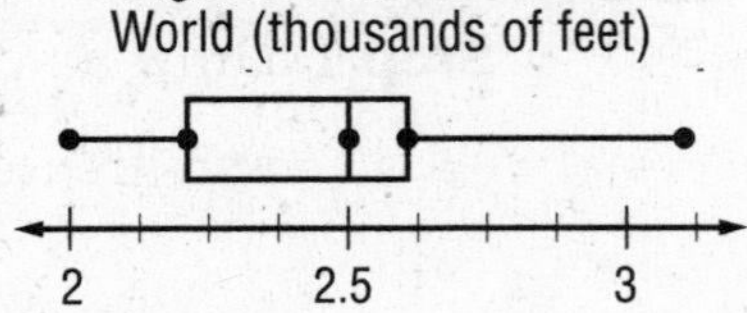

Source: *Time Almanac*

Data in the ______ quartile are more spread out than the data in the ______ quartile. You can see that data in the ______ quartile are the most spread out because the whisker is ______ than other parts of the plot.

© Glencoe/McGraw-Hill

Your Turn What does the length of the box-and-whisker plot below tell you about the data?

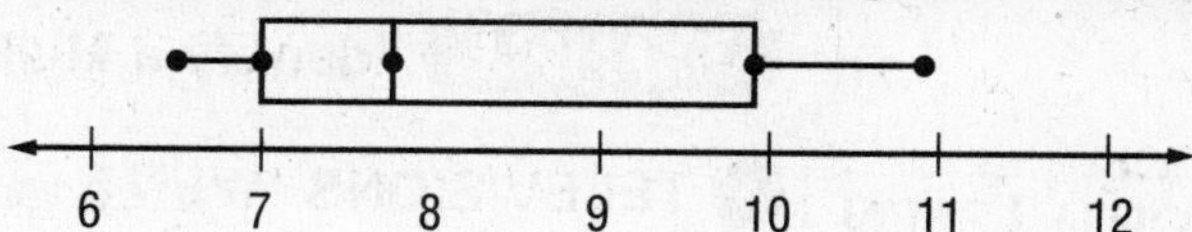

HOMEWORK ASSIGNMENT

Page(s):

Exercises:

© Glencoe/McGraw-Hill

9–7 Misleading Graphs and Statistics

WHAT YOU'LL LEARN

- Recognize when graphs and statistics are misleading.

FOLDABLES™

ORGANIZE IT

Under the tab for Lesson 9–7 record what you learn about recognizing misleading statistics or graphs. Try to collect an example of misleading statistics or graphs in print. Glue or tape them into your Foldable and explain how and why they are misleading.

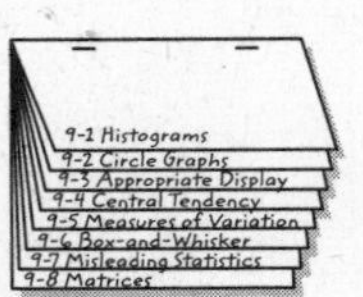

EXAMPLE Identify a Misleading Graph

1 TELEVISIONS **Which graph below could be used to indicate a greater difference in number of televisions? Explain.**

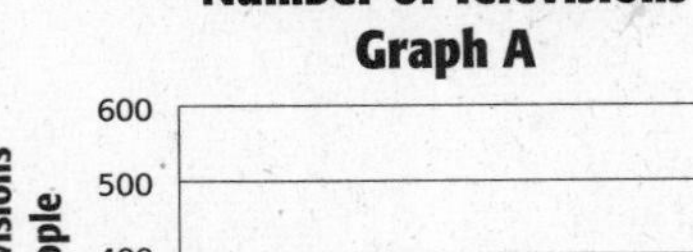

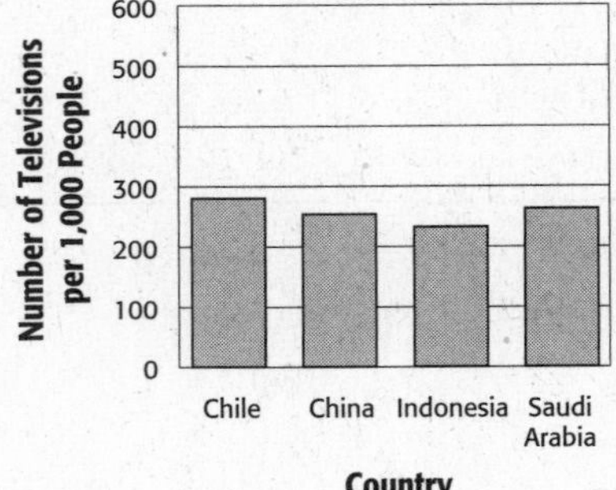

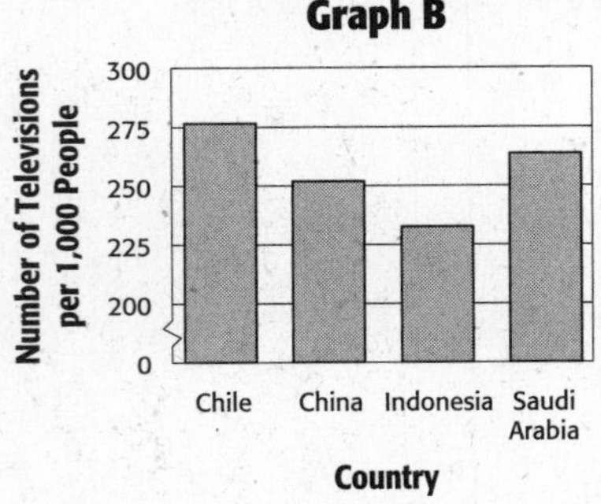

Both graphs show the order from greatest to least number of televisions per 1,000 people in Chile, Saudi Arabia, China, and Indonesia. However, the intervals in graph B represent ______ instead of ______ like graph A.

Graph B shows a greater difference in televisions.

Your Turn Which graph below could be used to show a greater difference in favorite classes?

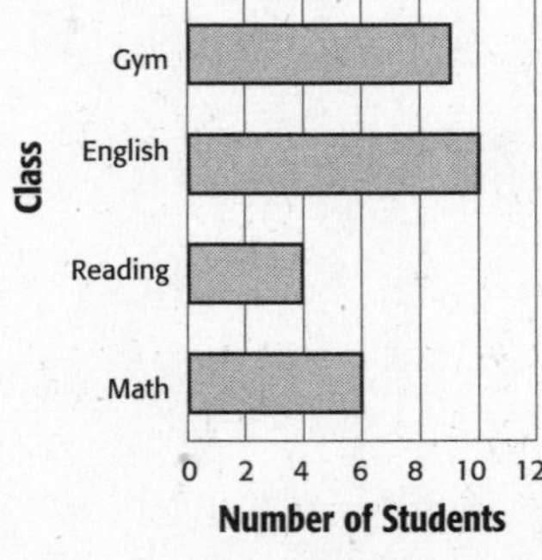

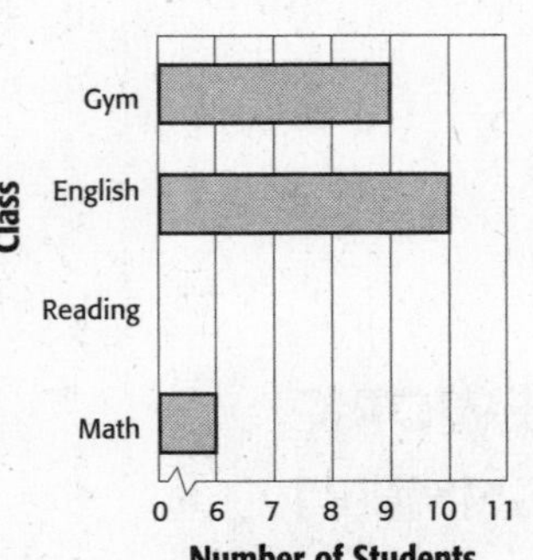

© Glencoe/McGraw-Hill

EXAMPLES Identify Different Uses of Statistics

GYMNASTICS The scores for girls on a team competing on vault at a meet are 8.3, 8.5, 8.5, 8.8, 9.0, and 9.2.

2 Find the mean, median, and mode of the vault scores.

Mean $\frac{\text{sum of values}}{\text{number of values}} =$ _____ or about _____

Median $\frac{8.5 + 8.8}{2} = \frac{17.3}{2}$ or _____

Mode _____

3 Which average would the team use to make its results look the best? Explain.

A gymnastics team would most likely want to show the highest average in scores. The _____ shows the highest event score, _____.

Your Turn **The scores for girls on a team competing in the short program are 5.2, 5.5, 5.5, 5.9, 5.8, and 6.0.**

a. Find the mean, median, and mode of the scores.

b. Which average would the team use to make its results look the best? Explain.

© Glencoe/McGraw-Hill

HOMEWORK ASSIGNMENT

Page(s):

Exercises:

9–8 Matrices

BUILD YOUR VOCABULARY (pages 222–223)

WHAT YOU'LL LEARN

- Use matrices to organize data.

A ______ arrangement of numerical data is called a **matrix**.

$$\begin{bmatrix} 7 & -9 & 2 \\ -4 & 0 & 1 \\ 5 & 6 & -8 \end{bmatrix}$$

In a matrix, the numbers side by side ______ form a **row**.

In a matrix, the numbers ______ under one another form a **column**.

Each ______ in a matrix is called an **element**.

A matrix is described by its **dimensions**, or the number of ______ and ______.

FOLDABLES™

ORGANIZE IT

Under the tab for Lesson 9–8, list 2 or 3 matrices. Label the row and columns and give the dimensions.

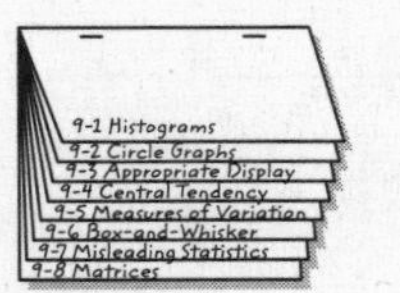

EXAMPLE Identify Dimensions and Elements

1 **State the dimensions of $\begin{bmatrix} -3 & 2 & 8 \\ \textcircled{1} & -6 & 5 \\ 6 & -1 & 12 \end{bmatrix}$. Then identify the position of the circled element.**

The matrix has ______ rows and ______ columns. The dimensions of the matrix are ______.

The circled element is in the ______ row and the ______ column.

© Glencoe/McGraw-Hill

EXAMPLES Add and Subtract Matrices

Add or subtract. If there is no sum or difference, write *impossible*.

2 $\begin{bmatrix} -5 & -2 & -1 \\ 8 & 1 & -3 \end{bmatrix} + \begin{bmatrix} 2 & 3 & -2 \\ -2 & 7 & 9 \end{bmatrix}$

$$\begin{bmatrix} -5 & -2 & -1 \\ 8 & 1 & -3 \end{bmatrix} + \begin{bmatrix} 2 & 3 & -2 \\ -2 & 7 & 9 \end{bmatrix} = \begin{bmatrix} -5 + 2 & -2 + 3 & -1 + (-2) \\ 8 + (-2) & \square & \square \end{bmatrix}$$

$$= \begin{bmatrix} -3 & 1 & -3 \\ 6 & \square & \square \end{bmatrix}$$

3 $\begin{bmatrix} -3 & -3 \\ 8 & 2 \end{bmatrix}$ **[96]**

The first matrix has ☐ rows and ☐ columns. The second matrix has ☐ row and ☐ column. Since the matrices do not have the same ☐, it is ☐ to subtract them.

4 $\begin{bmatrix} 3 & 2 & -8 & 6 \\ 5 & 12 & 4 & -2 \\ 7 & -3 & 8 & 2 \\ 6 & 1 & 1 & -6 \end{bmatrix} - \begin{bmatrix} 2 & 1 & 7 & 2 \\ -2 & -3 & 2 & 5 \\ 4 & 1 & 6 & -2 \\ 2 & 9 & 12 & -4 \end{bmatrix}$

$$= \begin{bmatrix} 3 - 2 & 2 - 1 & \square & 6 - 2 \\ 5 - (-2) & \square & 4 - 2 & \square \\ 7 - 4 & -3 - 1 & \square & 2 - (-2) \\ \square & 1 - 9 & 1 - 12 & \square \end{bmatrix}$$

$$= \begin{bmatrix} \square & 1 & \square & 4 \\ 7 & 15 & \square & \square \\ 3 & \square & 2 & 4 \\ 4 & -8 & -11 & \square \end{bmatrix}$$

© Glencoe/McGraw-Hill

Your Turn

a. State the dimensions of $\begin{bmatrix} -2 & 5 & \textcircled{0} \\ 3 & -6 & -4 \end{bmatrix}$. Then identify the position of the circled element.

Add or subtract. If there is no sum or difference, write *impossible*.

b. $\begin{bmatrix} -2 & -1 & -4 \\ 5 & 1 & 2 \end{bmatrix} + \begin{bmatrix} 3 & 5 & 2 \\ 3 & -2 & 0 \end{bmatrix}$

c. $\begin{bmatrix} 2 & 2 \\ 5 & 1 \end{bmatrix} - \begin{bmatrix} -4 & 6 & -7 \\ 3 & 0 & 11 \end{bmatrix}$

d. $\begin{bmatrix} 2 & -1 & -8 & 3 \\ 1 & -6 & 4 & -2 \\ 4 & -3 & 7 & 2 \\ 6 & 2 & 0 & -6 \end{bmatrix} - \begin{bmatrix} 2 & 1 & 3 & 2 \\ -2 & -3 & 2 & 6 \\ 8 & 2 & 7 & -2 \\ 6 & 4 & 11 & 0 \end{bmatrix}$

HOMEWORK ASSIGNMENT

Page(s):

Exercises:

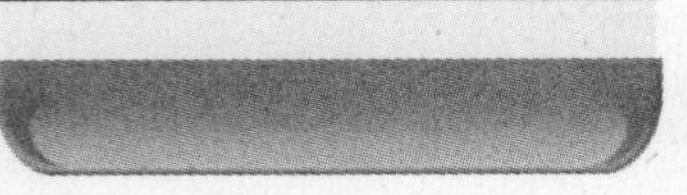

© Glencoe/McGraw-Hill

CHAPTER 9

BRINGING IT ALL TOGETHER

STUDY GUIDE

FOLDABLES™	VOCABULARY PUZZLEMAKER	BUILD YOUR VOCABULARY
Use your **Chapter 9 Foldable** to help you study for your chapter test.	To make a crossword puzzle, word search, or jumble puzzle of the vocabulary words in Chapter 9, go to: www.glencoe.com/sec/math/t_resources/free/index.php	You can use your completed **Vocabulary Builder** *(pages 222–223)* to help you solve the puzzle.

9-1 Histograms

1. Explain the difference between a bar graph and a histogram.

Use the histogram at the right.

2. How many months have less than two days of rain?

3. How many months had between two and seven days of rain?

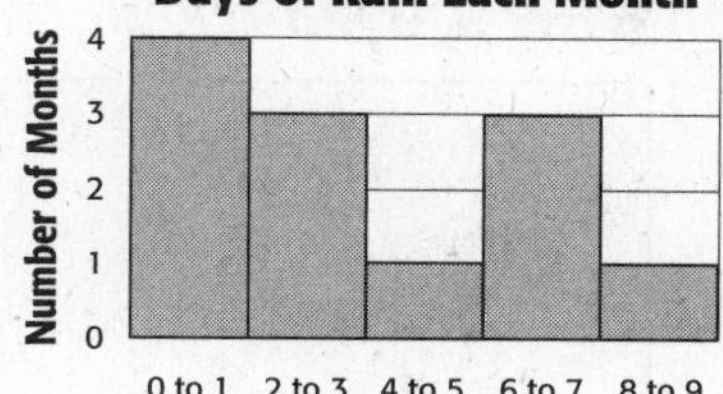

9-2 Circle Graphs

Use the circle graph at the right.

4. What percent of her time does Luisa spend studying?

5. How many degrees are in the section that represents sports?

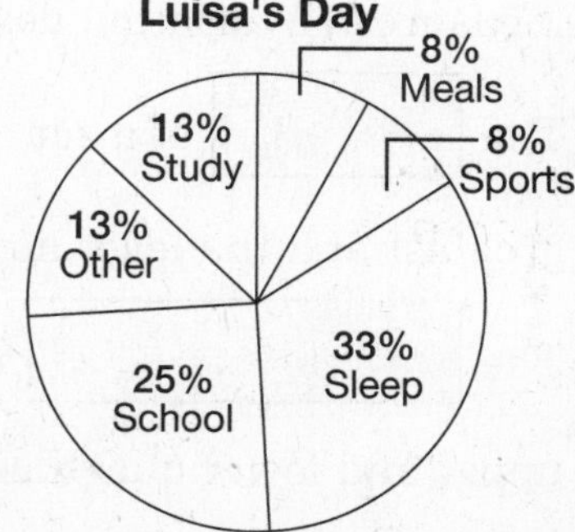

© Glencoe/McGraw-Hill

9-3 Choosing an Appropriate Display

Choose the letter that best matches the type of display to its use.

6. Line Graph

7. Bar Graph

8. Histogram

9. Line Plot

a. shows the frequency of data that has been organized into equal intervals

b. shows the number of items in specific categories in the data using bars

c. shows change over a period of time

d. shows how many times each number occurs in the data

9-4 Measures of Central Tendency

10. Name the three most common measures of central tendency.

11. Which measure of central tendency best represents the data? Why? 9, 9, 20, 22, 25, 27

9-5 Measures of Variation

Complete.

12. Measures of variation describe the ______ of data.

13. The ______ of a set of data is the difference between the greatest and the least numbers in the set.

14. The ______ range is the difference between the upper and lower quartiles.

© Glencoe/McGraw-Hill

9-6

Box-and-Whisker Plots

15. Draw a box-and-whisker plot for the data. 1, 1, 1, 2, 3, 3, 4, 5, 5

9-7

Misleading Graphs and Statistics

16. When writing an employment ad for an automotive dealership, would it be best to use the mean, median, or mode of the number of cars sold to encourage a commissioned salesperson to apply for the job?

Wagner Automotive Sales

Month	Cars Sold	Month	Cars Sold
Jan.	16	July	44
Feb.	5	Aug.	40
March	34	Sept.	38
April	49	Oct.	45
May	47	Nov.	48
June	79	Dec.	38

9-8

Matrices

17. Describe the matrix $\begin{bmatrix} 26 & 45 & 36 & 14 \\ 5 & 4 & 9 & 3 \end{bmatrix}$ in terms of its dimensions.

18. Before you can add or subtract two matrices, what must be the same? Explain.

19. What is each number of a matrix called?

20. Find $\begin{bmatrix} 2 & -5 & 3 \\ 4 & 1 & -9 \end{bmatrix} + \begin{bmatrix} 0 & 7 & -8 \\ 1 & -6 & -9 \end{bmatrix}$.

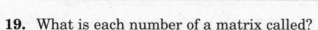

© Glencoe/McGraw-Hill

ARE YOU READY FOR THE CHAPTER TEST?

Visit **msmath3.net** to access your textbook, more examples, self-check quizzes, and practice tests to help you study the concepts in Chapter 9.

Check the one that applies. Suggestions to help you study are given with each item.

☐ **I completed the review of all or most lessons without using my notes or asking for help.**

- You are probably ready for the Chapter Test.
- You may want to take the Chapter 9 Practice Test on page 461 of your textbook as a final check.

☐ **I used my Foldable or Study Notebook to complete the review of all or most lessons.**

- You should complete the Chapter 9 Study Guide and Review on pages 458–460 of your textbook.
- If you are unsure of any concepts or skills, refer back to the specific lesson(s).
- You may also want to take the Chapter 9 Practice Test on page 461.

☐ **I asked for help from someone else to complete the review of all or most lessons.**

- You should review the examples and concepts in your Study Notebook and Chapter 9 Foldable.
- Then complete the Chapter 9 Study Guide and Review on pages 458-460 of your textbook.
- If you are unsure of any concepts or skills, refer back to the specific lesson(s).
- You may also want to take the Chapter 9 Practice Test on page 461.

Student Signature

Parent/Guardian Signature

Teacher Signature

© Glencoe/McGraw-Hill

Algebra: More Equations and Inequalities

FOLDABLES™

Use the instructions below to make a Foldable to help you organize your notes as you study the chapter. You will see Foldable reminders in the margin of this Interactive Study Notebook to help you in taking notes.

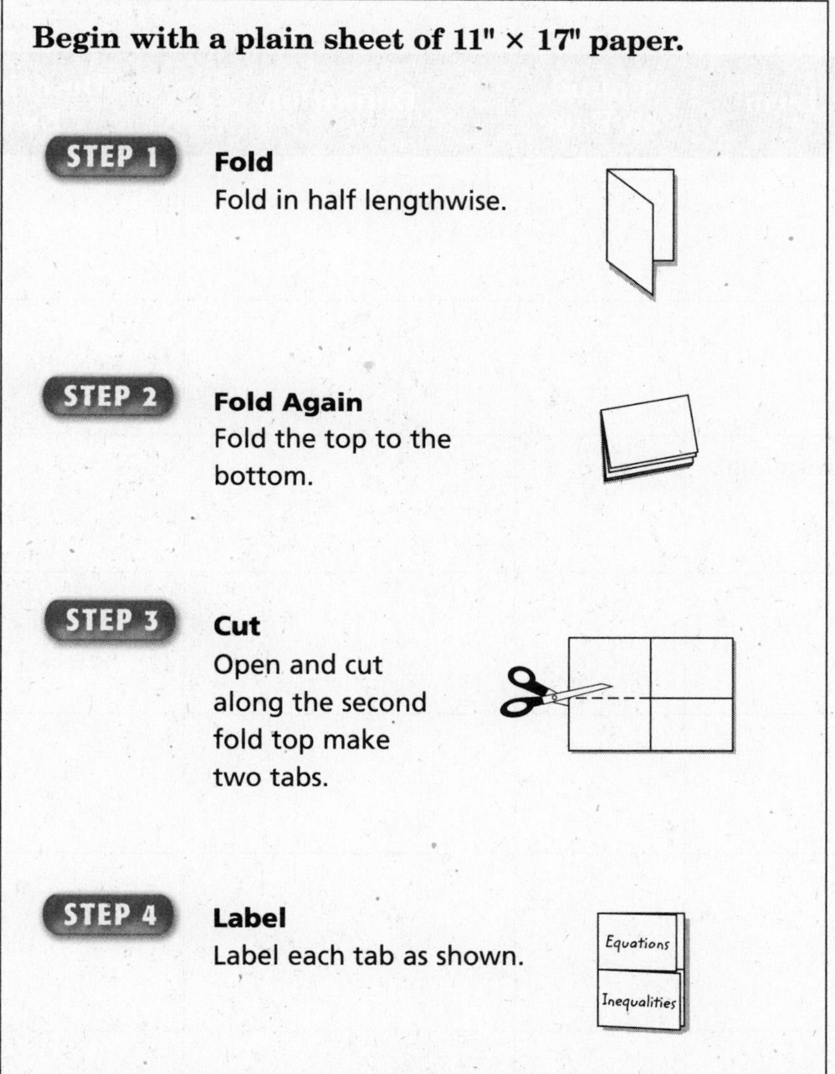

Begin with a plain sheet of 11" × 17" paper.
STEP 1
Fold
Fold in half lengthwise.
STEP 2
Fold Again
Fold the top to the bottom.
STEP 3
Cut
Open and cut along the second fold top make two tabs.
STEP 4
Label
Label each tab as shown.
Equations
Inequalities

NOTE-TAKING TIP: When you take notes, define new terms and write about the new concepts you are learning in your own words. Write your own examples that use the new terms and concepts.

© Glencoe/McGraw-Hill

BUILD YOUR VOCABULARY

This is an alphabetical list of new vocabulary terms you will learn in Chapter 10. As you complete the study notes for the chapter, you will see Build Your Vocabulary reminders to complete each term's definition or description on these pages. Remember to add the textbook page number in the second column for reference when you study.

Vocabulary Term	Found on Page	Definition	Description or Example
coefficient			
constant			
equivalent expressions			
like terms			
simplest form			
simplifying the expression			
term			
two-step equation			

© Glencoe/McGraw-Hill

10-1 Simplifying Algebraic Expressions

BUILD YOUR VOCABULARY (page 252)

Equivalent expressions are expressions that have the ______ regardless of the value of the variable.

WHAT YOU'LL LEARN

- Use the Distributive Property to simplify algebraic expressions.

EXAMPLE Write Equivalent Expressions

1 **Use the Distributive Property to rewrite $3(x + 5)$.**

$3(x + 5) = 3(x) + 3(5)$

$= 3x +$ ______ Simplify.

Your Turn **Use the Distributive Property to rewrite each expression.**

a. $2(x + 6)$

b. $(a + 6)3$

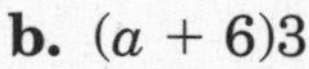

EXAMPLES Write Expressions with Subtraction

Use the Distributive Property to rewrite each expression.

2 $(q - 3)9$

$(q - 3)\,9 = [q + (-3)]\,9$ Rewrite $q - 3$ as $q + (-3)$

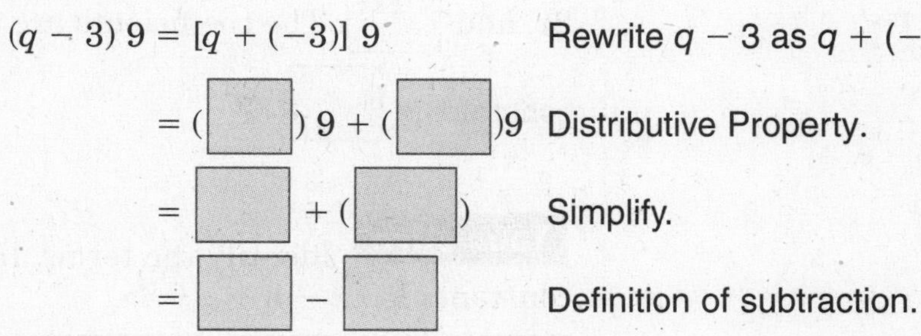

$= (\ \ )\,9 + (\ \ )9$ Distributive Property.

$= \ \ + (\ \ )$ Simplify.

$= \ \ - \ \ $ Definition of subtraction.

REVIEW IT

What is the sign of the product when you multiply two integers with different signs? with the same sign? *(Lesson 1-6)*

3 $-3(z - 7)$

$-3(z - 7) = -3\,[z + (-7)]$ Rewrite $z - 7$ as $z + (-7)$.

$= -3(z) + (-3)(-7)$ Distributive Property

$= -3z +$ ______ Simplify.

© Glencoe/McGraw-Hill

Your Turn **Use the Distributive Property to rewrite each expression.**

a. $(q - 2)\,8$

b. $-2(z - 4)$

BUILD YOUR VOCABULARY (page 252)

When a plus sign separates an algebraic expression into parts, each part is called a **term**.

The numeric factor of a term that contains a [ ] is called the **coefficient** of the variable.

Like terms are terms that contain the [ ] variables, such as 2*x* and *x*.

A term without a [ ] is called a **constant.**

EXAMPLE Identify Parts of an Expression

4 **Identify the terms, like terms, coefficients, and constants in $3x - 5 + 2x - x$.**

$3x - 5 + 2x - x$

$= 3x + (\ \ ) + 2x + (\ \ )$ Definition of Subtraction

$= 3x + (-5) + 2x + (-1x)$ Identity Property; $-x = -1x$

The terms are $3x$, [ ], $2x$, and $-x$. The like terms are $3x$, $2x$, and [ ]. The coefficients are 3, [ ], and –1. The constant is [ ].

Your Turn Identify the terms, like terms, coefficients, and constants in $6x - 2 + x - 4x$.

© Glencoe/McGraw-Hill

BUILD YOUR VOCABULARY (page 252)

An algebraic expression is in **simplest form** if it has no

______ and no ______.

When you use properties to ______ like terms you are **simplifying the expression**.

EXAMPLES Simplify Algebraic Expressions

Simplify each expression.

5 $\mathbf{6n - n}$

$6n$ and n are ______ terms.

$6n - n = 6n -$ ______ Identity Property; $n =$ ______

$= (6-1)n$ Distributive Property

$=$ ______ Simplify.

6 $\mathbf{8z + z - 5 - 9z + 2}$

$8z$, z, and ______ are like terms. -5 and ______ are also like terms.

$8z + z - 5 - 9z + 2$

$= 8z + z + ($ ______ $) + ($ ______ $) + 2$ Definition of subtraction.

$= 8z + z + (-9z) + (-5) + 2$ Commutative Property

$= [8 + 1 + (-9)]$ ______ $+ [(-5) + 2]$ Distributive Property

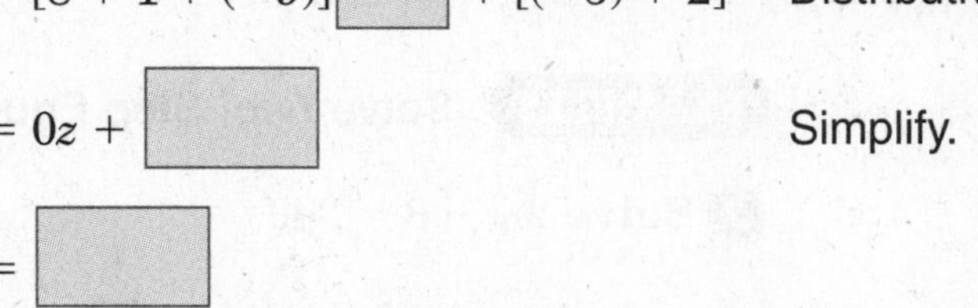

$= 0z +$ ______ Simplify.

$=$ ______

Your Turn **Simplify each expression.**

a. $7n + n$

______

b. $6s + 2 - 10s$

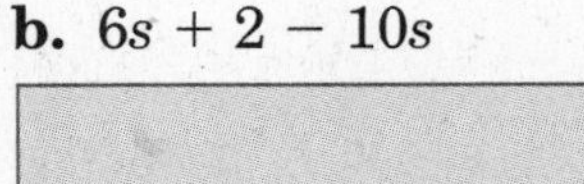

______

c. $6z + z - 2 - 8z + 2$

______

HOMEWORK ASSIGNMENT

Page(s):

Exercises:

© Glencoe/McGraw-Hill

10–2 Solving Two-Step Equations

BUILD YOUR VOCABULARY (page 252)

A **two-step equation** contains ______.

WHAT YOU'LL LEARN

- Solve two-step equations.

REMEMBER IT

Two-step equations can also be solved using models. Refer to page 474 of your textbook.

EXAMPLE Solve a Two-Step Equation

1 **Solve $5x + 1 = 26$.**

Use the Subtraction Property of Equality.

$5x + 1 = 26$ Write the equation.

______ ______ Subtract ______ from each side.

$5x = 25$

Use the Division Property of Equality.

$5x = 25$

$\frac{5x}{\square} = \frac{25}{\square}$ Divide each side by ______.

$x =$ ______ Simplify.

Your Turn Solve $3x + 2 = 20$.

______

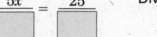

EXAMPLES Solve Two- Step Equations

2 **Solve $2n - 8 = 34$.**

$2n - 8 = 34$ Write the equation.

$2n - 8 + \square = 34 + \square$ Add ______ to each side.

$\frac{2n}{\square} = \frac{42}{\square}$ Simplify. Divide each side by ______.

$n = 21$ Simplify.

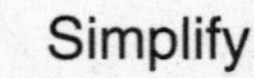

© Glencoe/McGraw-Hill

FOLDABLES™

ORGANIZE IT

Under the "Equations" tab, include examples of how to solve a two step equation. You can use your notes later to tell someone else what you learned in this lesson.

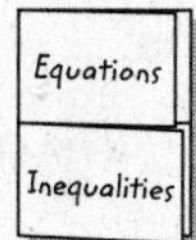

3 Solve $-4 = \frac{1}{3}z + 2$.

$-4 = \frac{1}{3}z + 2$ — Write the equation.

$-4 - \_\_\_ = \frac{1}{3}z + 2 - \_\_\_$ — Subtract ___ from each side.

$-6 = \frac{1}{3}z$ — Simplify.

$\_\_\_(-6) = \_\_\_ \cdot \frac{1}{3}z$ — Multiply each side by ___.

$-18 = z$ — Simplify.

Your Turn **Solve each equation.**

a. $5n - 6 = 39$

b. $-5 = \frac{1}{2}z + 8$

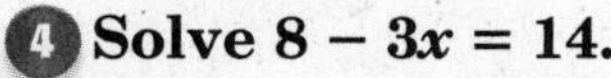

EXAMPLE **Equations with Negative Coefficients**

REMEMBER IT

When you are solving an equation, watch for the negative signs. In Example 4, the coefficient of the variable, *x*, is −3, not +3. So, divide each side by −3 to solve for *x*.

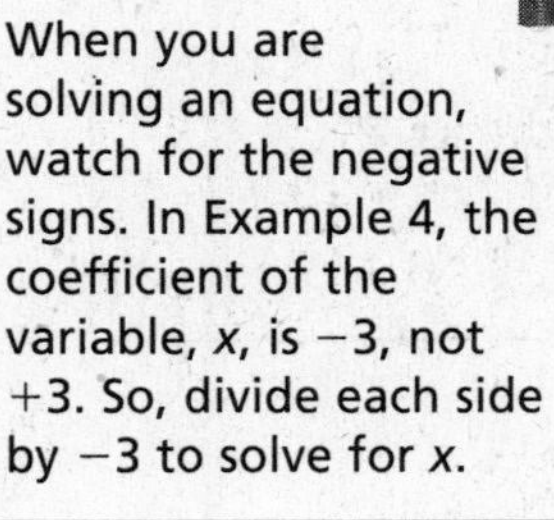

4 Solve $8 - 3x = 14$.

$8 - 3x = 14$ — Write the equation.

$8 + (\_\_\_) = 14$ — Definition of subtraction.

$8 - 8 + (\_\_\_) = 14 - 8$ — Subtract 8 from each side.

$-3x = 6$ — Simplify.

$\frac{-3x}{\_\_\_} = \frac{6}{\_\_\_}$ — Divide each side by ___.

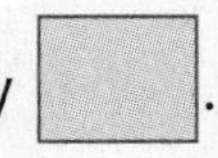

$x = -2$ — Simplify.

Your Turn Solve $5 - 2x = 11$.

© Glencoe/McGraw-Hill

REVIEW IT

Simplify $-c + 4c$.

______

______

______

______

______

______

EXAMPLE Combine Like Terms Before Solving

5 **Solve $14 = -k + 3k - 2$.**

$14 = -k + 3k - 2$	Write the equation.
$14 = -1k + 3k - 2$	______ Property; $-k = 1k$
$14 = \square - 2$	Combine like terms; $-1k + 3k = (-1 + 3)k$ *or* $2k$.
$14 + \square = 2k - 2 + \square$	Add ______ to each side.
$16 = 2k$	Simplify.
$\frac{16}{\square} = \frac{2k}{\square}$	Divide each side by ______.
$8 = k$	Simplify.

Your Turn Solve $10 = -n + 4n - 5$.

HOMEWORK ASSIGNMENT

Page(s):

Exercises:

© Glencoe/McGraw-Hill

10–3 Writing Two-Step Equations

EXAMPLES Translate Sentences Into Equations

WHAT YOU'LL LEARN

- Write two-step equations that represent real-life situations.

Translate each sentence into an equation.

Sentence	Equation
1 Three more than a half a number is 15.	$\frac{1}{2}n + \square = 15$
2 Nineteen is two more than five times a number.	 $19 = \square + 2$
3 Eight less that twice a number is −35.	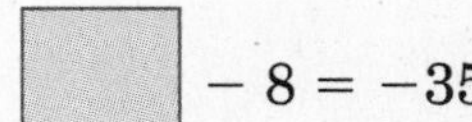 $\square - 8 = -35$

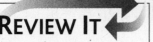

REVIEW IT

What are at least two words that will tell you that a sentence can be written as an equation? *(Lesson 1-7)*

____

____

____

____

____

____

EXAMPLE Translate and Solve an Equation

4 **Two more than $\frac{1}{3}$ of a number is 6. Find the number.**

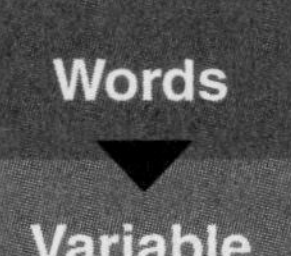

Words Two more than $\frac{1}{3}$ of a number is 6.

Variable Let n = the number.

Equation $\frac{1}{3}n + 2 = \square$

$\frac{1}{3}n + 2 = 6$ Write the equation.

$\frac{1}{3}n + 2 - \square = 6 - \square$ Subtract $\square$ from each side.

$\frac{1}{3}n = \square$ Simplify.

$n = \square$ Mentally multiply each side by  $\square$.

The number is $\square$.

© Glencoe/McGraw-Hill

FOLDABLES™

ORGANIZE IT

Record the main ideas, definitions of vocabulary words, and other notes as you learn how to write two-step equations. Write your notes under the "Equations" tab.

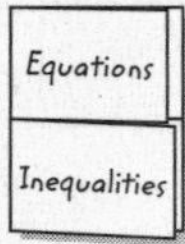

EXAMPLE Write and Solve a Two-Step Equation

5 **TRANSPORTATION A taxi ride costs \$3.50 plus \$2 for each mile traveled. If Jan pays \$11.50 for the ride, how many miles did she travel?**

Her costs start at \$3.50 and adds \$2 until it reaches \$11.50. Organize tha data for the first few miles into a table and look for a pattern.

Miles	Cost
0	$3.50 + 2(0) = 3.50$
1	$3.50 + 2(1) = 5.50$
2	$3.50 + 2(2) = 7.50$
3	$3.50 + 2(3) = 9.50$

Write an equation to represent the situation. Let m represent the number of miles.

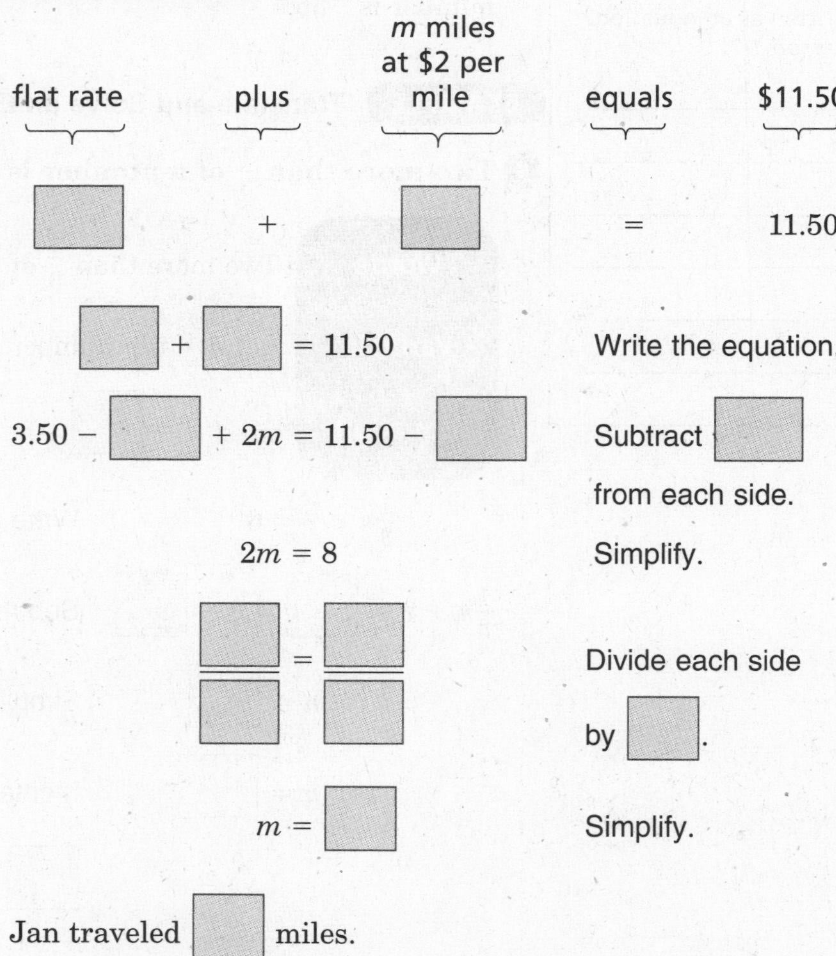

Jan traveled ______ miles.

© Glencoe/McGraw-Hill

Your Turn **Translate each sentence into an equation.**

a. Five more than one third a number is 7.

b. Fifteen is three more than six times a number.

c. Six less that three times a number is −22.

d. Three more than six times a number is 15. Find the number.

e. A rental car costs $100 plus $0.25 for each mile traveled. If Kaya pays $162.50 for the car, how many miles did she travel?

© Glencoe/McGraw-Hill

HOMEWORK ASSIGNMENT

Page(s):

Exercises:

10–4 Solving Equations with Variables on Each Side

WHAT YOU'LL LEARN

- Solve equations with variables on each side.

EXAMPLE Equations with Variables on Each Side

1 **Solve $7x + 4 = 9x$.**

$7x + 4 = 9x$ — Write the equation.

$7x - \square + 4 = 9x - \square$ — Subtract $\square$ from each side.

$\square = \square$ — Simplify by combining like terms.

$\square = \square$ — Divide each side by $\square$.

Your Turn Solve $3x + 6 = x$.

EXAMPLE Equations with Variables on Each Side

FOLDABLES™

ORGANIZE IT

Describe in your own words the steps to follow when you solve an equation with variables on both sides. Write an example of such an equation and solve it.

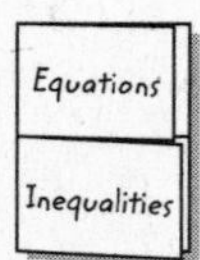

2 **Solve $3x - 2 = 8x + 13$.**

$3x - 2 = 8x + 13$ — Write the equation.

$3x - \square - 2 = 8x - \square + 13$ — Subtract $\square$ from each side.

$-5x - 2 = 13$ — Simplify.

$-5x - 2 + \square = 13 + \square$ — Add $\square$ to each side

$\square = \square$ — Simplify.

$x = \square$ — Divide each side by $\square$.

Your Turn Solve $4x - 3 = 5x + 7$.

© Glencoe/McGraw-Hill

EXAMPLE Use an Equation to Solve a Problem

3 Find the value of x so that the polygons have the same perimeter.

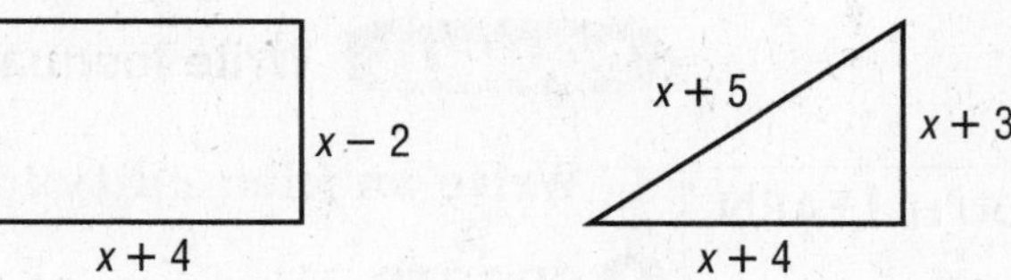

Triangle

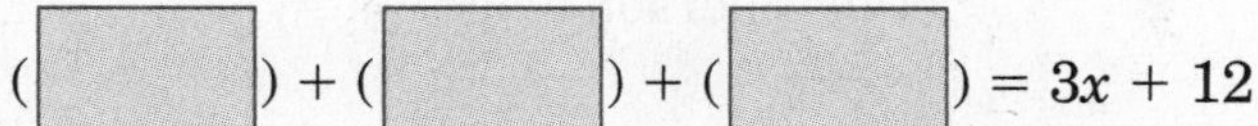

$(\quad) + (\quad) + (\quad) = 3x + 12$

Rectangle

$(\quad) + (\quad) + (\quad) + (\quad) = 4x + 4$

Perimeter of Triangle = Perimeter of Rectangle

$3x + 12 = 4x + 4$

$3x - \square + 12 = 4x - \square + 4$

$12 = \square + 4$

$12 - \square = x + 4 - \square$

$\square = x$

Your Turn Find the value of x so that the polygons have the same perimeter.

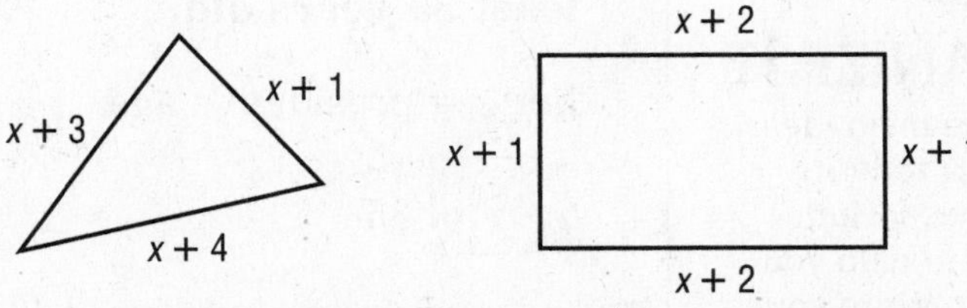

HOMEWORK ASSIGNMENT

Page(s):

Exercises:

© Glencoe/McGraw-Hill

10–5 Inequalities

WHAT YOU'LL LEARN

- Write and graph inequalities.

EXAMPLES Write Inequalities with < or >

Write an inequality for each sentence.

1 SPORTS **Members of the little league team must be under 14 years old.**

Let a = person's age.

a  14

2 CONSTRUCTION **The ladder must be over 30 feet tall to reach the top of the building.**

Let h = ladder's height.

h ☐ 30

Your Turn **Write an inequality for each sentence.**

a. Members of the peewee football team must be under 10 years old.

b. The new building must be over 300 feet tall.

EXAMPLES Write Inequalities with ≤ or ≥

Write an equality for each sentence.

3 POLITICS **The president of the United States must be at least 35 years old.**

Let a = president's age.

a  35

4 CAPACITY **A theater can hold a maximum of 300 people.**

Let p = theater's capacity.

p ☐ 300

FOLDABLES™

ORGANIZE IT

Record the main ideas about how to write inequalities. Include examples to help you remember. Write your notes under the "Inequalities" tab.

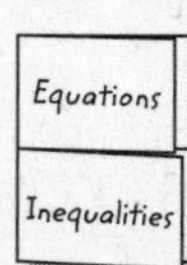

© Glencoe/McGraw-Hill

Your Turn **Write an inequality for each sentence.**

a. To vote, you must be at least 18 years old.

b. A football stadium can hold a maximum of 10,000 people.

EXAMPLES **Determine the Truth of an Inequality**

For the given value, state whether the inequality is true or false.

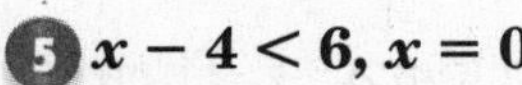

5 $x - 4 < 6, x = 0$

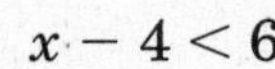

$x - 4 < 6$ Write the inequality.

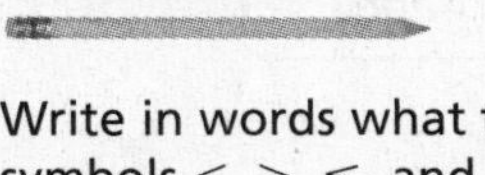

$\square - 4 \overset{?}{<} 6$ Replace *x* with $\square$.

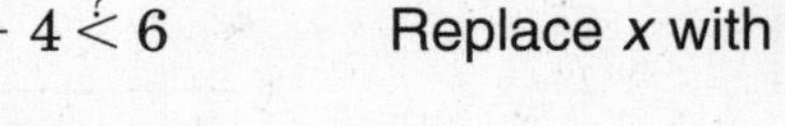

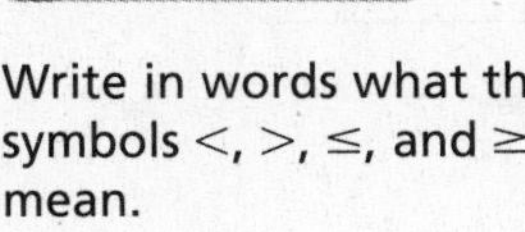

$\square < 6$ Simplify.

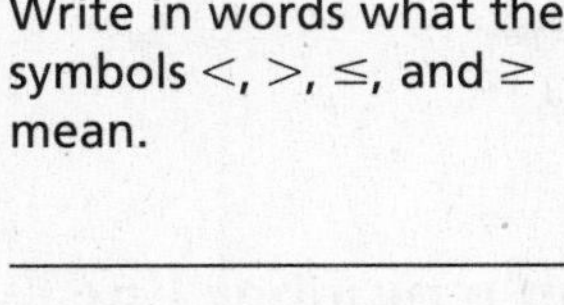

Since $\square$ is less than $\square$, $\square < \square$ is $\square$.

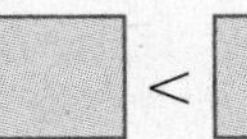

WRITE IT

Write in words what the symbols <, >, ≤, and ≥ mean.

____

____

____

____

6 $3x \geq 4, x = 1$

$3x \geq 4$ Write the inequality.

$3\square \overset{?}{\geq} 4$ Replace *x* with 1.

$\square \not\geq 4$ Simplify.

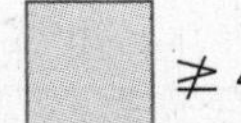

Since $\square$ is not greater than or equal to 4, the sentence is $\square$.

Your Turn **For the given value, state whether the inequality is *true* or *false*.**

a. $x - 5 < 8, x = 16$

b. $2x \geq 9, x = 5$

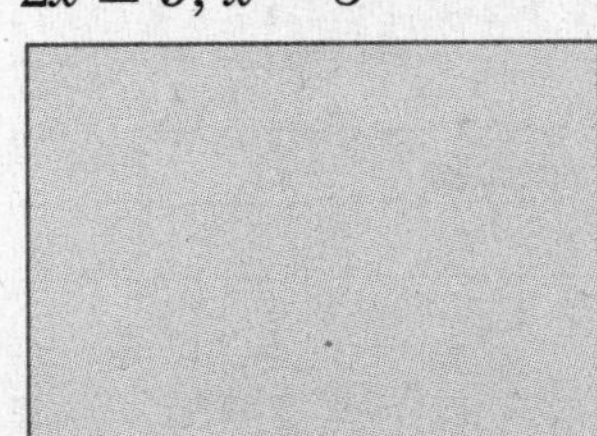

© Glencoe/McGraw-Hill

EXAMPLES Graph an Inequality

Graph each inequality on a number line.

7 $n \leq -1$

Place a ______ circle at -1. Then draw a line and an arrow to the ______.

The closed circle means the number -1 is included in the graph.

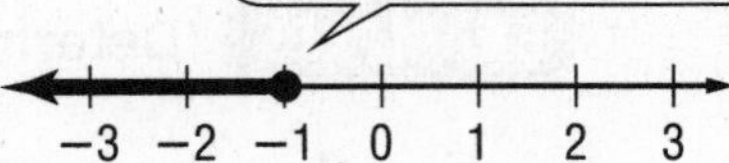

8 $n > -1$

Place an ______ circle at -1. Then draw a line and an arrow to the ______.

The open circle means -1 is *not* included in the graph.

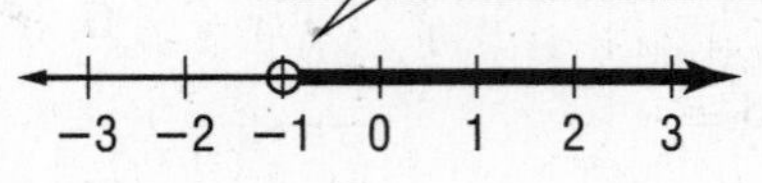

Your Turn **Graph each inequality on a number line.**

a. $n \leq -3$

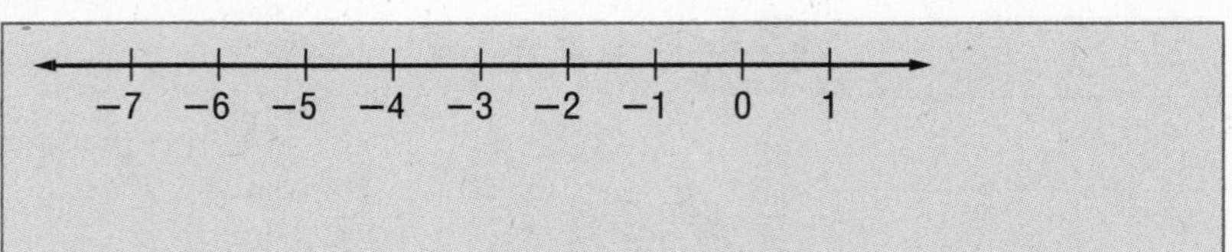

b. $n > -3$

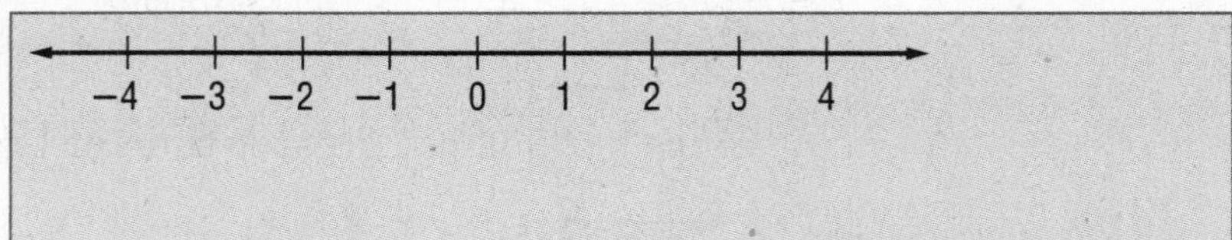

HOMEWORK ASSIGNMENT

Page(s):

Exercises:

© Glencoe/McGraw-Hill

10–6 Solving Inequalities by Adding or Subtracting

WHAT YOU'LL LEARN

- Solve inequalities by using the Addition or Subtraction Properties of Inequality.

KEY CONCEPT

Addition and Subtraction Properties of Inequality. When you add or subtract the same number from each side of an inequality, the inequality remains true.

EXAMPLE Solve an Inequality Using Addition

1 Solve $n - 4 > 6$. Check your solution.

$n - 4 > 6$	Write the inequality.
$n - 4 +$ ___ $> 6 +$ ___	Add ___ to each side.
$n >$ ___	Simplify.

Check

$n - 4 > 6$	Write the inequality.
$11 - 4 \overset{?}{>} 6$	Replace n with a number  than 10, such as 11.
___ > 6 ✔	The statement is true.

Any number greater than 10 will make the statement true, so the solution is ___.

EXAMPLE Solve an Inequality Using Subtraction

2 Solve $x + 8 \leq -7$. Check your solution.

$x + 8 \leq -7$	Write the inequality.
$x + 8 -$ ___ $\leq -7 -$ ___	Subtract ___ from each side.
$x \leq$ ___	Simplify.

Check Replace x in the original inequality with -15 and then with a number 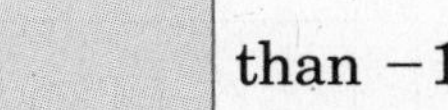 than -15.

$x + 8 \leq -7$	$x + 8 \leq -7$
$-15 + 8 \overset{?}{\leq} -7$	$-16 + 8 \overset{?}{\leq} -7$
___ ≤ -7 ✔	___ ≤ -7 ✔

The solution is ___.

© Glencoe/McGraw-Hill

Your Turn **Solve each equation. Check your solution.**

a. $n - 5 > 8$

b. $x + 2 \leq -3$

EXAMPLE Graph the Solution of an Inequality

FOLDABLES

ORGANIZE IT

Write the main ideas, definitions and other notes as you learn how to solve inequalities by adding or subtracting. Be sure to include examples. Write your notes under the "Inequalities" tab.

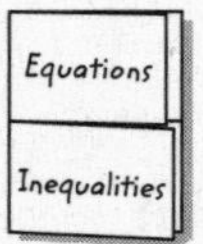

3 **Solve $s - \frac{3}{4} < 2$. Then graph the solution on a number line.**

$$s - \frac{3}{4} < 2$$

$$s - \frac{3}{4} + \underline{\qquad} < 2 + \underline{\qquad}$$

$$s < \underline{\qquad}$$

The solution is $s <$ ______. Graph the solution.

Place an ______ circle at ______. Draw a line and an arrow to the ______.

Your Turn Solve $s - \frac{1}{4} < 4$. Then graph the solution on a number line.

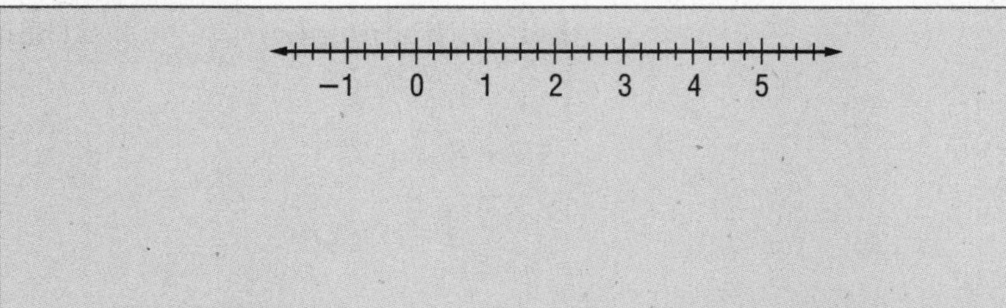

© Glencoe/McGraw-Hill

EXAMPLE Use an Inequality to Solve a Problem

4 TOWING CAPACITY **A pickup truck is towing a trailer that weighs 3,525 pounds. The maximum towing capacity of the truck is 4,700 pounds. Determine how much more weight can be added to the trailer and still be towed by the truck.**

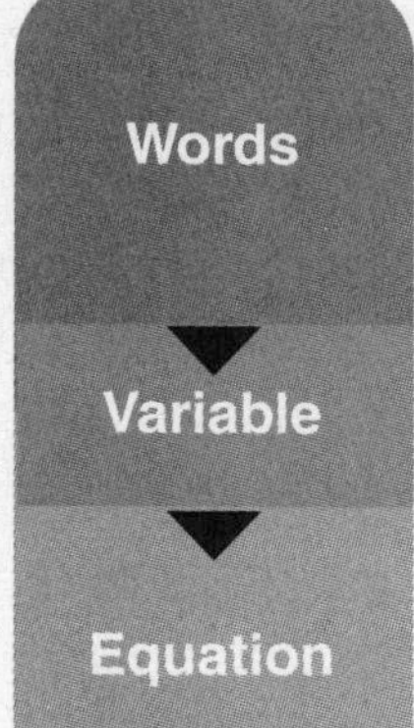

Words: The phrase *maximum capacity* means *less than or equal to*. So, the current weight being towed plus any more weight must be less than or equal to 4,700 pounds.

Variable: Let w = more weight added.

Equation:

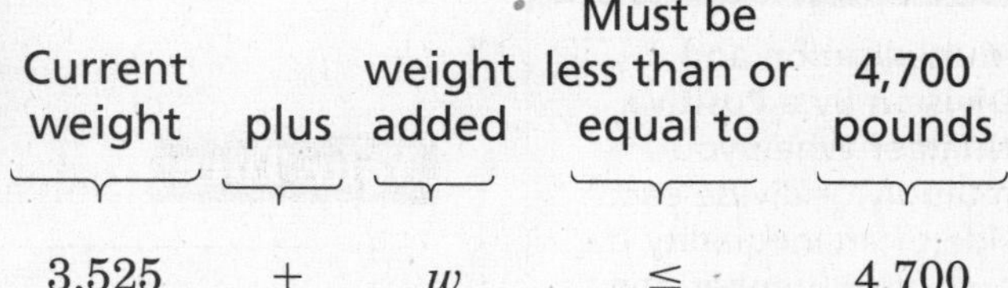

Current weight	plus	weight added	Must be less than or equal to	4,700 pounds
3,525	$+$	w	$\leq$	4,700

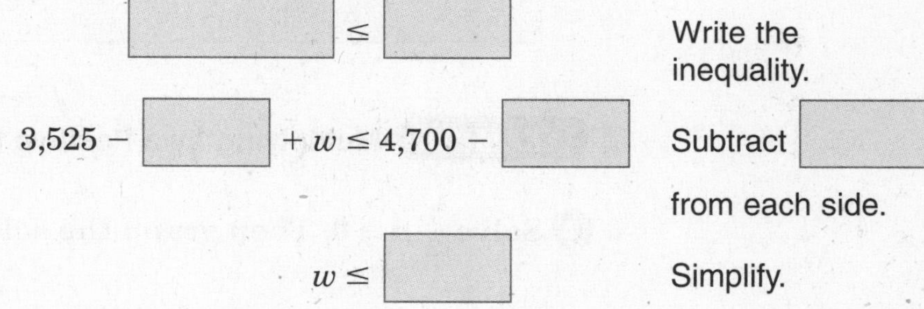

$\boxed{} \leq \boxed{}$ Write the inequality.

$3{,}525 - \boxed{} + w \leq 4{,}700 - \boxed{}$ Subtract $\boxed{}$ from each side.

$w \leq \boxed{}$ Simplify.

Up to ______ more pounds can be added to the trailer.

Your Turn A weightlifter can lift up to 375 pounds. He is currently lifting 255 pounds. Determine how much more weight can be added and still be lifted by the weightlifter.

© Glencoe/McGraw-Hill

HOMEWORK ASSIGNMENT

Page(s):

Exercises:

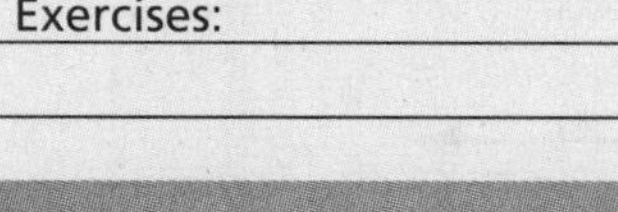

10–7 Solving Inequalities by Multiplying or Dividing

What You'll Learn

- Solve inequalities by using the Multiplication or Division Properties of Inequality.

Key Concept

Multiplication and Division By a Positive Number When you multiply or divide each side of an inequality by a positive number, the inequality remains true.

EXAMPLE Divide by a Positive Number

1 **Solve $6x \leq -30$.**

$6x < -30$	Write the inequality.
$\frac{6x}{\square} < \frac{-30}{\square}$	Divide each side by $\square$.
$x < \square$	Simplify.

Your Turn Solve $4x < -24$.

EXAMPLE Multiplying by a Positive Number

2 **Solve $\frac{1}{2}p \geq 9$. Then graph the solution.**

$\frac{1}{2}p \geq 9$	Write the inequality.
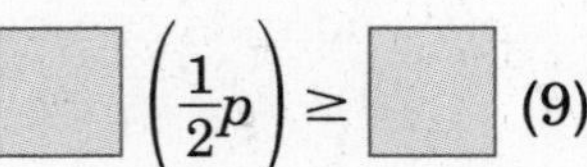 $\square\left(\frac{1}{2}p\right) \geq \square(9)$	Multiply each side by $\square$.
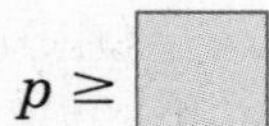 $p \geq \square$	Simplify.

Graph the solution, $p \geq \square$.

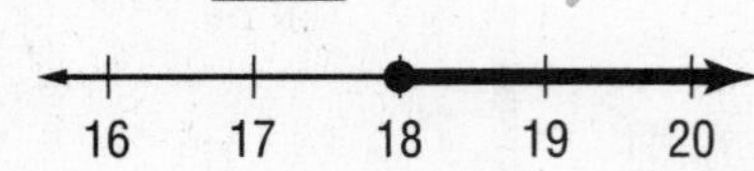

Your Turn Solve $\frac{1}{2}p \geq 5$. Then graph the solution.

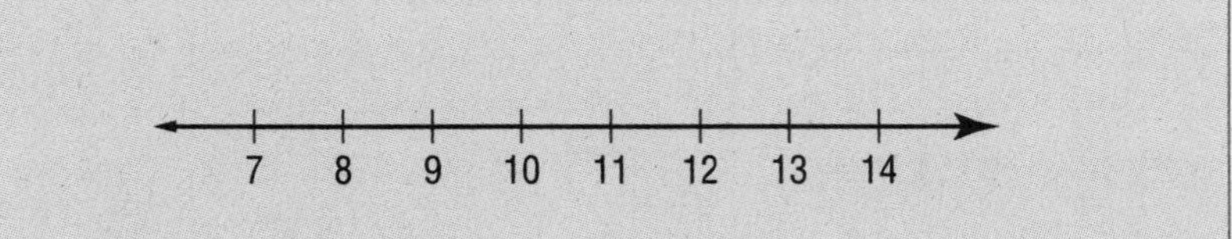

© Glencoe/McGraw-Hill

EXAMPLES **Multiply or Divide by a Negative Number**

KEY CONCEPT

Multiplication and Division By a Negative Number When you multiply or divide each side of an inequality by a negative number, the direction of the inequality symbol must be reversed for the inequality to remain true.

3 Solve $\frac{b}{-4} \leq 5$.

$\frac{b}{-4} \leq 5$ Write the inequality.

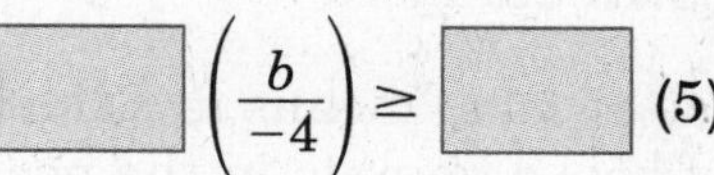

Multiply each side by ____ and reverse the inequality symbol.

$b \geq$ ____ Simplify.

4 Solve $-4n > -60$. Then graph the solution.

$-4n > -60$ Write the inequality.

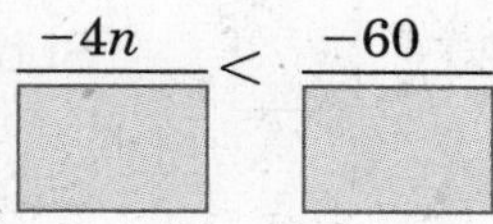

Divide each side by ____ and reverse the inequality symbol.

Check this result.

Graph the solution, $n <$ ____.

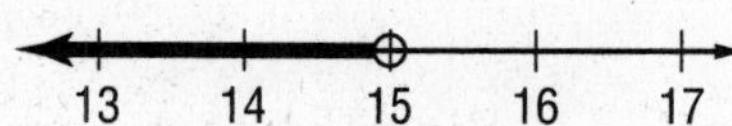

Your Turn

a. Solve $\frac{n}{-2} \leq 7$.

b. Solve $-3n > -18$. Then graph the solution.

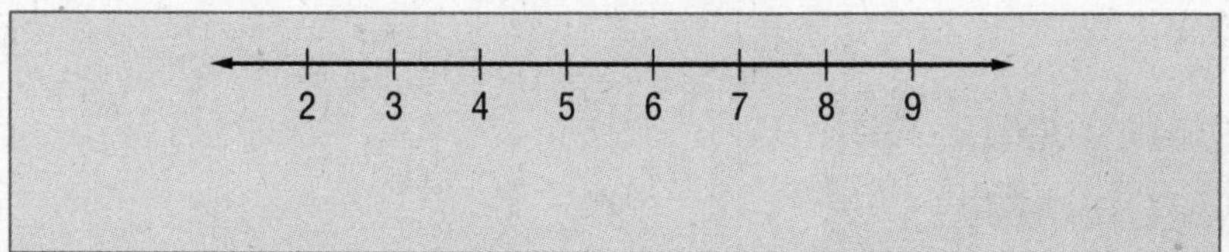

© Glencoe/McGraw-Hill

EXAMPLE Solve a Two-Step Inequality

5 **PACKAGES A box weighs 1 pound. It is filled with books that weigh 2 pounds each. Jesse can carry at most 20 pounds. Assuming space is not an issue, write and solve an inequality to find how many books he can put in the box and still carry it.**

The phrase *at most* means *less than or equal to.* Let p = the number of books he puts in the box. Then write an inequality.

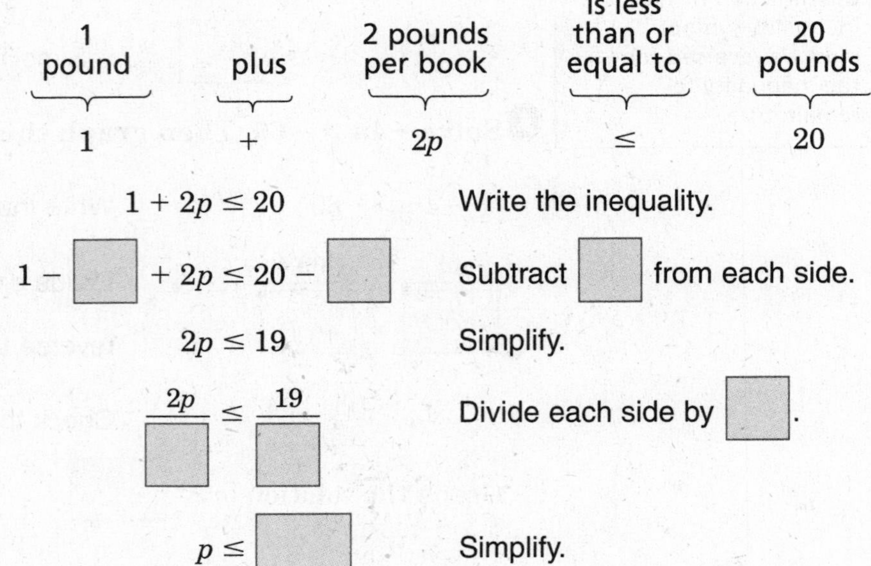

1 pound
plus
2 pounds per book
is less than or equal to
20 pounds
1
+
2p
≤
20

$1 + 2p \leq 20$ Write the inequality.

$1 - \square + 2p \leq 20 - \square$ Subtract $\square$ from each side.

$2p \leq 19$ Simplify.

$\frac{2p}{\square} \leq \frac{19}{\square}$ Divide each side by $\square$.

$p \leq \square$ Simplify.

Since he cannot put half a book in the box, Jesse can put at most $\square$ books in the box.

Your Turn A box weighs 2 pounds. It is filled with toys that weigh 1 pound each. Danielle can carry at most 30 pounds. Assuming space is not an issue, write and solve an inequality to find how many toys she can put in the box and still carry it.

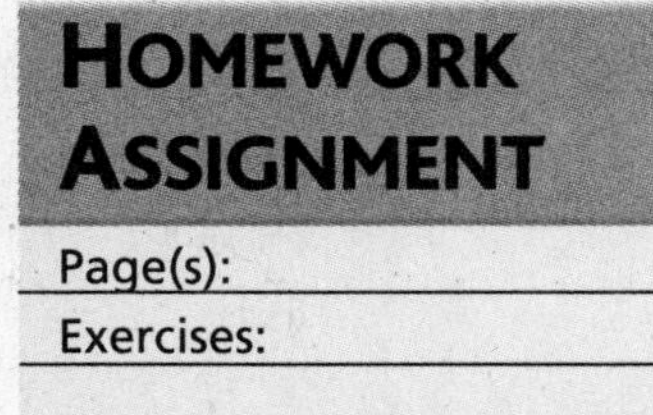

© Glencoe/McGraw-Hill

CHAPTER 10

BRINGING IT ALL TOGETHER

STUDY GUIDE

FOLDABLES™	VOCABULARY PUZZLEMAKER	BUILD YOUR VOCABULARY
Use your **Chapter 10 Foldable** to help you study for your chapter test.	To make a crossword puzzle, word search, or jumble puzzle of the vocabulary words in Chapter 10, go to: www.glencoe.com/sec/math/t_resources/free/index.php	You can use your completed **Vocabulary Builder** *(page 252)* to help you solve the puzzle.

10-1 Simplifying Algebraic Expressions

1. Simplify the expression $3x - 4 - 8x + 2$ by writing the missing information:

______ and ______ are like terms. ______ and ______ are also like terms.

$3x - 4 - 8x + 2 = 3x +$ ______ $- 8x + 2$	Definition of subtraction
$= 3x +$ ______ $+ (-4) + 2$	Commutative Property
$=$ ______ $x + (-4) + 2$	Distributive Property
$=$ ______	Simplify

10-2 Solving Two-Step Equations

2. Define *two-step equation.*

What is the first step in solving each equation?

3. $3y - 2 = 16$ **4.** $5 - 6x = -19$ **5.** $32 = 4b + 6 - b$

© Glencoe/McGraw-Hill

10-3 Writing Two-Step Equations

Write each sentence as an algebraic equation.

6. Four less than six times a number is -40.

7. The quotient of a number and 9, decreased by 3 is equal to 24.

8. Jennifer bought 3 CD's, each having the same price. Her total for the purchase was $51.84, which included $3.84 in sales tax. Find the price of each CD.

Let p represent

Equation: Price of 3 CD's + ______ = ______

______ + ______ $= 51.84$

$3p + 3.84 -$ ______ $= 51.84 -$ ______

______ = ______

______ $= \frac{48}{3}$

$p =$ ______

10-4 Frequency Tables

9. What is the first step in solving equations with variables on each side?

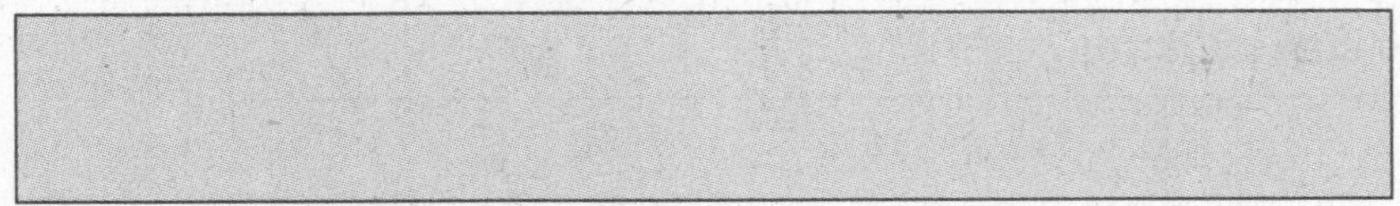

Solve each equation.

10. $3x + 2 = 2x + 5$

11. $6x - 2 = 3x$

12. $7x - 2 = 9x + 6$

© Glencoe/McGraw-Hill

10-5 Inequalities

Write an inequality for each sentence using the symbol <, >, ≤, or ≥.

13. Children under the age of 2 fly free.

14. You must be at least 12 years old to go on the rocket ride.

15. He can spend no more than $15.00 for a gift.

Write the solution shown by each graph.

16. 

17.

10-6 Solving Inequalities by Adding or Subtracting

18. Explain what solving an inequality means.

Solve each inequality.

19. $n - 6 < 21$

20. $-10 \geq 7 + b$

10-7 Solving Inequalities by Multiplying or Dividing

21. Describe the conditions under which you reverse the direction of the inequality symbol when solving an inequality.

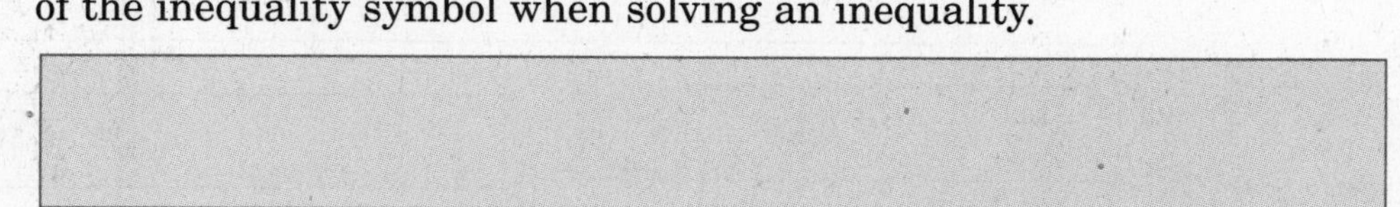

Solve each inequality.

22. $4c - 10 < -26$

23. $\frac{d}{2} + 6 \leq -4$

© Glencoe/McGraw-Hill

ARE YOU READY FOR THE CHAPTER TEST?

Visit **msmath3.net** to access your textbook, more examples, self-check quizzes, and practice tests to help you study the concepts in Chapter 10.

Check the one that applies. Suggestions to help you study are given with each item.

☐ **I completed the review of all or most lessons without using my notes or asking for help.**

- You are probably ready for the Chapter Test.
- You may want to take the Chapter 10 Practice Test on page 507 of your textbook as a final check.

☐ **I used my Foldable or Study Notebook to complete the review of all or most lessons.**

- You should complete the Chapter 10 Study Guide and Review on pages 505–506 of your textbook.
- If you are unsure of any concepts or skills, refer back to the specific lesson(s).
- You may also want to take the Chapter 10 Practice Test on page 507.

☐ **I asked for help from someone else to complete the review of all or most lessons.**

- You should review the examples and concepts in your Study Notebook and Chapter 10 Foldable.
- Then complete the Chapter 10 Study Guide and Review on pages 505–506 of your textbook.
- If you are unsure of any concepts or skills, refer back to the specific lesson(s).
- You may also want to take the Chapter 10 Practice Test on page 507.

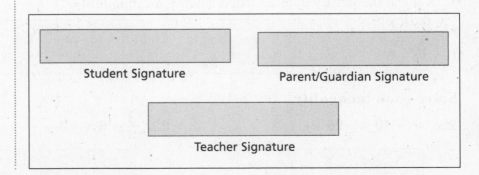

Student Signature

Parent/Guardian Signature

Teacher Signature

© Glencoe/McGraw-Hill

Algebra: Linear Functions

Use the instructions below to make a Foldable to help you organize your notes as you study the chapter. You will see Foldable reminders in the margin of this Interactive Study Notebook to help you in taking notes.

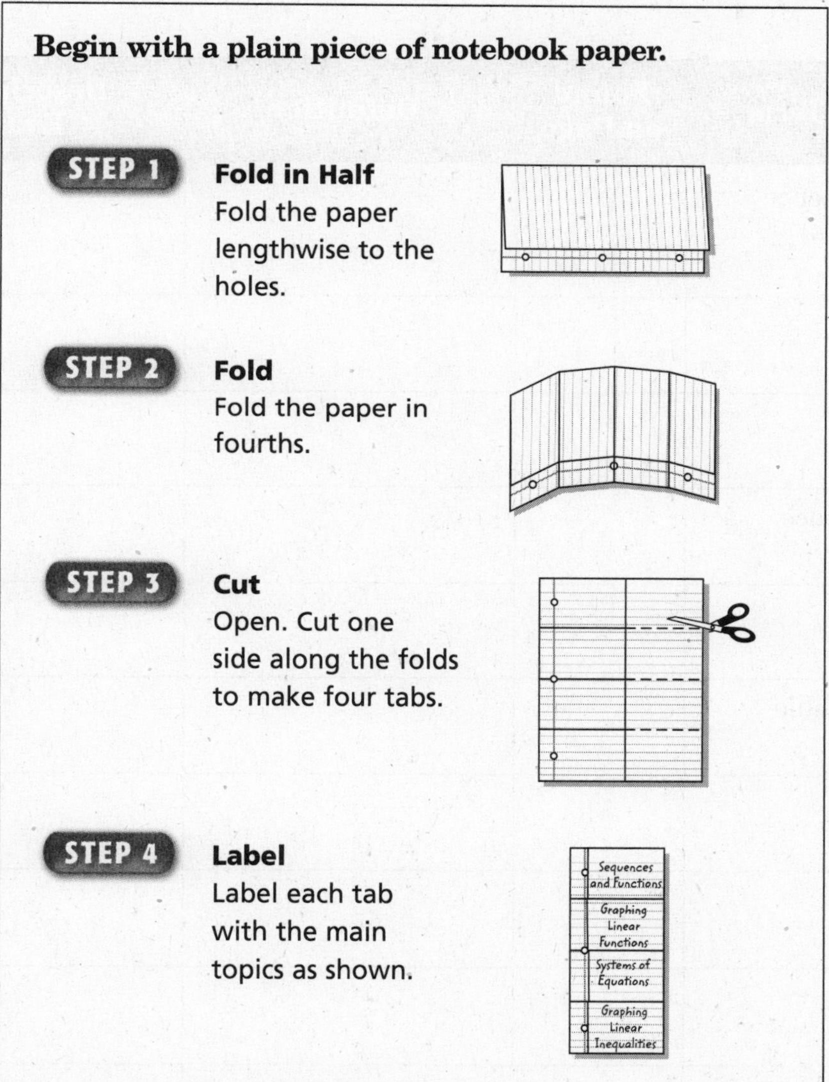

Begin with a plain piece of notebook paper.
STEP 1
Fold in Half
Fold the paper lengthwise to the holes.
STEP 2
Fold
Fold the paper in fourths.
STEP 3
Cut
Open. Cut one side along the folds to make four tabs.
STEP 4
Label
Label each tab with the main topics as shown.
Sequences and Functions
Graphing Linear Functions
Systems of Equations
Graphing Linear Inequalities

NOTE-TAKING TIP: When you begin studying a chapter in a textbook, first skim through the chapter to become familiar with the topics. As you skim, write questions about what you don't understand and what you'd like to know. Then, as you read the chapter, write answers to your questions.

© Glencoe/McGraw-Hill

BUILD YOUR VOCABULARY

This is an alphabetical list of new vocabulary terms you will learn in Chapter 11. As you complete the study notes for the chapter, you will see Build Your Vocabulary reminders to complete each term's definition or description on these pages. Remember to add the textbook page number in the second column for reference when you study.

Vocabulary Term	Found on Page	Definition	Description or Example
arithmetic sequence [air-ith-MEH-tik]			
best-fit line			
boundary			
common difference			
common ratio			
dependent variable			
domain			
function			
function table			
geometric sequence [je-o-MET-rik]			
half plane			

© Glencoe/McGraw-Hill

Vocabulary Term	Found on Page	Definition	Description or Example
independent variable			
linear function			
range			
scatter plot			
sequence			
slope formula			
slope-intercept form			
substitution			
system of equations			
term			
x-intercept			
y-intercept			

© Glencoe/McGraw-Hill

11–1 Sequences

WHAT YOU'LL LEARN

- Recognize and extend arithmetic and geometric sequences.

FOLDABLES

ORGANIZE IT

Under the "Sequences and Functions" tab of your Foldable, explain the difference between arithmetic and geometric sequences.

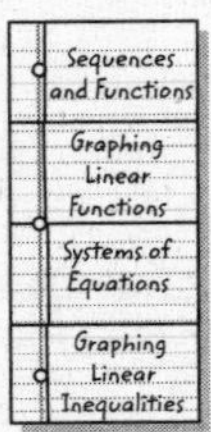

BUILD YOUR VOCABULARY (pages 278–279)

A **sequence** is an ________ of numbers.

Each number in a ________ is called a **term.**

An **arithmetic sequence** is a sequence in which the ________ between any two consecutive terms is the same.

The difference between any two ________ ________ in an ________ sequence is called the **common difference.**

EXAMPLE Identify Arithmetic Sequences

1 **State whether the sequence 23, 15, 7, −1, −9 . . . is arithmetic. If it is, state the common difference. Write the next three terms of the sequence.**

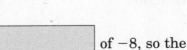

23, 15, 7, −1, −9 (each step −8)

Notice that 15 − 23 = −8, 7 − 15 = −8, and so on.

The terms have a common ________ of −8, so the sequence is ________.

Continue the pattern to find the next three terms.

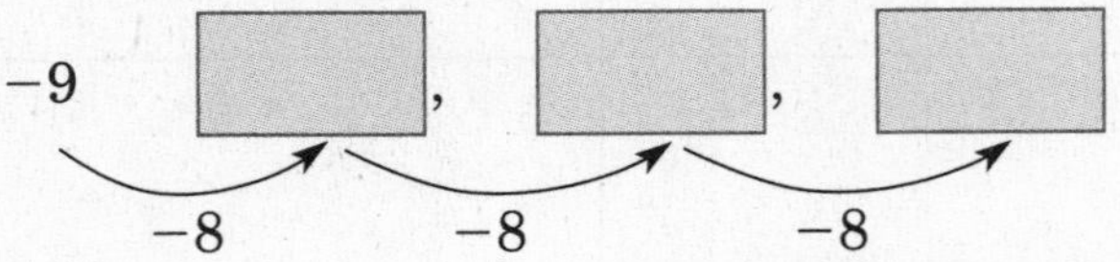

The next three terms are ________, ________, and ________.

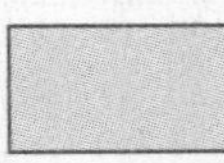

© Glencoe/McGraw-Hill

Your Turn State whether the sequence 29, 27, 25, 23, 21, . . . is arithmetic. If it is, state the common difference. Write the next three terms of the sequence.

BUILD YOUR VOCABULARY (pages 278–279)

A **geometric sequence** is a sequence in which the ______ between any two consecutive terms is the same.

The quotient between any two ______ terms in a ______ sequence is called the **common ratio.**

EXAMPLES Identify Geometric Sequences

State whether each sequence is geometric. If it is, state the common ratio. Write the next three terms of the sequence.

2 **−1, 3, −9, 27, −81, . . .**

Notice that 3 ÷ (−1) = ______, −9 ÷ 3 = ______, and so on.

−1, 3, −9, 27, −81

× ______ × ______ × ______ × ______

The terms have a common ratio of ______, so the sequence is ______.

Continue the pattern to find the next three terms.

−81 ______, ______, ______

× ______ × ______ × ______

The next three terms are ______, ______, and ______.

© Glencoe/McGraw-Hill

3 **6, 12, 13, 26, 27, . . .**

6, 12, 13, 26 27

×2 +1 ×2 +1

Since there is no 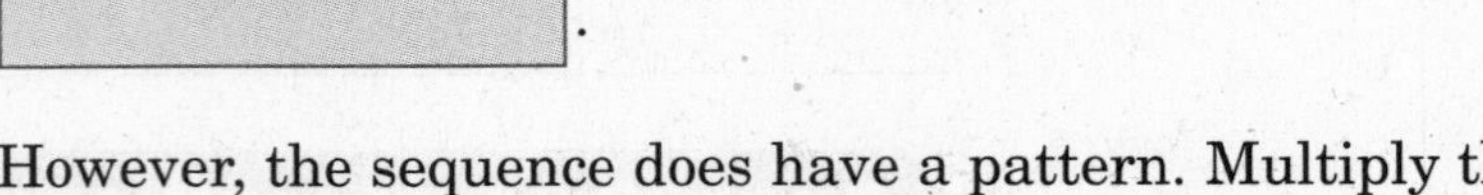, the sequence is

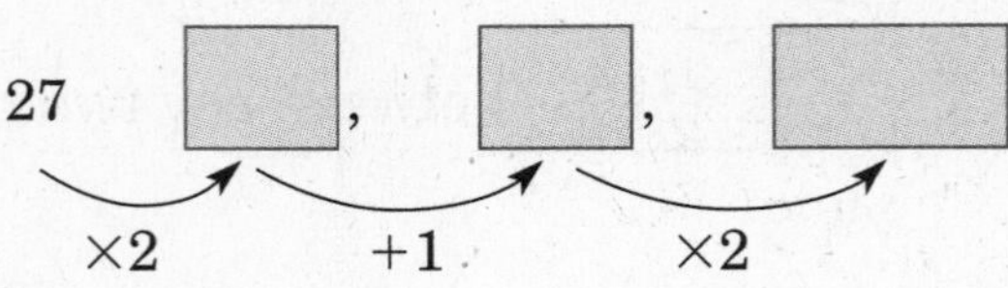.

However, the sequence does have a pattern. Multiply the last term by ___, and then add ___ to get the next term.

27 ___, ___, ___

×2 +1 ×2

The next three terms are ___, ___, and ___.

Your Turn **State whether each sequence is geometric. If it is, state the common ratio. Write the next three terms of the sequence.**

a. 1, −2, 4, −8, 16, . . .

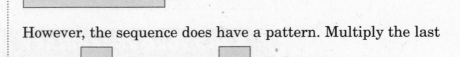

b. 1, 3, 7, 15, 31, . . .

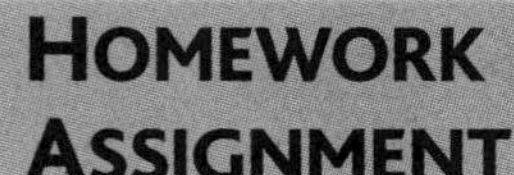

HOMEWORK ASSIGNMENT

Page(s):

Exercises:

© Glencoe/McGraw-Hill

11–2 Functions

WHAT YOU'LL LEARN

- Complete function tables.

BUILD YOUR VOCABULARY (pages 278–279)

A ______ where one thing ______ another is called a **function.**

FOLDABLES

ORGANIZE IT

Under the "Sequences and Functions" tab of your Foldable, write how you would find the value of a function. You may wish to include an example.

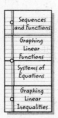

Sequences and Functions | Graphing Linear Functions | Systems of Equations | Graphing Linear Inequalities

EXAMPLES Find a Function Value

Find each function value.

1 $f(4)$ **if** $f(x) = x - 8$

$f(x) = x - 8$

$f(\_\_) = \_\_ - 8$ — Substitute __ for x into the function rule.

$= \_\_$ — Simplify.

So, $f(4) = \_\_$.

2 $f(-6)$ **if** $f(x) = 3x + 4$

$f(x) = 3x + 4$

$f(\_\_) = 3(\_\_) + 4$ — Substitute __ for x into the function rule.

$f(\_\_) = \_\_ + 4$ — Multiply.

$= \_\_$ — Simplify.

So, $f(-6) = \_\_$.

Your Turn **Find each function value.**

a. $f(2)$ if $f(x) = x - 7$

b. $f(-2)$ if $f(x) = 2x + 6$

© Glencoe/McGraw-Hill

BUILD YOUR VOCABULARY (pages 278–279)

You can organize the ________, rule, and output of a function into a **function table.**

The variable for the ________ of a function is called the **independent variable.**

The variable for the ________ of a function is called the **dependent variable.**

The set of ________ values in a function is called the **domain.**

The set of ________ values in a function is called the **range.**

EXAMPLE Make a Function Table

3 Complete the function table for $f(x) = 4x - 1$.

Substitute each value of x, or ________, into the function rule.

Then simplify to find the ________.

Input x	Rule $4x - 1$	Output $f(x)$
−3		
−2		
−1		
0		
1		

$f(x) = 4x - 1$

$f(-3) =$ ________ or ________

$f(-2) =$ ________ or ________

$f(-1) =$ ________ or ________

$f(0) =$ ________ or ________

$f(1) =$ ________ or ________

Input x	Rule $4x - 1$	Output $f(x)$
−3	________	________
−2	________	________
−1	________	________
0	________	________
1	________	________

© Glencoe/McGraw-Hill

Your Turn Complete the function table for $f(x) = 3x - 2$.

Input x	Rule $3x - 2$	Output $f(x)$
−3		
−2		
−1		
0		
1		

EXAMPLES Functions with Two Variables

PARKING FEES The price for parking at a city lot is $3.00 plus $2.00 per hour.

4 Write a function using two variables to represent the price of parking for *h* hours.

Words — Cost of parking equals $3.00 plus $2.00 per hour.

Function — $p =$ ______ + ______

The function $p =$ ______ represents the situation.

5 How much would it cost to park at the lot for 2 hours?

Substitute ______ for h into the function rule.

$p =$ ______ + ______

$p = 3 + 2$ ______ or ______

It will cost ______ to park for 2 hours.

Your Turn **TAXI The price of a taxi ride is $5.00 plus $4.00 per hour.**

a. Write a function using two variables to represent the price of riding a taxi for h hours.

______

b. How much would it cost for a 3-hour taxi ride?

______

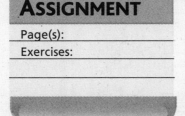

HOMEWORK ASSIGNMENT

Page(s):

Exercises:

© Glencoe/McGraw-Hill

11–3 Graphing Linear Functions

EXAMPLE Graph a Function

WHAT YOU'LL LEARN

- Graph linear functions by using function tables and plotting points.

FOLDABLES

ORGANIZE IT

Under the "Graphing Linear Functions" tab of your Foldable, include a linear function and its graph.

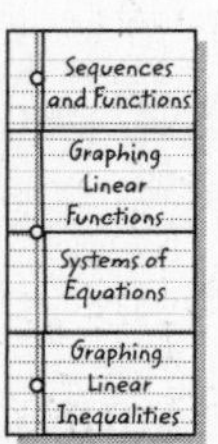

1 **Graph $y = x - 3$.**

Step 1 Choose some values for x. Make a function table. Include a column of ordered pairs of the form (x, y).

x	$x - 3$	y	(x, y)
0	___ − 3		
1	___ − 3		
2	___ − 3		
3	___ − 3		

Step 2 Graph each ordered pair.

Draw a line that passes through each point. Note that the ordered pair for any point on this line is a solution of $y = x - 3$. The line is the complete graph of the function.

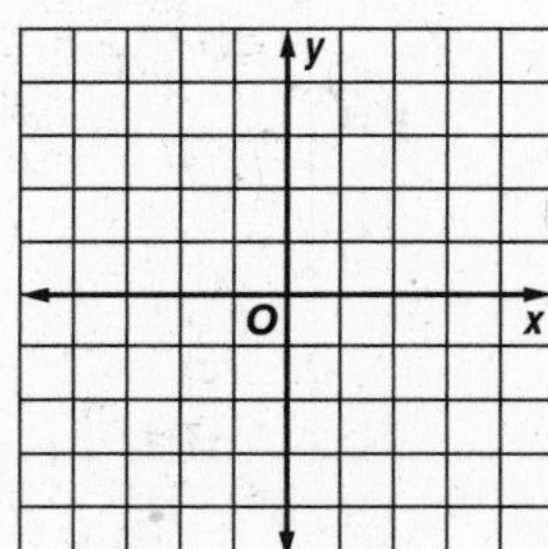

Check It appears from the graph that $(-1, -4)$ is also a solution. Check this by substitution.

$y = x - 3$ Write the function.

___ $\stackrel{?}{=}$ ___ $- 3$ Replace x and y.

___ $=$ ___ Simplify.

Your Turn Graph $y = x - 2$.

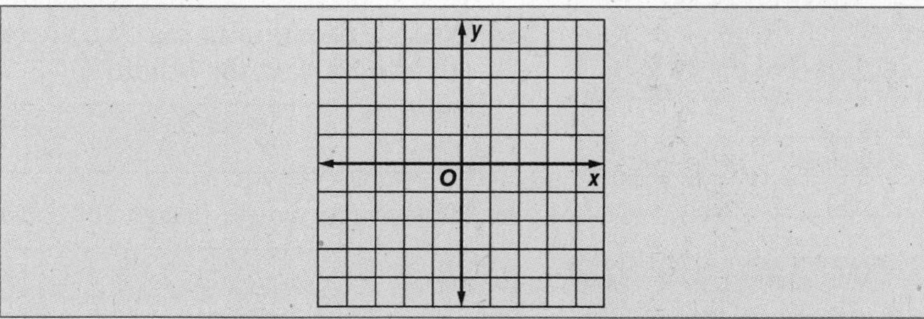

© Glencoe/McGraw-Hill

BUILD YOUR VOCABULARY (pages 278–279)

A function in which the graph of solutions forms a ____ is called a linear function.

The value of x where the graph crosses the  is called the ***x*-intercept.**

The value of y where the graph crosses the ____ is called the ***y*-intercept.**

EXAMPLE Use *x*- and *y*-intercepts

2 Which graph represents $y = 2x + 1$?

a.

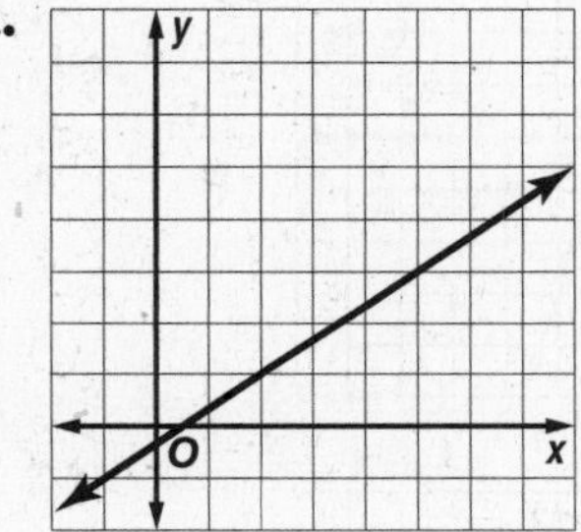

b.

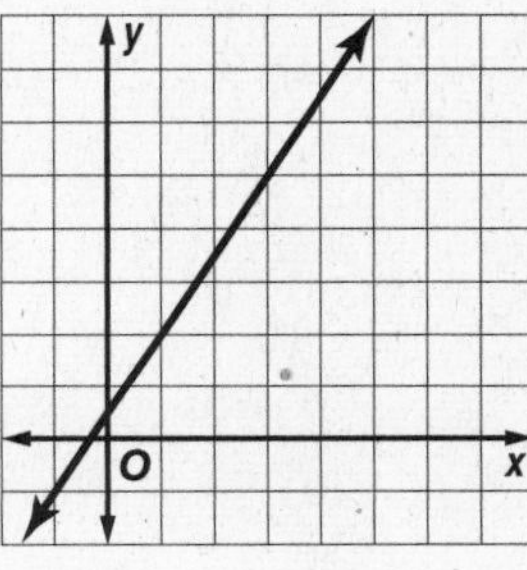

c.

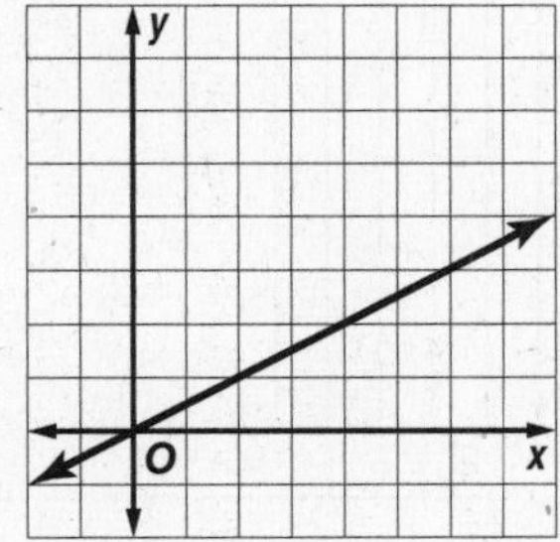

d.

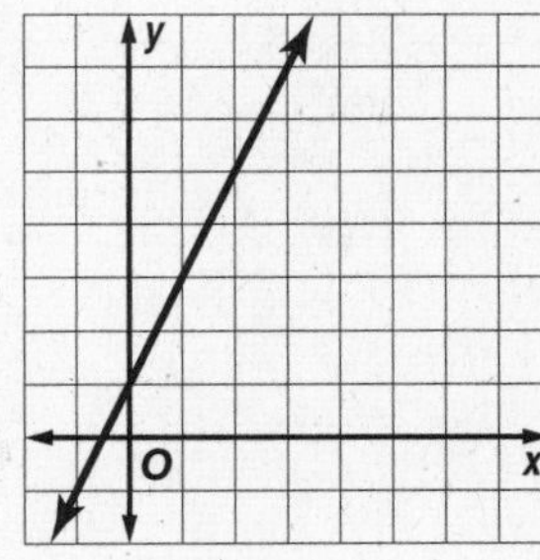

The graph will cross the x-axis when $y =$ ____.

____ $= 2x + 1$	Replace y with ____.
____ $=$ ____	Subtract ____. Then simplify.
____ $= x$	Divide by ____. Then simplify.

© Glencoe/McGraw-Hill

The graph will cross the y-axis when $x =$ ☐.

$y = 2$ ☐ $+ 1$ Replace x with ☐.

$y =$ 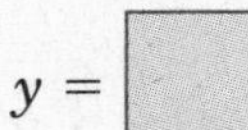 Simplify.

The x-intercept is  and the y-intercept is .

Graph ☐ is the only graph with both of these intercepts.

Your Turn Which graph represents $y = 3x + 3$?

a.

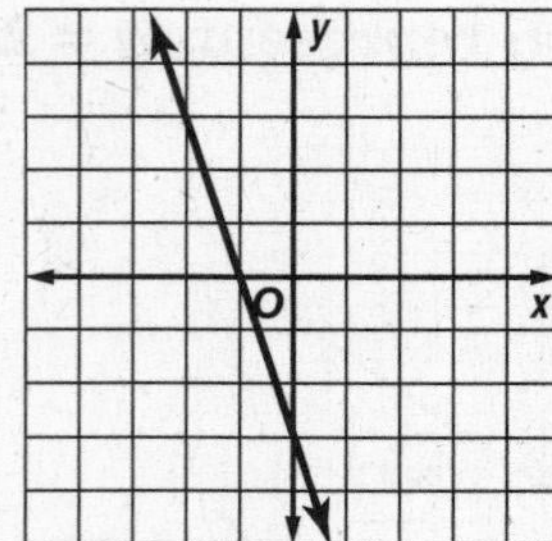

b.

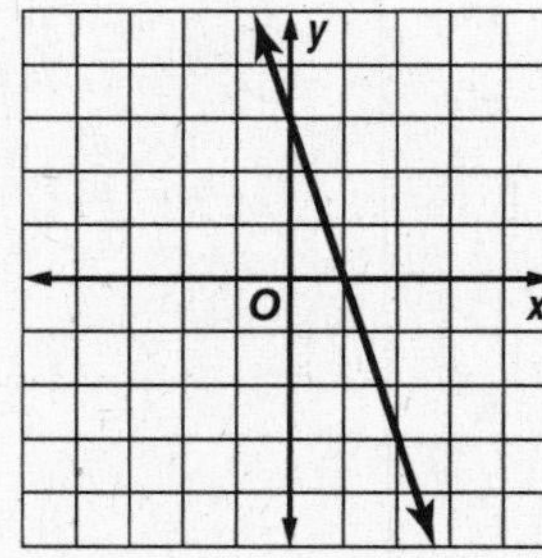

c.

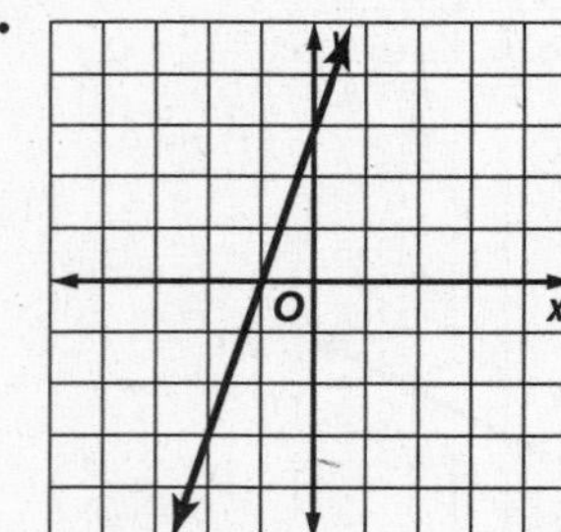

d.

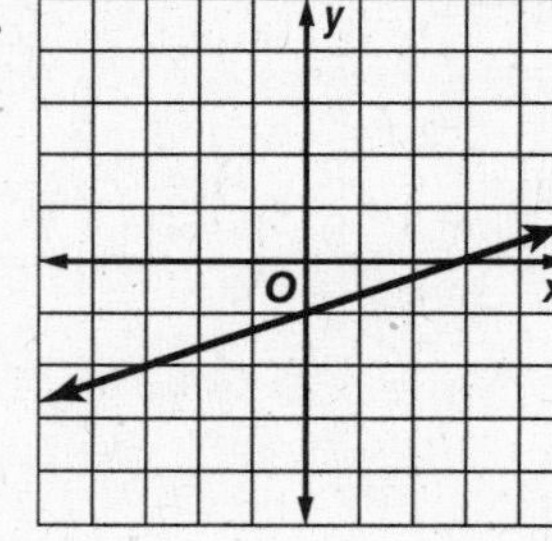

HOMEWORK ASSIGNMENT

Page(s):

Exercises:

© Glencoe/McGraw-Hill

11-4 The Slope Formula

WHAT YOU'LL LEARN

- Find the slope of a line using the slope formula.

KEY CONCEPTS

Slope Formula The slope *m* of a line passing through points (x_1, y_1) and (x_2, y_2) is the ratio of the difference in the *y*-coordinates to the corresponding difference in the *x*-coordinates.

EXAMPLE Positive Slope

1 **Find the slope of the line that passes through $A(3, 3)$ and $B(2, 0)$.**

$m = \frac{y_2 - y_1}{x_2 - x_1}$ Definition of slope

$m = \frac{0 - 3}{2 - 3}$ $(x_1, y_1) = (3, 3)$, $(x_2, y_2) = (2, 0)$

$m = \frac{-3}{-1}$ or 3 Simplify.

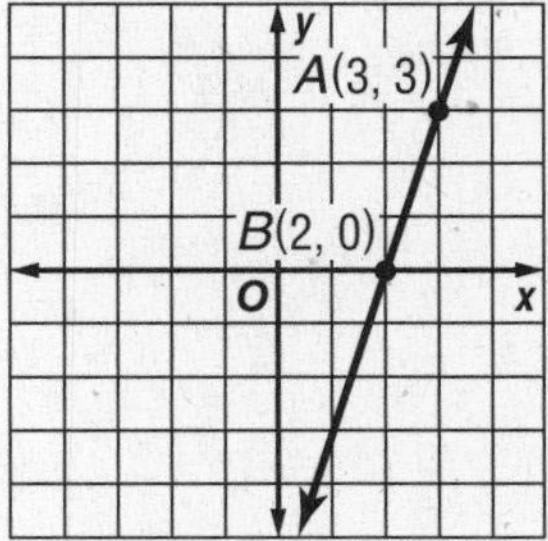

EXAMPLE Negative Slope

2 **Find the slope of the line that passes through $X(-2, 3)$ and $Y(3, 0)$.**

$m = \frac{y_2 - y_1}{x_2 - x_1}$ Definition of slope

$m =$ 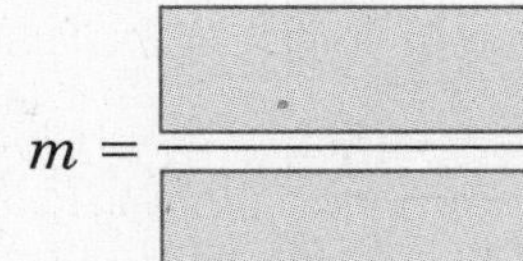 $(x_1, y_1) = (-2, 3)$, $(x_2, y_2) = (3, 0)$

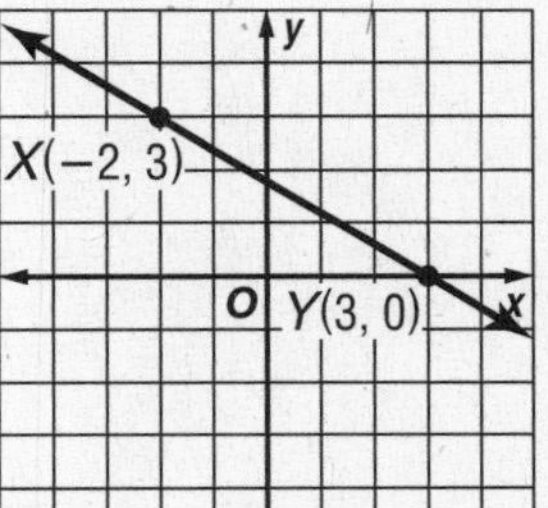

$m = \frac{-3}{5}$ or $-\frac{3}{5}$ Simplify.

EXAMPLE Zero Slope

3 **Find the slope of the line that passes through $P(6, 5)$ and $Q(2, 5)$.**

$m = \frac{y_2 - y_1}{x_2 - x_1}$ Definition of slope

$m =$ 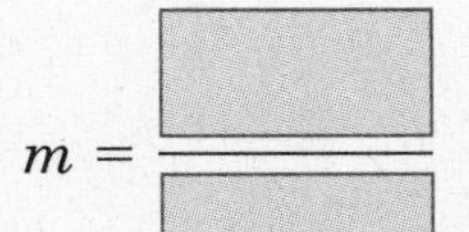 $(x_1, y_1) = (6, 5)$, $(x_2, y_2) = (2, 5)$

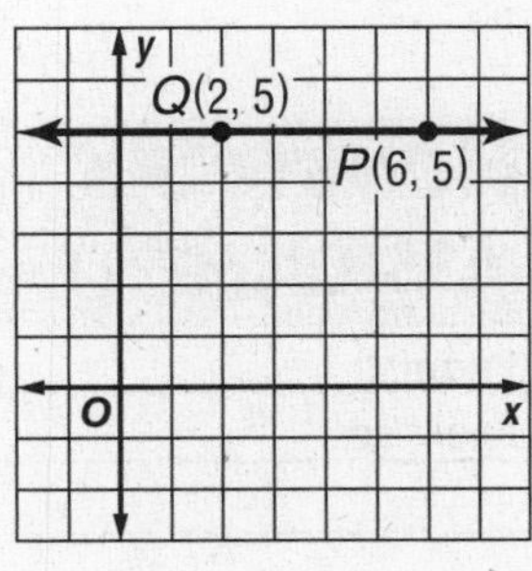

$m = \frac{0}{-4}$ or 0 Simplify.

The slope is 0. The slope of any horizontal line is .

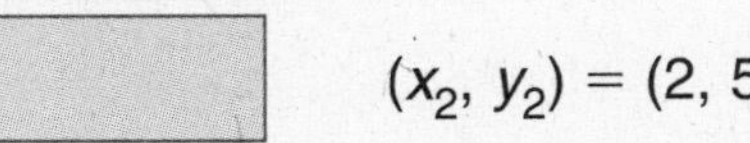

© Glencoe/McGraw-Hill

EXAMPLE Undefined Slope

4 **Find the slope of the line that passes through each pair of points.**

$m = \frac{y_2 - y_1}{x_2 - x_1}$ Definition of slope

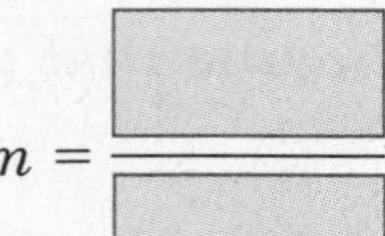

$m = \frac{\square}{\square}$ $(x_1, y_1) = (2, 4)$, $(x_2, y_2) = (2, 6)$

$m = \frac{2}{0}$ Simplify.

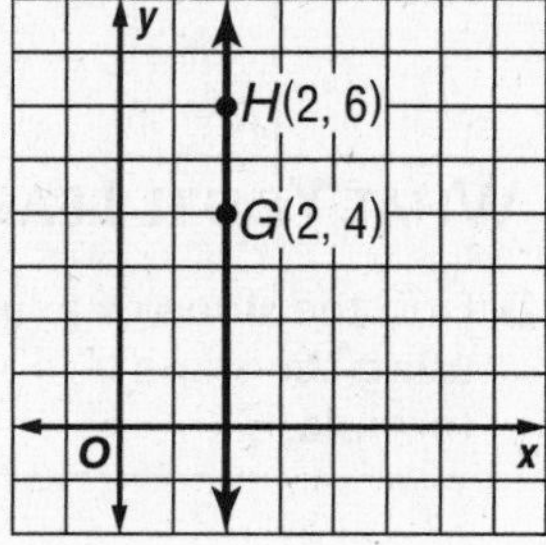

Division by 0 is not defined. So, the slope is undefined. The slope of any ______ line is undefined.

Your Turn **Find the slope of the line that passes through each pair of points.**

a. $A(4, 3)$, $B(1, 0)$

b. $X(-3, 3)$, $Y(1, 0)$

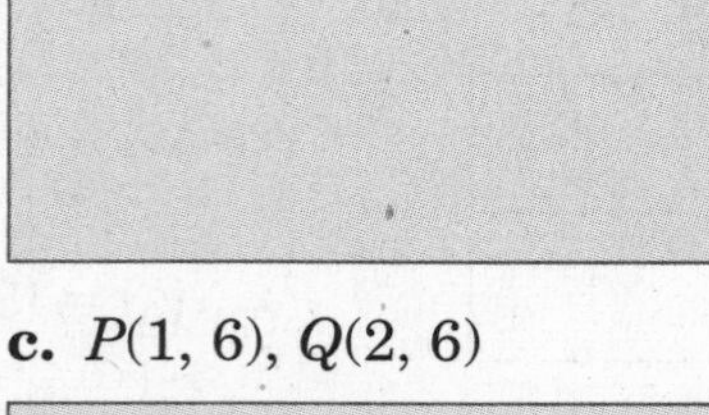

c. $P(1, 6)$, $Q(2, 6)$

d. $G(-2, 1)$, $H(-2, 0)$

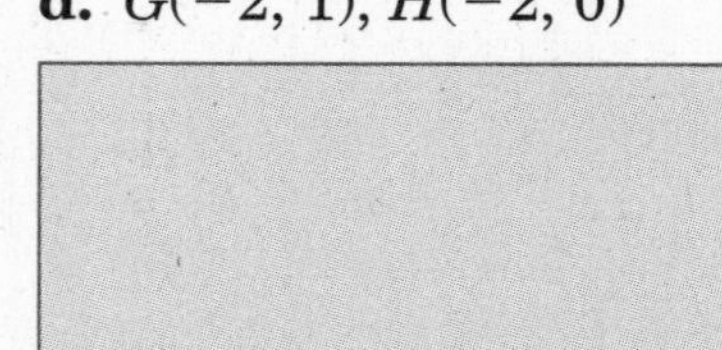

HOMEWORK ASSIGNMENT

Page(s):

Exercises:

© Glencoe/McGraw-Hill

11–5 Slope-Intercept Form

What You'll Learn

- Graph linear equations using the slope and y-intercept.

Build Your Vocabulary (page 279)

Slope-intercept form is when an equation is written in the form ______, where m is the ______ and b is the ______.

EXAMPLES Find the Slopes and y-intercepts of Graphs

State the slope and the y-intercept of the graph of each equation.

1 $y = \frac{3}{4}x - 5$

$y = \frac{3}{4}x + (\ \_\_\_\ )$ Write the equation in the form $y = mx + b$.

$y = mx + b$ $m = \frac{3}{4}$ $b =$ ______

The slope of the graph is ______, and the y-intercept is ______.

2 $2x + y = 8$

$2x + y = 8$ Write the original equation.

$-$ ______ $-$ ______ Subtract ______ from each side.

$y =$ ______ Simplify.

$y =$ ______ Write the equation in the form $y = mx + b$.

$y = mx + b$ $m =$ ______, $b =$ ______

The slope of the graph is ______ and the y-intercept is ______.

© Glencoe/McGraw-Hill

Your Turn **State the slope and the y-intercept of the graph of each equation.**

a. $y = \frac{1}{4}x - 2$

b. $3x + y = 5$

EXAMPLE Graph an Equation

3 **Graph $y = \frac{2}{3}x + 2$ using the slope and y-intercept.**

Step 1 Find the slope and y-intercept.

$$y = \frac{2}{3}x + 2$$

slope $= \frac{2}{3}$

y-intercept $= 2$

Step 2 Graph the y-intercept ______.

Step 3 Use the slope to locate a second point on the line.

$m = \frac{2}{3}$ change in y: up 2 units; change in x: right 3 units

Step 4 Draw a line through the two points.

Your Turn Graph $y = \frac{1}{3}x + 3$ using the slope and y-intercept.

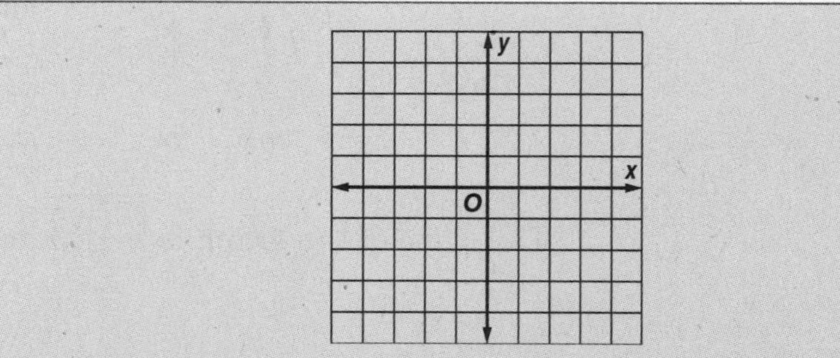

© Glencoe/McGraw-Hill

11–6 Scatter Plots

WHAT YOU'LL LEARN

- Construct and interpret scatter plots.

BUILD YOUR VOCABULARY (pages 278–279)

A **scatter plot** is a graph that shows the ________ between ________ sets of data.

A **best-fit line** is a line that is very close to ________ of the data points in a scatter plot.

EXAMPLES Identify a Relationship

Determine whether a scatter plot of the data for the following might show a *positive*, *negative*, or *no relationship*.

1. **cups of hot chocolate sold at a concession stand and the outside temperature**

   As the temperature decreases, the number of cups of hot chocolate sold ________. Therefore, the scatter plot might show a ________ relationship.

2. **birthday and number of sports played**

   The number of sports played does not depend on your birthday. Therefore, the scatter plot shows ________ relationship.

Your Turn **Determine whether a scatter plot of the data for the following might show a *positive*, *negative*, or *no relationship*.**

a. number of cups of lemonade sold at a concession stand and the outside temperature

____

b. age and the color of your hair

____

© Glencoe/McGraw-Hill

EXAMPLES Draw a Best-Fit Line

GROWTH A boy's heights, measured from age 2 to age 9, are given.

Age (years)	Height (cm)
2	88
3	92
4	103
5	106
6	117
7	121
8	125
9	136

3 Make a scatter plot using the data. Then draw a line that seems to best represent the data.

Boy's Height, Ages 2 Through 9

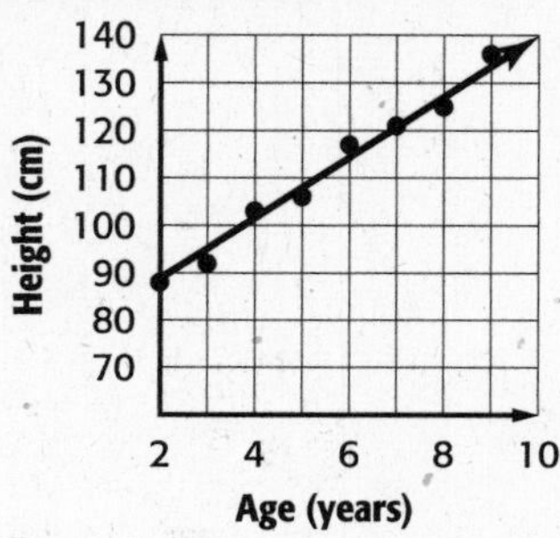

Graph each of the data points.

Draw a ______ that best fits the data.

4 Write an equation for the best-fit line.

The line passes through the points at (4, 103) and (6, 117). Use these points to find the slope of the line.

$m = \frac{y_2 - y_1}{x_2 - x_1}$ Definition of slope

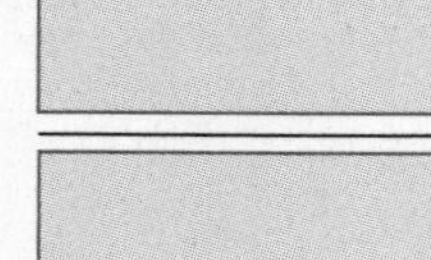

$(x_1, y_1) = (4, 103), (x_2, y_2) = (6, 117)$

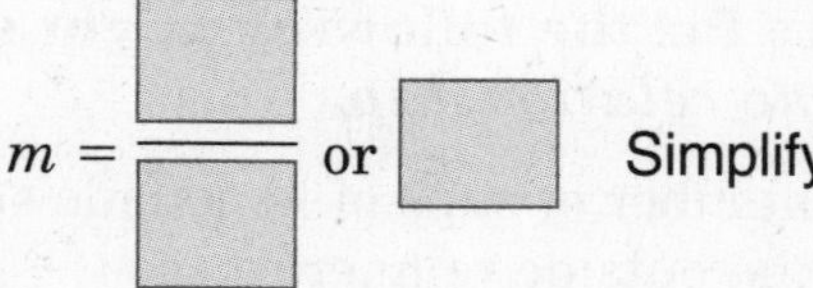

$m =$ ______ or ______ Simplify.

The slope is ______ and the y-intercept is ______.

© Glencoe/McGraw-Hill

Use the slope and the y-intercept to write the equation.

$y = mx + b$ Slope-intercept form

$y = \square c + \square$ $m = \square, b = \square$

The equation for the best-fit line is $\square$.

5 Use the equation to predict the boy's height at age 11.

$y = 7x + 75$ Equation for the best-fit line

$y = 7(11) + 75$ or $\square$

At age 11, the boy's height is about $\square$ centimeters.

Your Turn **The table shows the average hourly earnings of U.S. production workers since 1995.**

a. Make a scatter plot using the data.

b. Write an equation for the best-fit line using points (0, 11.43) and (5, 13.76).

c. Use the equation to predict the average hourly earnings of U.S. production workers in 2004.

U.S. Production Workers Earnings	
Year Since 1995	Average Hourly Earnings
0	\$11.43
1	\$11.82
2	\$12.28
3	\$12.78
4	\$13.24
5	\$13.76
6	\$14.32

Source: *The World Almanac*

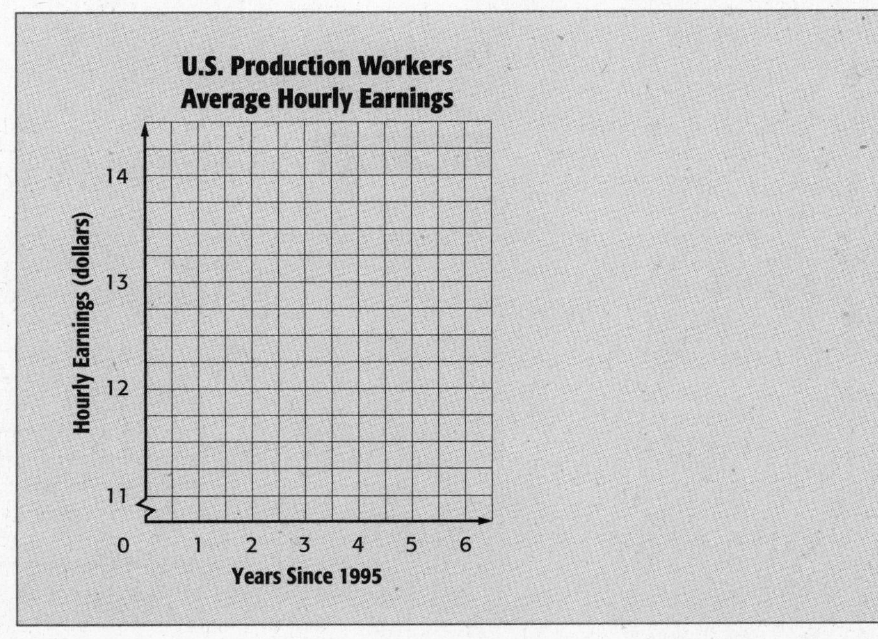

© Glencoe/McGraw-Hill

11-7 Graphing Systems of Equations

BUILD YOUR VOCABULARY (pages 278–279)

A ______ of two or more equations ______ ______ is called a **system of equations.**

WHAT YOU'LL LEARN

- Solve systems of linear equations by graphing.

FOLDABLES

ORGANIZE IT

Under the "Systems of Equations" tab of your Foldable, explain how you know whether a system of equations has one solution, infinitely many solutions, or no solution.

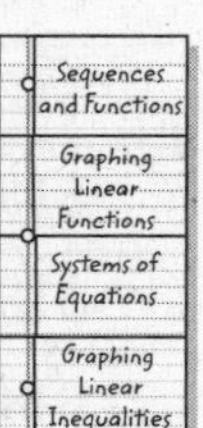

EXAMPLE One Solution

1 **Solve the system $y = x - 1$ and $y = 2x - 2$ by graphing.**

The graphs of the equations appear to intersect at ______. Check this estimate.

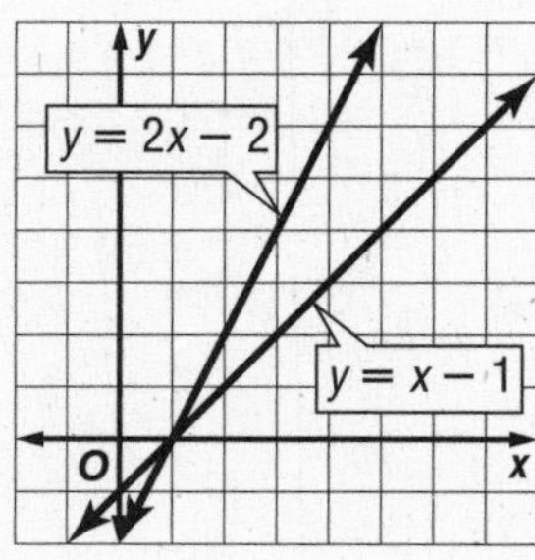

Check $y = x - 1$ $\quad\quad$ $y = 2x - 2$

$\square \stackrel{?}{=} 1 - \square$ $\quad\quad$ $\square \stackrel{?}{=} 2\square - 2$

$\square = \square$ $\quad\quad$ $\square = \square$

The solution is ______.

Your Turn Solve the system $y = x + 3$ and $y = 3x - 3$ by graphing.

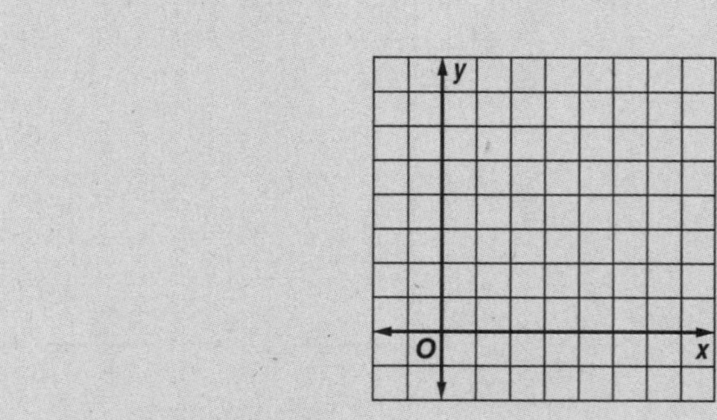

© Glencoe/McGraw-Hill

EXAMPLE Infinitely Many Solutions

2 Solve the system $y = \frac{1}{2}x + 3$ and $y - \frac{1}{2}x = 3$ by graphing.

Write $y - \frac{1}{2}x = 3$ in slope-intercept form.

$y - \frac{1}{2}x = 3$ Write the equation.

$y - \frac{1}{2}x +$ ______ $= 3 +$ ______ Add ______ to each side.

$y =$ 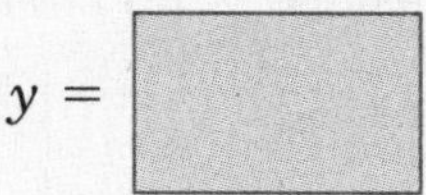 Both equations are the same.

The solution is ______ the coordinates of points on the graph of $y = \frac{1}{2}x + 3$.

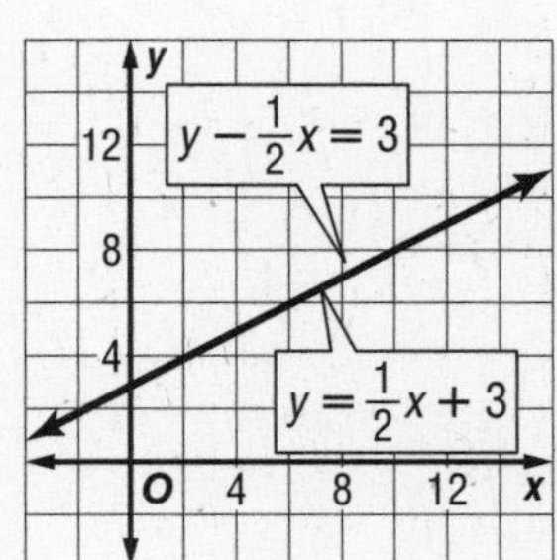

Your Turn Solve the system $y = 2x + 4$ and $y - 2x = 4$ by graphing.

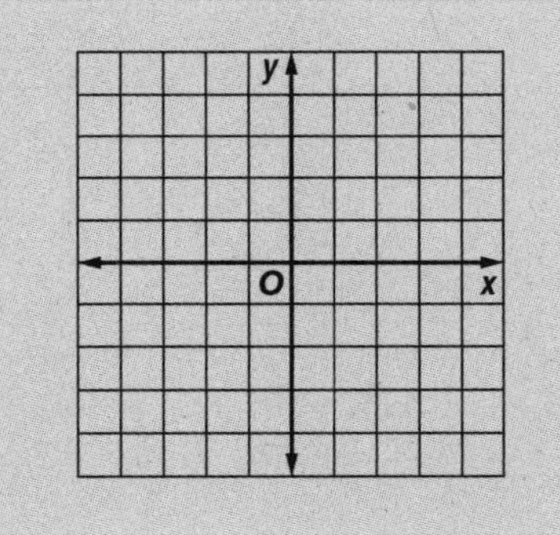

EXAMPLE No Solution

3 PHONE RATES One phone company charges \$3 a month plus 10 cents a minute for long-distance calls. Another company charges \$5 a month plus 10 cents a minute for long-distance calls. For how many minutes will the total monthly charges of the two companies be the same?

Let x equal the number of minutes.

Let y equal the total monthly cost.

Write an equation to represent each company's charge for long-distance phone calls.

1st company: $y =$ ______

2nd company: $y =$ ______

© Glencoe/McGraw-Hill

Graph the system of equations.

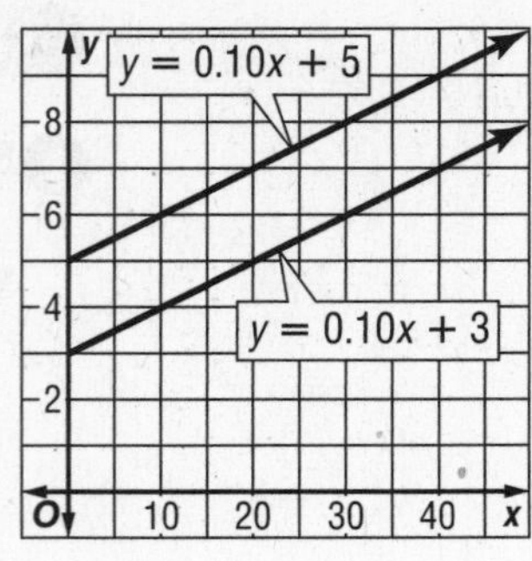

$y = 0.10x + 3$

$y = 0.10x + 5$

The graphs appear to be

 lines.

Since there is no coordinate pair that is a solution of both equations, there is ______ solution of this system of equations.

For ______ amount of minutes, the first company will charge less than the second company.

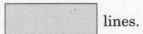

Your Turn One phone company charges $20 a month plus 15 cents a minute for long-distance calls. Another company charges $25 a month plus 15 cents a minute for long-distance calls. For how many minutes will the total monthly charges of the two companies be the same?

BUILD YOUR VOCABULARY (pages 278–279)

A more accurate way to solve a system of ______ than by graphing, is by using a method called **substitution**.

© Glencoe/McGraw-Hill

EXAMPLE Solve by Substitution

REMEMBER IT

There is exactly one solution when the graphs of a system of equations have different slopes.

There is no solution when the graphs have the same slope and different *y*-intercepts.

There are infinitely many solutions when the graphs have the same slope and the same *y*-intercepts.

4 Solve the system $y = 3x - 4$ and $y = 5$ by substitution.

Since y must have the ______ value in both equations, you can replace y with ______ in the first equation.

$y = 3x - 4$	Write the first equation.
$\square = 3x - 4$	Replace y with ______.
$5 + \square = 3x - 4 + \square$	Add ______ to each side.
$\square = 3x$	Simplify.
$\frac{9}{\square} = \frac{3x}{\square}$	Divide each side by ______.
$\square = x$	Simplify.

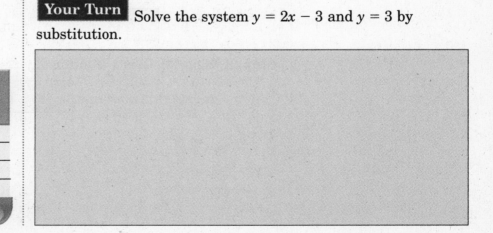

The solution of this system of equations is ______. You can check the solution by graphing. The graphs appear to intersect at ______, so the solution is correct.

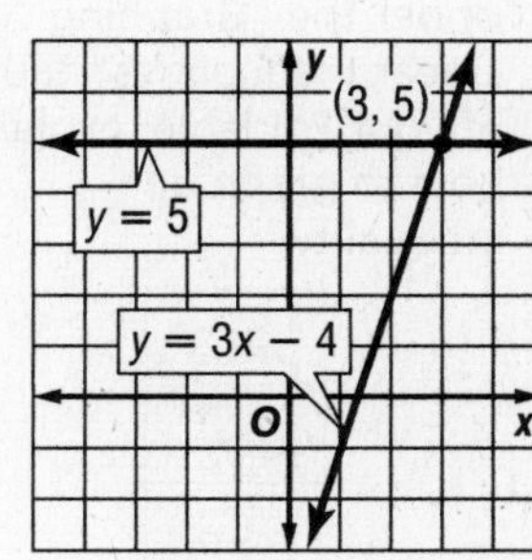

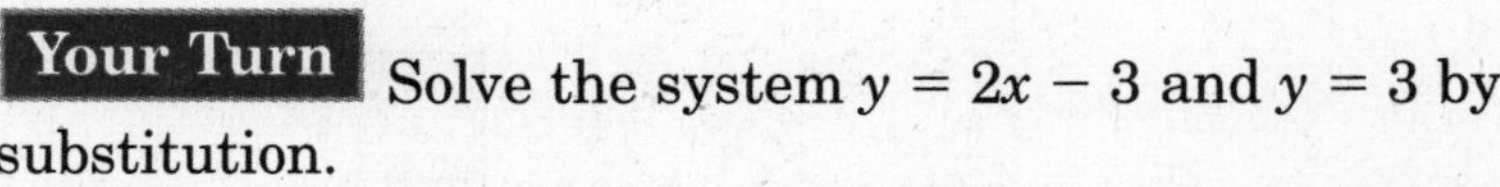

Your Turn Solve the system $y = 2x - 3$ and $y = 3$ by substitution.

HOMEWORK ASSIGNMENT

Page(s):

Exercises:

© Glencoe/McGraw-Hill

11-8 Graphing Linear Inequalities

WHAT YOU'LL LEARN

- Graph linear inequalities.

BUILD YOUR VOCABULARY (pages 278–279)

The line which separates the ______ from the points that are ______ solutions in the graph of a ______ inequality is called the **boundary**.

The ______ which contains the ______ in the graph of a linear inequality is called the **half plane**.

FOLDABLES ORGANIZE IT

Under the "Graphing Linear Inequalities" tab of your Foldable, explain how to graph an inequality.

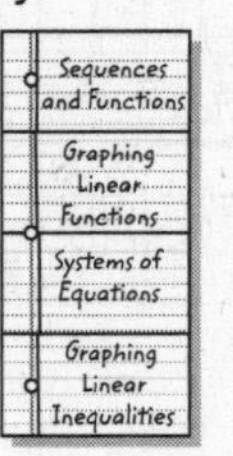

EXAMPLE Graph an Inequality

1 **Graph $y \leq -x + 3$.**

Step 1 Graph the boundary line $y = -x + 3$. Since $\leq$ is used in the inequality, make the boundary line a ______ line.

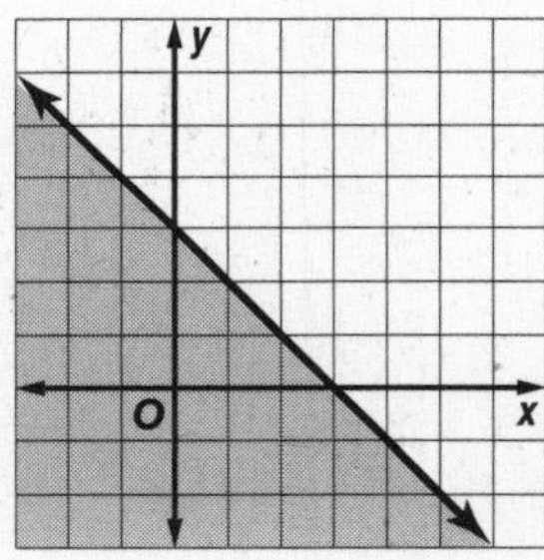

Step 2 Test a point not on the boundary line, such as (0, 0).

$y \overset{?}{\leq} -x + 3$	Write the inequality.
$\square \leq -(\square) + 3$	Replace *x* and *y*.
$0 \leq \square$ ✔	Simplify.

Step 3 Since (0, 0) is a solution of $y \leq -x + 3$, shade the region that contains ______.

© Glencoe/McGraw-Hill

Your Turn Graph $y \leq 2x - 2$.

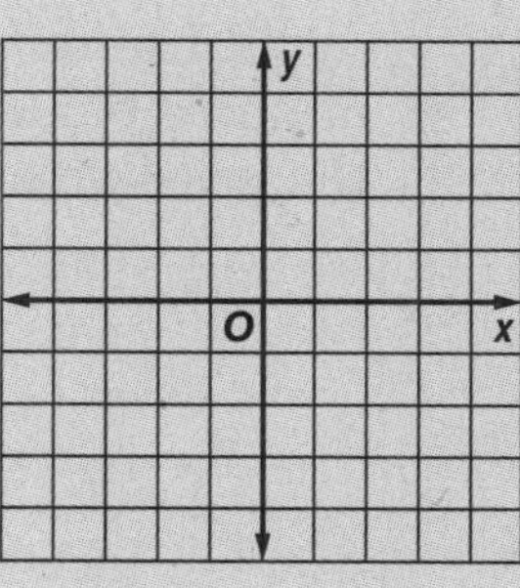

EXAMPLE **Graph an Inequality to Solve a Problem**

2 CRAFTS Sylvia is making items to sell at the school craft sale. Each flower basket takes 15 minutes, and each flower pin takes 1 minute to make. She can spend, at most, 60 minutes making crafts one afternoon. Graph the inequality showing the possible numbers of each item she can make that afternoon.

Let x represent the number of flower baskets and y represent the number of flower pins. Write the inequality $15x + 1y \leq 60$.

The related equation is $15x + y = 60$.

$15x + y = 60$	Write the equation.
$15x + y - 15x = 60 - 15x$	Subtract $15x$ from each side.
$y =$ ☐	Slope-intercept form

Graph $y = -15x + 60$. Test $(0, 0)$ in the original inequality.

$15x + 1y \leq 60$	Write the inequality.
$15(\square) + 1(\square) \overset{?}{\leq} 60$	Replace x and y with ☐.
$\square \leq 60$	Simplify.

Since Sylvia cannot make a negative number or a fractional number of flower baskets or flower pins, the answer is any pair of integers represented in the shaded region. For example, she could make 1 basket and 30 pins.

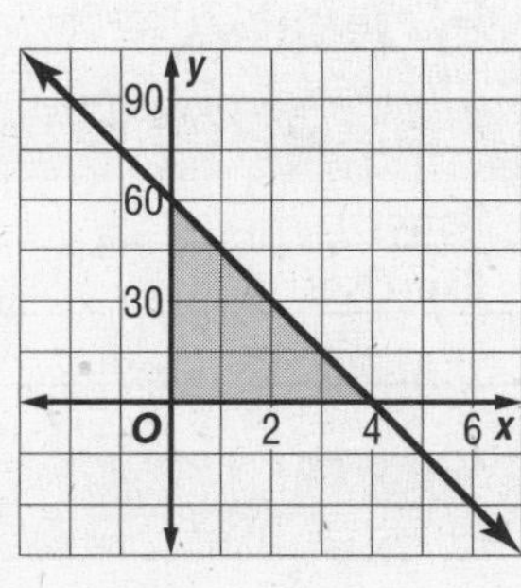

© Glencoe/McGraw-Hill

Your Turn Josef is baking items to sell at the school bake sale. Each cake takes 40 minutes, and each pre-made brownie takes 1 minute to unwrap. He can spend, at most, 120 minutes baking one afternoon. Graph the inequality showing the possible numbers of each item he can make that afternoon.

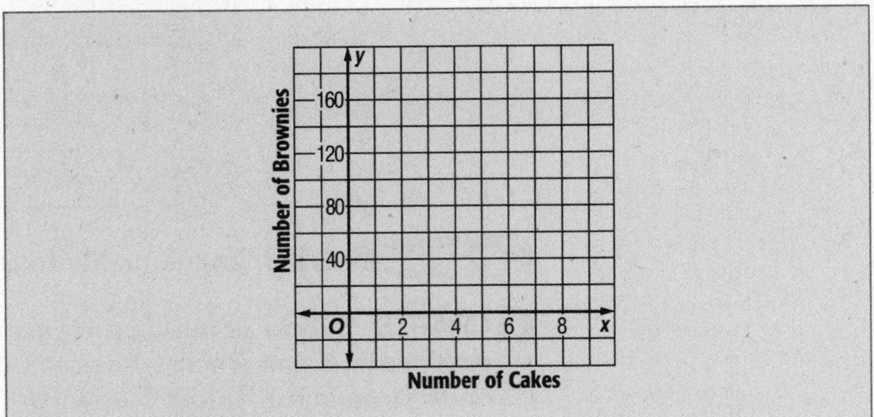

HOMEWORK ASSIGNMENT

Page(s):

Exercises:

© Glencoe/McGraw-Hill

CHAPTER 11

BRINGING IT ALL TOGETHER

STUDY GUIDE

FOLDABLES™	VOCABULARY PUZZLEMAKER	BUILD YOUR VOCABULARY
Use your **Chapter 11 Foldable** to help you study for your chapter test.	To make a crossword puzzle, word search, or jumble puzzle of the vocabulary words in Chapter 11, go to: www.glencoe.com/sec/math/t_resources/free/index.php	You can use your completed **Vocabulary Builder** *(pages 278–279)* to help you solve the puzzle.

11–1 Sequences

State whether each sequence is *arithmetic*, *geometric*, or *neither*. If it is arithmetic or geometric, state the common difference or common ratio. Write the next three terms of each sequence.

1. 14, 19, 24, 29, 34, . . .

2. 2, 0, −3, −7, −12, . . .

3. −1, 3, −9, 27, −81, . . .

11–2 Functions

Match each description with the word it describes.

4. an output value of a function

5. the set of values of the dependent variable

6. the underlined letter in $f(x) = 2\underline{x} + 5$

a. independent variable
b. dependent variable
c. domain
d. range

7. Complete the function table for $f(x) = 2x + 2$. Then give the domain and range.

Domain:

Range:

x	$2x + 2$	$f(x)$
−2		
0		
1		
3		

© Glencoe/McGraw-Hill

11–3 Graphing Linear Functions

8. Complete the function table. Then graph $y = -x + 2$.

x	$-x + 2$	y	(x, y)
-2			
0			
2			
3			

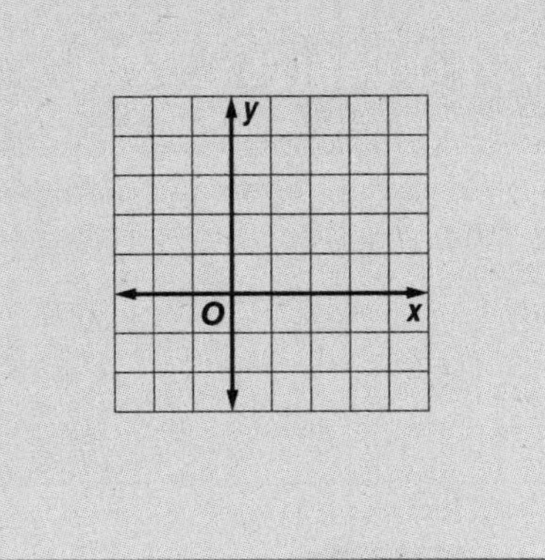

9. Explain how to find the x-intercept of the graph of a linear function. Then find the x-intercept of $y = 2x + 8$.

10. Explain how to find the y-intercept of the graph of a linear function. Then find the y-intercept of $y = 2x + 8$.

11–4 The Slope Formula

Find the slope of the line that passes through each pair of points.

11. $A(1, -2), B(4, 4)$ **12.** $C(1, 2), D(3, -2)$ **13.** $E(-1, 2), F(2, 2)$

11–5 Slope-Intercept Form

State the slope and the y-intercept for the graph of each equation.

14. $y = -3x + 4$ **15.** $y = \frac{2}{3}x - 7$ **16.** $\frac{1}{2}x + y = 8$

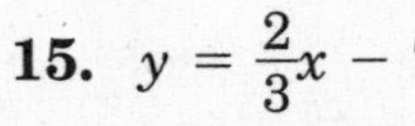

© Glencoe/McGraw-Hill

11–6

Scatter Plots

17. Complete. A scatter plot that shows a negative relationship will have a pattern of data points that go ________.

Write whether a scatter plot of the data for the following might show a *positive*, *negative*, or *no relationship*.

18. favorite color and type of pet

19. the amount of rain and the water level of a pond

11–7

Graphing Systems of Equations

Solve each system of equations by substitution.

20. $y = -3x + 2$
$y = -1$

21. $y = x + 5$
$y = x + 7$

22. $y = 4x + 1$
$y = -x + 6$

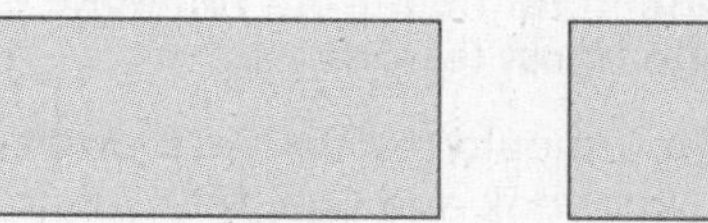

11–8

Graphing Linear Inequalities

Write *true* or *false* beside each statement. If the statement is false, write the correct words in place of the underlined words.

23. The <u>non-shaded</u> region contains all the solutions of the inequality.

24. A <u>dashed</u> boundary line means that points on the line are solutions of the inequality.

25. If the inequality uses the symbol <u>$<$ or $>$</u>, the boundary line will be a solid line.

© Glencoe/McGraw-Hill

ARE YOU READY FOR THE CHAPTER TEST?

Visit **msmath3.net** to access your textbook, more examples, self-check quizzes, and practice tests to help you study the concepts in Chapter 11.

Check the one that applies. Suggestions to help you study are given with each item.

☐ **I completed the review of all or most lessons without using my notes or asking for help.**

- You are probably ready for the Chapter Test.
- You may want to take the Chapter 11 Practice Test on page 555 of you textbook as a final check.

☐ **I used my Foldable or Study Notebook to complete the review of all or most lessons.**

- You should complete the Chapter 11 Study Guide and Review on pages 552–554 of your textbook.
- If you are unsure of any concepts or skills, refer to the specific lesson(s).
- You may also want to take the Chapter 11 Practice Test on page 555.

☐ **I asked for help from someone else to complete the review of all or most lessons.**

- You should review the examples and concepts in your Study Notebook and Chapter 11 Foldable.
- Then complete the Chapter 11 Study Guide and Review on pages 552–554 of your textbook.
- If you are unsure of any concepts or skills, refer to the specific lesson(s).
- You may also want to take the Chapter 11 Practice Test on page 555.

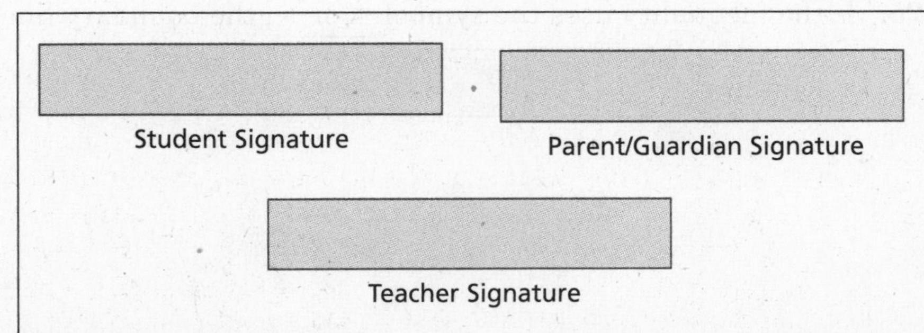

© Glencoe/McGraw-Hill

Algebra: Nonlinear Functions and Polynomials

Use the instructions below to make a Foldable to help you organize your notes as you study the chapter. You will see Foldable reminders in the margin of this Interactive Study Notebook to help you in taking notes.

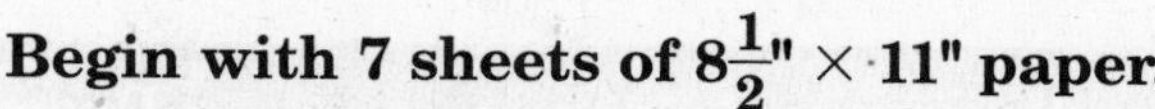

Begin with 7 sheets of $8\frac{1}{2}$" × 11" paper.

STEP 1 **Fold and Cut**
Fold a sheet of paper in half lengthwise. Cut a 1" tab along the left edge through one thickness.

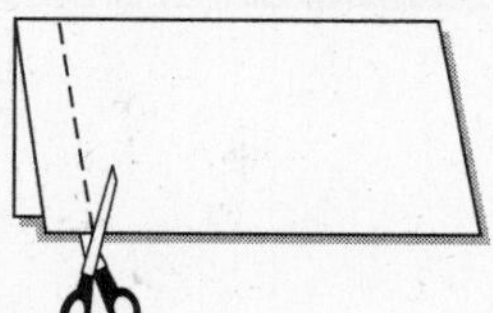

STEP 2 **Glue and Label**
Glue the 1" tab down. Write the title of the lesson on the front tab.

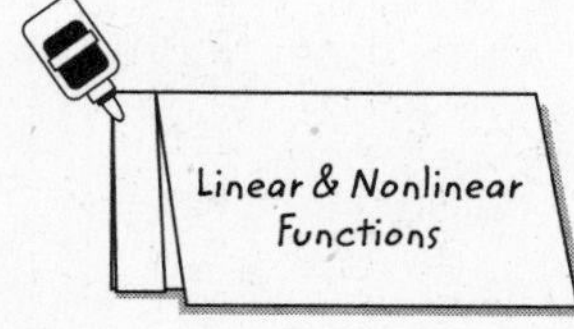

STEP 3 **Repeat and Staple**
Repeat Steps 1–2 for the remaining sheets of paper. Staple together to form a booklet.

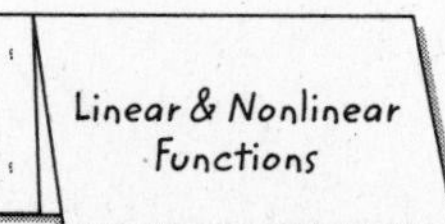

NOTE-TAKING TIP: When you take notes, define new terms and write about the new concepts you are learning in your own words. Write your own examples that use the new terms and concepts.

© Glencoe/McGraw-Hill

BUILD YOUR VOCABULARY

This is an alphabetical list of new vocabulary terms you will learn in Chapter 12. As you complete the study notes for the chapter, you will see Build Your Vocabulary reminders to complete each term's definition or description on these pages. Remember to add the textbook page number in the second column for reference when you study.

Vocabulary Term	Found on Page	Definition	Description or Example
monomial			
nonlinear function			
polynomial			
quadratic function			

© Glencoe/McGraw-Hill

12–1 Linear and Nonlinear Functions

What You'll Learn

- Determine whether a function is linear or nonlinear.

Build Your Vocabulary (page 308)

Nonlinear functions do not have ________ rates of change. Therefore, their graphs are not straight lines.

Foldables

Organize It

Explain how to identify linear and nonlinear functions using graphs, equations, and tables on the Lesson 12-1 section of your Foldable journal.

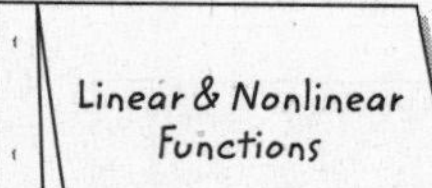

EXAMPLES Identify Functions Using Graphs

Determine whether each graph represents a *linear* or *nonlinear* function. Explain.

1

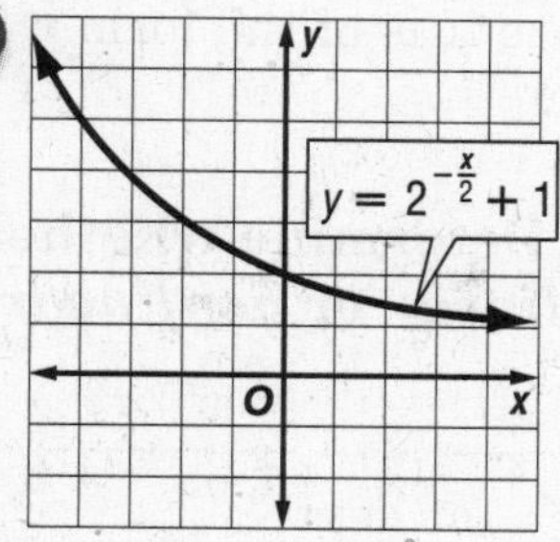

The graph is a curve, not a straight line. So it represents a ________ function.

2

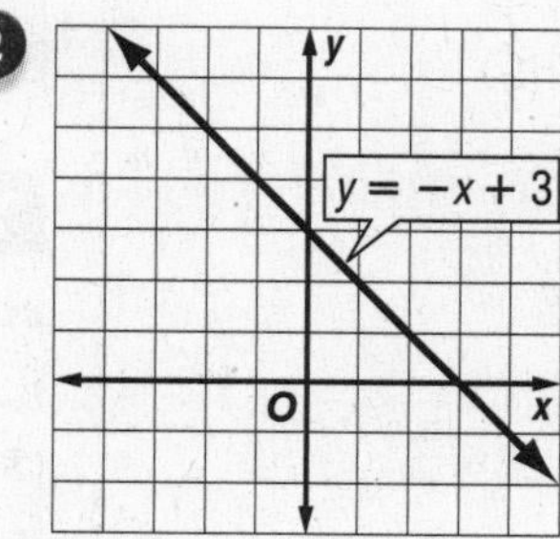

The graph is a straight line. So it represents a ________ function.

Your Turn **Determine whether each graph represents a *linear* or *nonlinear* function. Explain.**

a.

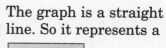

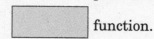

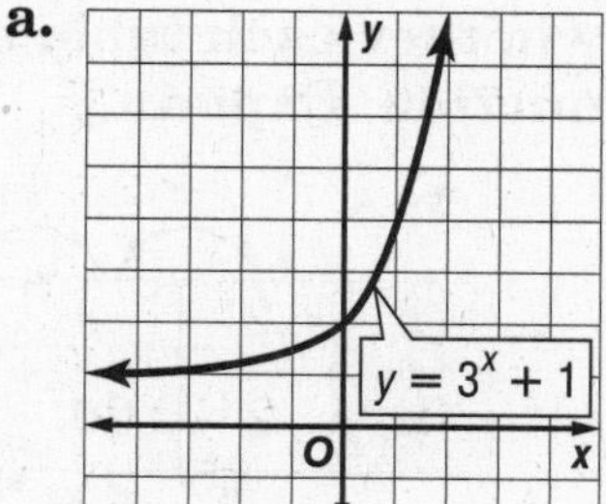

b.

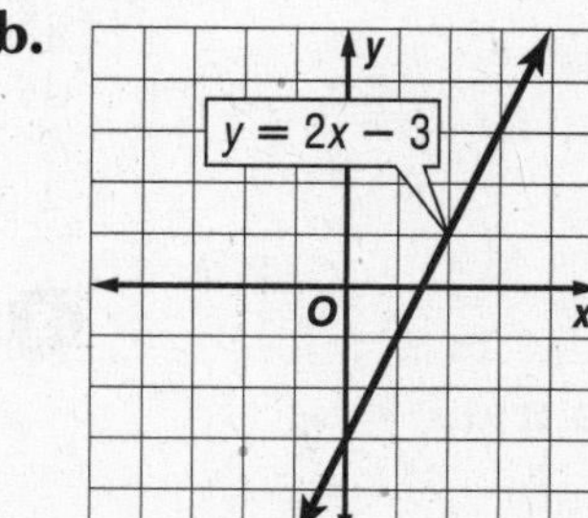

© Glencoe/McGraw-Hill

EXAMPLES Identify Functions Using Equations

Determine whether each equation represents a *linear* or *nonlinear* function. Explain.

3 $y = 5x^2 + 3$

Since x is raised to the ________ power, the equation cannot be written in the form $y = mx + b$. So, this function is ________.

4 $y - 4 = 5x$

Rewrite the equation as $y =$ ________. This equation is ________ since it is of the form $y = mx + b$.

Your Turn **Determine whether each equation represents a *linear* or *nonlinear* function. Explain.**

a. $y = x^2 - 1$

____

b. $y = x$

____

EXAMPLES Identify Functions Using Tables

Determine whether each table represents a *linear* or *nonlinear* function. Explain.

5

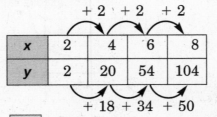

+ 2 + 2 + 2

x	2	4	6	8
y	2	20	54	104

+ 18 + 34 + 50

As x increases by ____, y increases by a greater amount each time. The rate of change is not ________, so this function is ________.

© Glencoe/McGraw-Hill

6

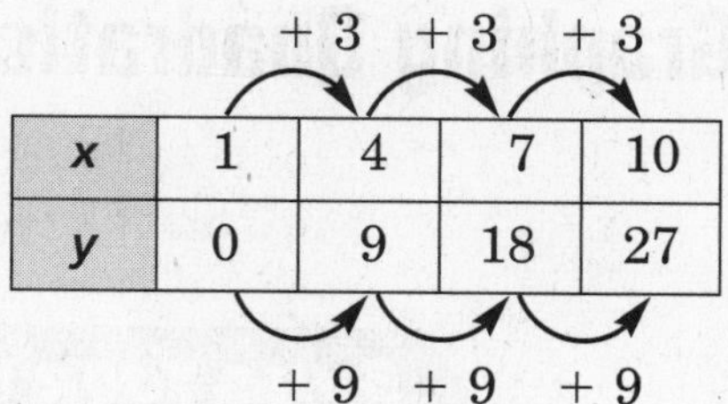

x	1	4	7	10
y	0	9	18	27

As x increases by ____, y increases by ____ each time. The

rate of change is ____, so this function is ____.

Your Turn **Determine whether each table represents a *linear* or *nonlinear* function. Explain.**

a.

x	1	3	5	7
y	3	7	11	15

b.

x	1	2	3	4
y	1	8	27	64

HOMEWORK ASSIGNMENT

Page(s):

Exercises:

© Glencoe/McGraw-Hill

12-2 Graphing Quadratic Functions

WHAT YOU'LL LEARN

- Graph quadratic functions.

BUILD YOUR VOCABULARY (page 308)

A **quadratic function** is a function in which the ______ power of the ______ is ______.

FOLDABLES

ORGANIZE IT

Record what you learn about graphing quadratic functions and using the graphs to solve problems on the Lesson 12-2 section of your Foldable.

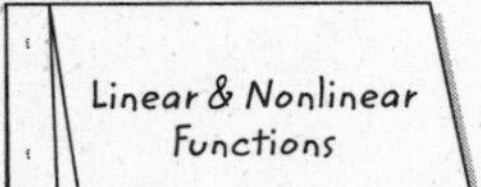

EXAMPLES Graph Quadratic Functions: $y = ax^2$

1 **Graph $y = 5x^2$.**

To graph a quadratic function, make a table of values, plot the ordered pairs, and connect the points with a smooth curve.

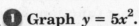

x	$5x^2$	y	(x, y)
−2	$5(-2)^2 =$		$(-2,\ \ )$
−1	$5(-1)^2 =$		$(-1,\ \ )$
0	$5(0)^2 =$		$(0,\ \ )$
1	$5(1)^2 =$		$(1,\ \ )$
2	$5(2)^2 =$		$(2,\ \ )$

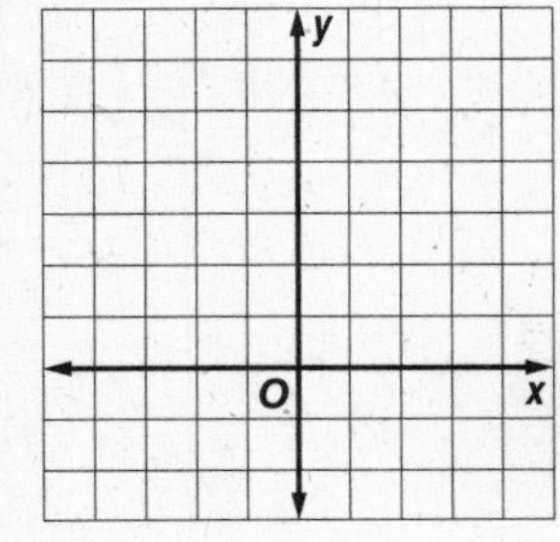

Your Turn Graph $y = -3x^2$.

x	$-3x^2$	y	(x, y)

© Glencoe/McGraw-Hill

EXAMPLE Graph Quadratic Functions: $y = ax^2 + c$

2 **Graph $y = 3x^2 + 1$.**

x	$3x^2 + 1$	y	(x, y)
−2	$3(-2)^2 + 1 =$ ___	___	(−2, ___)
−1	$3(-1)^2 + 1 = 4$	4	(−1, 4)
0	$3(0)^2 + 1 =$ ___	___	(0, ___)
1	$3(1)^2 + 1 = 4$	4	(1, 4)
2	$3(2)^2 + 1 = 13$	13	(2, 13)

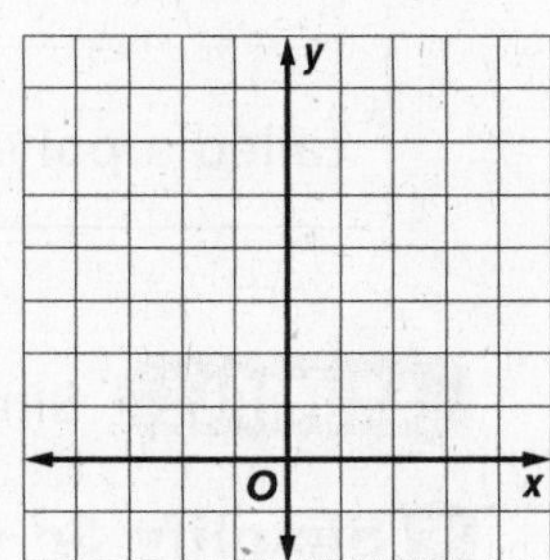

Your Turn Graph $y = -2x^2 - 1$.

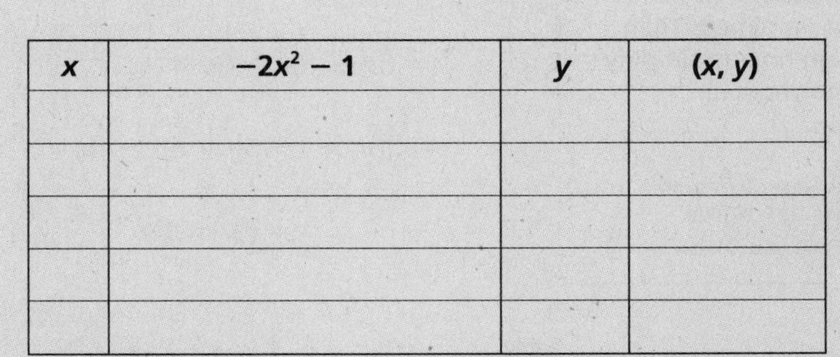

x	$-2x^2 - 1$	y	(x, y)

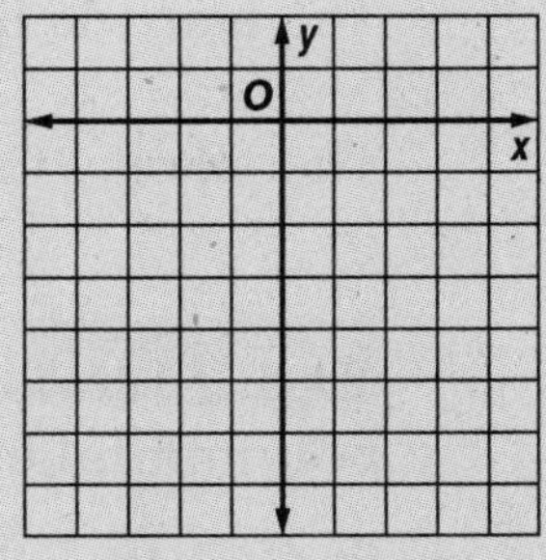

HOMEWORK ASSIGNMENT

Page(s):

Exercises:

© Glencoe/McGraw-Hill

12–3 Simplifying Polynomials

What You'll Learn

- Simplify polynomials.

Build Your Vocabulary (page 308)

A **monomial** is a number, a variable, or a ______ of a number and one or more variables.

An algebraic expression that is the ______ or ______ of one or more ______ is called a **polynomial**.

Foldables™

Organize It

In the Lesson 12-3 section of your Foldable, include an example of a polynomial that needs to be simplified. Then explain how to simplify the polynomial.

Linear & Nonlinear Functions

EXAMPLE Simplify Polynomials

1 **Simplify $3r + 8p - 6q - r$.**

The like terms in this expression are ______ and ______.

$3r + 8p - 6q - r$	Write the polynomial.
$= 3r + 8p - 6q +$ ______	Definition of subtraction
$= [3r + (-r)] + 8p - 6q$	Group ______.
$=$ ______ $+ 8p - 6q$	Simplify by combining like terms.

2 **Simplify $-6x^2 + 14 + 3x$.**

There are no like terms in the expression.

Therefore, $-6x^2 + 14 + 3x$ is in ______ form.

Your Turn **Simplify each polynomial. If the polynomial cannot be simplified, write *simplest form*.**

a. $2r + 7p - 3q - 5r$

b. $-2x^2 + 4 + x$

© Glencoe/McGraw-Hill

3 **Simplify $2x + 8x^2 - 9x + 3 - 2x^2$.**

$2x + 8x^2 - 9x + 3 - 2x^2$ is equal to $2x + 8x^2 +$ ______ $+$

$3 +$ ______.

Write the polynomial. Then group and add ______ terms.

$2x + 8x^2 + (-9x) + 3 + (-2x^2)$

$= [8x^2 + (\ \_\_\_\_\ )] + [2x + (\ \_\_\_\_\ )] + 3$

$= \_\_\_\_ + (\ \_\_\_\_\ ) + 3$

$= \_\_\_\_ - 7x + 3$

Thus, $2x + 8x^2 - 9x + 3 - 2x^2 =$ ______.

Your Turn Simplify $3x + 2x^2 - 6x + 2 - 3x^2$.

REMEMBER IT

To be consistent, write the results of simplifying polynomials in standard form, with the powers of the variable decreasing from left to right.

HOMEWORK ASSIGNMENT

Page(s):

Exercises:

© Glencoe/McGraw-Hill

12-4 Adding Polynomials

EXAMPLES Add Polynomials

WHAT YOU'LL LEARN

- Add polynomials.

FOLDABLES

ORGANIZE IT

Record what you learn about adding polynomials in the Lesson 12-4 section of your Foldable.

Linear & Nonlinear Functions

1 **Find $(9x + 2) + (7x + 12)$.**

Method 1 Add vertically.

$$\begin{array}{rr} & 9x + 2 \\ (+) & 7x + 12 \\ \hline & \_\_\_\_ \end{array}$$

Align ____ terms.

Add.

Method 2 Add horizontally.

$(9x + 2) + (7x + 12)$

$= (9x + 7x) + (2 + 12)$ Associative and Commutative Properties

$=$ ____ $+$ ____

The sum is ____.

2 **Find $(4x^2 + 11x - 3) + (-2x^2 + 5x - 7)$.**

Method 1 Add vertically.

$$\begin{array}{rrrr} & 4x^2 + & 11x & - 3 \\ (+) & -2x^2 + & 5x & - 7 \\ \hline & \_\_\_\_ & + 16x & \_\_\_\_ \end{array}$$

Method 2 Add horizontally.

$(4x^2 + 11x - 3) + (-2x^2 + 5x - 7)$

$= (4x^2 - 2x^2) + ($ ____ $+$ ____ $) + (-3 - 7)$

$=$ ____ $+ 16x$ ____

The sum is ____.

© Glencoe/McGraw-Hill

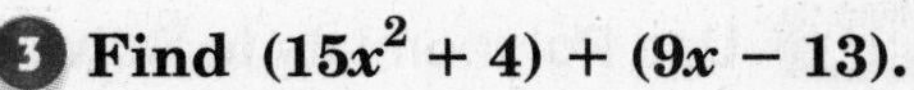

WRITE IT

Explain why it is helpful when adding polynomials to leave a space when there is no other term like another.

____

____

____

____

3 **Find $(15x^2 + 4) + (9x - 13)$.**

$(15x^2 + 4) + (9x - 13) = 15x^2 + \square + (4 - 13)$ Group like terms.

$= 15x^2 + 9x - \square$ Simplify.

The sum is $15x^2 + 9x - \square$.

4 **Find $(3x^2 + 14x - 9) + (-6x + 1)$.**

$$\begin{array}{rrrr} & 3x^2 & + 14x & - \quad 9 \\ (+) & & - \quad 6x & + \quad 1 \\ \hline & \square & + \quad 8x & \square \end{array}$$

Leave a space because there is no other term like $3x^2$.

The sum is $\square$.

Your Turn **Add.**

a. $(5x + 1) + (3x + 10)$

b. $(x^2 + 3x - 6) + (-3x^2 + 4x - 3)$

c. $(5x^2 + 2) + (2x - 9)$

d. $(2x^2 + 4x - 7) + (-3x + 5)$

© Glencoe/McGraw-Hill

EXAMPLE Use Polynomials to Solve a Problem

5 Find the measure of $\angle A$ in the figure below.

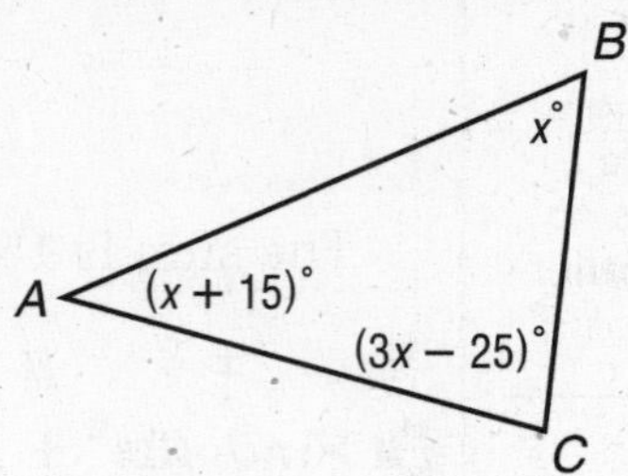

Write an equation to find the value of x.

The sum of the measures of the angles	equals	180	
$(x + 15) + x + (3x - 25)$	$=$	180	Write the equation.
(☐) + (☐)	$=$	180	Group like terms.
$5x - 10$	$=$	180	Simplify.
+ ☐		+ ☐	Add ☐ to each side.
$5x$	$=$	190	Simplify.
$\frac{5x}{5}$	$=$	$\frac{190}{5}$	Divide each side by 5.
x	$=$	☐	Simplify.

Find the measure of $\angle A$.

$m\angle A = x + 15$ Write the expression for the measure of angle A.

$=$ ☐ $+ 15$ Replace x with ☐.

$=$ 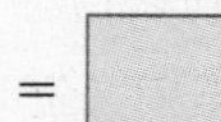 Simplify.

The measure of $\angle A$ is ☐.

HOMEWORK ASSIGNMENT

Page(s):

Exercises:

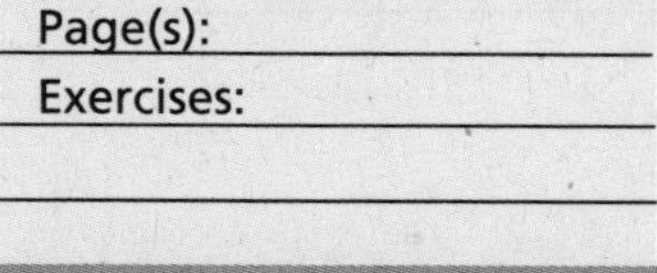

Your Turn Find the measure of $\angle A$ in the figure below.

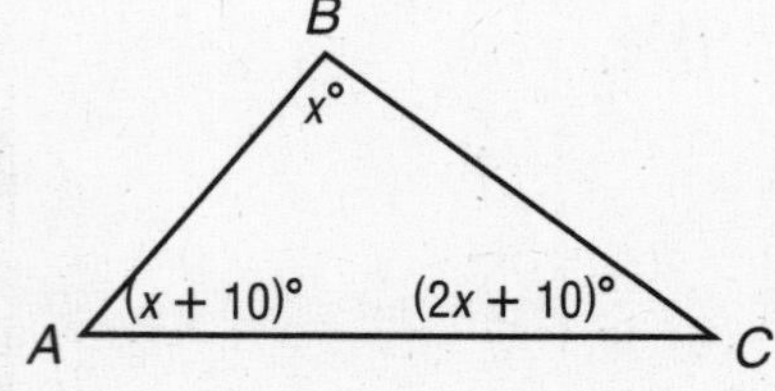

© Glencoe/McGraw-Hill

12–5 Subtracting Polynomials

WHAT YOU'LL LEARN

- Subtract polynomials.

FOLDABLES™

ORGANIZE IT

Record what you learn about subtracting polynomials in the Lesson 12-5 section of your Foldable journal.

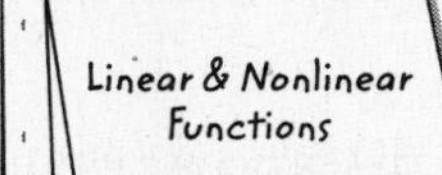

EXAMPLES Subtract Polynomials

Subtract.

1 $(8c + 3) - (6c + 2)$

$$\begin{array}{rl} 8c + 3 & \\ (-)\ 6c + 2 & \text{Align the terms.} \\ \hline 2c + 1 & \text{Subtract.} \end{array}$$

The difference is ______.

2 $(-2d^2 + 6d - 11) - (-3d + 4)$

$$\begin{array}{rl} -2d^2 + 6d - 11 & \\ (-)\qquad - 3d + 4 & \text{Align the terms.} \\ \hline -2d^2 + 9d - 15 & \text{Subtract.} \end{array}$$

The difference is ______.

EXAMPLES Subtract Using the Additive Inverse

3 **Find $(6z + 1) - (2z - 5)$.**

The additive inverse of $2z - 5$ is ______.

$(6z + 1) - (2z - 5)$

$= (6z + 1) + (-2z + 5)$ — To subtract $(2z - 5)$, add $(-2z + 5)$.

$= (6z - 2z) + (1 + 5)$ — Group like terms.

$= 4z + 6$ — Simplify by combining like terms.

The difference is ______.

4 **Find $(10f^2 - 15) - (-5f + 3)$.**

The additive inverse of $-5f + 3$ is ______.

$$\begin{array}{r} 10f^2 \qquad -15 \\ (-)\qquad -5f + 3 \\ \hline \end{array} \longrightarrow \begin{array}{r} 10f^2 \qquad - 15 \\ (+)\qquad 5f - 3 \\ \hline 10f^2 + 5f - 18 \end{array}$$

The difference is ______.

© Glencoe/McGraw-Hill

Your Turn **Subtract.**

a. $(6c + 5) - (2c + 2)$.

b. $(5d^2 + 2d + 10) - (-5d + 8)$.

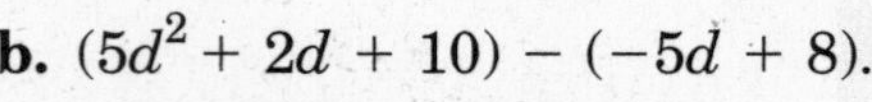

c. $(11z + 2) - (3z - 6)$.

d. $(12f^2 - 5) - (-2f + 4)$.

EXAMPLE Use Polynomials to Solve a Problem

5 EXPERIMENTS **Students are rolling identical marbles down two side-by-side ramps. The marble on ramp A rolls $3t^2 + 11t$ inches in t seconds. The marble on ramp B rolls $2t^2 + 4t$ inches in t seconds. How far apart are the marbles after 6 seconds?**

Write an expression for the difference of the distances traveled by each marble.

Words Marble A's distance minus marble B's distance.

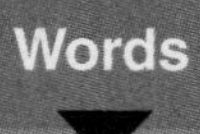

Variable t = the time in seconds

Expression $(3t^2 + 11t) - (2t^2 + 4t)$

$$\begin{array}{rr} & 3t^2 + 11t \\ (-) & 2t^2 + 4t \\ \hline \end{array} \quad \longrightarrow \quad \begin{array}{rr} & 3t^2 + 11t \\ (+) & -2t^2 - 4t \\ \hline & t^2 + 7t \end{array}$$

Now evaluate this expression for a time of 6 seconds.

$t^2 + 7t = (\square)^2 + 7(\square)$ Replace t with $\square$.

$= \square + 42$ or 78 Simplify.

After $\square$ seconds, the cars are $\square$ inches apart.

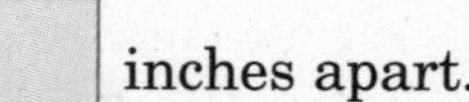

HOMEWORK ASSIGNMENT

Page(s):

Exercises:

Your Turn Students are rolling identical marbles down two side-by-side ramps. The marble on ramp A rolls $4t^2 + 12t$ inches in t seconds. The marble on ramp B rolls $t^2 + 2t$ inches in t seconds. How far apart are the marbles after 5 seconds?

© Glencoe/McGraw-Hill

12–6 Multiplying and Dividing Monomials

EXAMPLE Multiply Powers

WHAT YOU'LL LEARN

- Multiply and divide monomials.

KEY CONCEPT

Product of Powers To multiply powers with the same base, add their exponents.

FOLDABLES In the Lesson 12–6 section of your Foldable, record the product of powers rule.

1 Find $7^6 \cdot 7^2$. Express using exponents.

$7^6 \cdot 7^2 = 7^{6+2}$ The common base is ____.

$=$ ____  the exponents.

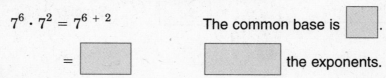

Check $7^6 \cdot 7^2 = (7 \cdot 7 \cdot 7 \cdot 7 \cdot 7 \cdot 7) \cdot (7 \cdot 7)$

$= 7 \cdot 7 \cdot 7 \cdot 7 \cdot 7 \cdot 7 \cdot 7 \cdot 7$

$=$ ____

Your Turn Find $2^5 \cdot 2^4$. Express using exponents.

EXAMPLE Multiply Monomials

2 Find $7x^2(11x^4)$. Express using exponents.

$7x^2(11x^4) = (7 \cdot 11)$ ____ Comm. and Assoc. Properties.

$=$ ____ (x^{2+4}) The common base is ____.

$=$ ____ ____ the exponents.

Your Turn Find $3x^2(-5x^5)$. Express using exponents.

© Glencoe/McGraw-Hill

EXAMPLE Divide Powers

Divide. Express using exponents.

3 $\dfrac{6^{12}}{6^2}$

$\dfrac{6^{12}}{6^2} = 6^{12-2}$ The common base is ____.

$= $ ____ Simplify.

4 $\dfrac{a^{14}}{a^8}$

$\dfrac{a^{14}}{a^8} = a^{14-8}$ The common base is ____.

$= $ ____ Simplify.

KEY CONCEPT

Quotient of Powers To divide powers with the same base, subtract their exponents.

Your Turn **Divide. Express using exponents.**

a. $\dfrac{3^{10}}{3^4}$

b. $\dfrac{x^{11}}{x^3}$

HOMEWORK ASSIGNMENT

Page(s):

Exercises:

© Glencoe/McGraw-Hill

12–7 Multiplying Monomials and Polynomials

EXAMPLES Use the Distributive Property

WHAT YOU'LL LEARN

- Multiply monomials and polynomials.

FOLDABLES

ORGANIZE IT

In the Lesson 12-7 section of your Foldable, explain the Distributive Property. Be sure to include an example.

Linear & Nonlinear Functions

1 Find $y(y + 12)$.

$y(y + 12) = y(\_\_\_) + y(\_\_\_)$ Distributive Property

$= \_\_\_ + \_\_\_$ $y \cdot y = \_\_\_$

The answer is ___.

2 Find $-6x(x + 3)$.

$-6x(x + 3) = -6x(\_\_\_) + (-6x)(\_\_\_)$ Distributive Property

$= -6x^2 + (-18x)$ $-6 \cdot x \cdot x = -6x^2$

$= \_\_\_$ Definition of subtraction

The answer is ___.

Your Turn **Multiply.**

a. $y(y + 3)$

___

b. $-2x(x + 6)$

___

EXAMPLES Use the Product of Powers Rule

3 Find $7w(w^2 + 6)$.

$7w(w^2 + 6)$

$= \_\_\_(w^2) + \_\_\_(6)$ Distributive Property

$= \_\_\_ + \_\_\_$ $7w(w^2) = 7w^{1+2}$ or ___

The answer is ___.

© Glencoe/McGraw-Hill

4 Find $9t(t^2 + 6t - 4)$.

$9t(t^2 + 6t - 4)$

$= 9t[t^2 + 6t + (-4)]$ Rewrite $t^2 + 6t - 4$ as $t^2 + 6t + (-4)$.

$= 9t(t^2) +$ ______ $+ 9t(-4)$ Distributive Property

$=$ ______ $+$ ______ $+$ ______ Simplify.

$=$ ______ Definition of subtraction

The answer is ______.

Your Turn **Multiply.**

a. $3w(w^2 - 8)$

b. $5t(t^2 + 7t - 6)$

HOMEWORK ASSIGNMENT

Page(s):

Exercises:

© Glencoe/McGraw-Hill

CHAPTER 12

BRINGING IT ALL TOGETHER

STUDY GUIDE

FOLDABLES	VOCABULARY PUZZLEMAKER	BUILD YOUR VOCABULARY
Use your **Chapter 12 Foldable** to help you study for your chapter test.	To make a crossword puzzle, word search, or jumble puzzle of the vocabulary words in Chapter 12, go to: www.glencoe.com/sec/math/t_resources/free/index.php	You can use your completed **Vocabulary Builder** *(page 308)* to help you solve the puzzle.

12-1 Linear and Nonlinear Functions

Write *linear* or *nonlinear* to name the kind of function described.

1. constant rate change

2. graph that is a curve

3. power of x may be greater than one

4. equation has the form $y = mx + b$

5. Name the kind of function represented. Explain your reasoning.

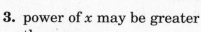

x	-3			
y	10	1	10	37

12-2 Graphing Quadratic Functions

Determine whether each equation represents a quadratic function. Write *yes* or *no*.

6. $y = 3x - 5$

7. $y = 6 - x^2$

8.

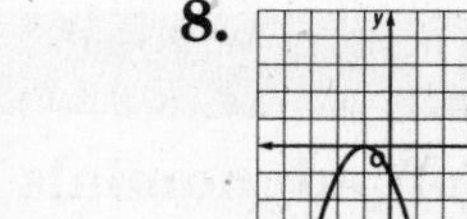

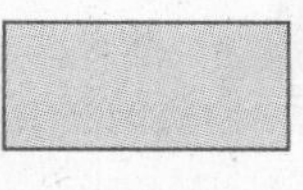

9. Explain how to graph a quadratic function.

© Glencoe/McGraw-Hill

12-3

Simplifying Polynomials

Write *true* or *false* beside each statement. If the statement is false, write the correct word in place of the underlined word.

10. The product of $4y$ and $-6y$ is a polynomial.

11. The expression $x + 5$ is a polynomial with two terms.

12. To simplify a polynomial, combine exponents.

Simplify each polynomial. If the polynomial cannot be simplified, write *simplest form*.

13. $2x^2 + 3x - x^2 + 4$

14. $6c - 4d - 1$

15. $8 + 3a - 10 - a + 2b$

12-4

Adding Polynomials

Determine whether each vertical addition can be performed as written. Explain.

16.

$$\begin{array}{r} x^2 + 5x - 6 \\ \underline{(+)\ 2x^2 + 3x + 4} \end{array}$$

17.

$$\begin{array}{r} x^2 + 6 \\ \underline{(+)\ 2x^2 - x} \end{array}$$

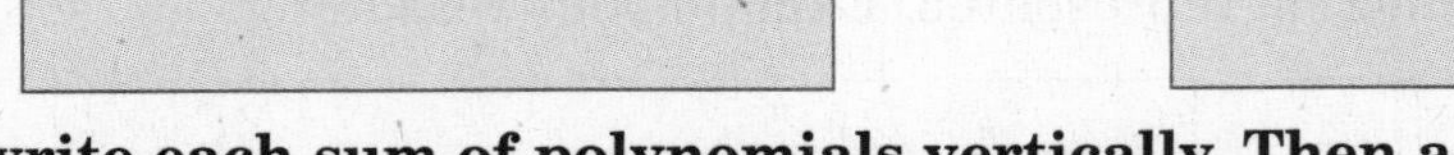

Rewrite each sum of polynomials vertically. Then add.

18. $(3d^2 + 14d - 2) + (-d^2 + 3d + 5)$

19. $(2n^2 + 3) + (n^2 - 5n + 1)$

12–5

Subtracting Polynomials

Rewrite each difference of polynomials as a sum of polynomials. Then add.

20. $(7x + 5) - (4x - 3)$

21. $(-3c^2 + 2c - 1) - (-c^2 - c - 2)$

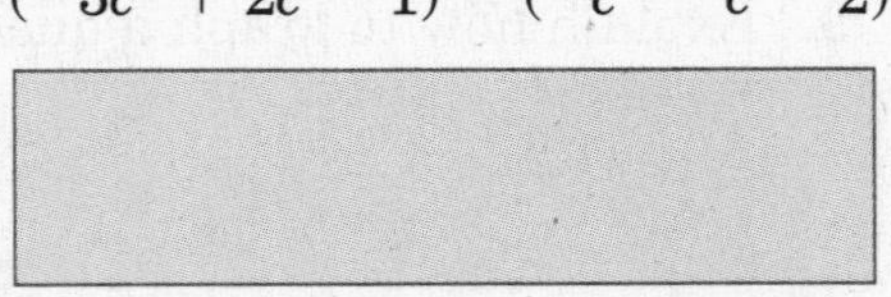

© Glencoe/McGraw-Hill

22. $(m^2 + 3m - 6) - (m^2 + 1)$

23. $(-6s + 9) - (4s^2 + 2s - 3)$

12-6

Multiplying and Dividing Monomials

Complete each sentence.

24. To multiply powers with the same base, ________ their exponents.

25. To divide powers with the same base, ________ their exponents.

26. Can $a^4 \cdot b^3$ be multiplied by adding exponents? Explain.

27. Can $\frac{8^3}{4^3}$ be divided by subtracting exponents? Explain.

Multiply or divide. Express using exponents.

28. $5^2 \cdot 5^6$ **29.** $(8x^3)(-3x^9)$ **30.** $\frac{2^5}{2^2}$ **31.** $\frac{18a^7}{6a^3}$

12–7

Multiplying Monomials and Polynomials

32. Explain what is done at each step in the following problem.

$3x(x^2 + 5x - 2)$

$= 3x[x^2 + 5x + (-2)]$

$= 3x(x^2) + 3x(5x) + 3x(-2)$

$= 3x^3 + 15x^2 + (-6x)$

$= 3x^3 + 15x^2 - 6x$

© Glencoe/McGraw-Hill

ARE YOU READY FOR THE CHAPTER TEST?

Visit **msmath3.net** to access your textbook, more examples, self-check quizzes, and practice tests to help you study the concepts in Chapter 12.

Check the one that applies. Suggestions to help you study are given with each item.

☐ **I completed the review of all or most lessons without using my notes or asking for help.**

- You are probably ready for the Chapter Test.
- You may want to take the Chapter 12 Practice Test on page 595 of your textbook as a final check.

☐ **I used my Foldable or Study Notebook to complete the review of all or most lessons.**

- You should complete the Chapter 12 Study Guide and Review on pages 593–594 of your textbook.
- If you are unsure of any concepts or skills, refer to the specific lesson(s).
- You may also want to take the Chapter 12 Practice Test on page 595.

☐ **I asked for help from someone else to complete the review of all or most lessons.**

- You should review the examples and concepts in your Study Notebook and Chapter 12 Foldable.
- Then complete the Chapter 12 Study Guide and Review on pages 593–594 of your textbook.
- If you are unsure of any concepts or skills, refer to the specific lesson(s).
- You may also want to take the Chapter 12 Practice Test on page 595.

Student Signature

Parent/Guardian Signature

Teacher Signature

© Glencoe/McGraw-Hill